Lecture Notes in Mathematics

Edited by A. Dold and B. Eckmann

706

Probability Measures on Groups

Proceedings of the Fifth Conference
Oberwolfach, Germany,
January 29th – February 4, 1978

Edited by H. Heyer

Springer-Verlag
Berlin Heidelberg New York 1979

Editor
Herbert Heyer
Mathematisches Institut
Universität Tübingen
Auf der Morgenstelle 10
D-7400 Tübingen

Library of Congress Cataloging in Publication Data

Main entry under title:

Probability measures on groups.

(Lecture notes in mathematics ; 706)
Bibliography: p.
Includes index.
1. Probabilities--Congresses. 2. Groups, Theory of--Congresses. 3. Measure theory--Congresses. 4. Random walks (Mathematics)--Congresses. I. Heyer, Herbert. II. Title. III. Series: Lecture notes in mathematics (Berlin) ; 706.
QA3.L28 no. 706 [QA273.A1] 510'.8s [519.2]
79-10236

AMS Subject Classifications (1970): primary: 43A05, 60K05, 60B15, 60J15 secondary: 60B05, 60J35, 60G50

ISBN 3-540-09124-6 Springer-Verlag Berlin Heidelberg New York
ISBN 0-387-09124-6 Springer-Verlag New York Heidelberg Berlin

Printed in Germany

Printing and binding: Beltz Offsetdruck, Hemsbach/Bergstr.
2141/3140-543210

INTRODUCTION

Probability theory on groups has become a field of considerable interest to analysts in recent years. Despite of the progress which still goes on, the degree of maturity of certain parts of the theory has encouraged a number of probabilists to write extended expositions on the subject. These publications prove both the breadth and the profundity of the problems and their solutions. We only mention the book [1] by C. Berg and G. Forst on the potential theory of continuous convolution semigroups on locally compact Abelian groups, the Lecture Notes on random walks on Lie groups by Y. Guivarc'h, M. Keane and B. Roynette [4], on continuous convolution semigroups and generating distributions by W. Hazod [5] and on convolution products of measures on arbitrary topological groups and semigroups by A. Mukherjea and N.A. Tserpes [10], finally the editor's monograph [6] containing a detailed presentation of the basic topics of the theory for general locally compact groups. Stimulated by the written work the specialists keep articulating the legitimate desire of a permanent exchange of ideas. Meetings and conferences help surveying the published results and initiate new directions of research. In this spirit the edition of the contributions collected from the fifth Oberwolfach conference on Probability Measures on Groups will be more than justified.

When looking into the actual trends of research within the theory of probability on groups one observes essentially two different aspects of extending the classical theory:
The one aspect consists in generalizing a given classical theorem valid for the real line or the torus group to possibly all locally compact groups. Here the generalization is mostly achieved at the costs of the strength of the given statement. The second aspect is based on the idea to preserve the strong formulation of the classical result as strictly as possible and still extend it to more general groups. While the first aspect gives rise to an enlargement of the fundamental concepts of the theory, the second aspect enriches the collection of examples which have to be available in order to test the general theory. Under the first aspect the theory has been generalized via Lie groups and Lie projective groups to arbitrary locally compact groups [6] and by various other methods to completely regular or metric groups and

semigroups [10]. The second aspect led to generalizations for the affine group, the motion group, general nilpotent and solvable Lie groups as well as their compact extensions [4].

The papers published in this volume represent fairly well the wide spectrum of the actual theory. We shall arrange the contributions under five different themes and comment briefly on the subjects treated.

1. *Infinite convolutions* of probability measures on groups and semigroups from a field of research which is basic for the theory. Interestingly enough the Itô-Kawada theorem still motivates further work in this direction. By a well-known result of Y. Derriennic and A. Mukherjea the powers of a probability measure generating a non-compact second countable locally compact group converge to the zero measure. Quantitative results on the rate of this convergence are presented for infinite discrete groups, especially for free groups by P. GERL using combinatorial methods, and by P. BOUGEROL using the method of concentration functions. The last author points out that it will be promising to generalize his method from groups to homogeneous spaces. A. MUKHERJEA's contribution based on the work of G. Brown, W. Moran and P. Hartman contains purity properties of the limits of infinite convolutions. The paper of V. LOSERT and H. RINDLER touches the relationship between random walks on compact groups and the theory of equidistribution.

2. *Continuous convolution semigroups*: After the analysis of continuous convolution semigroups of measures has reached its culmination with the general Lévy-Khintchine representation, new detail aspects have moved into the center of discussion. Th. DRISCH deals in his paper with the analyticity of convolution semigroups. W. HAZOD's main interest in his contribution is the problem of mixing convolution and operator semigroups with application to subordination. Aspects of particular interest are an order relation defined by the subordination procedure and the Bochner-stable semigroups introduced as subordinated semigroups. Convolution semigroups of matrix-valued measures correspond to contraction semigroups of invariant operators on the vector-valued continuous functions on the group. Their representation as contraction semigroups of positive measures and their generation by operators satisfying a maximum principle is the object of the paper by J.-P. ROTH.

The papers of C. BERG and of C. BERG and G. FORST pertain to the potential theory of continuous convolution semigroups on an Abelian group. C. Berg constructs continuous singular Hunt convolution kernels on the real line and on the infinite-dimensional torus. Jointly with G. Forst he reviews some results from the theory with special emphasis on similarities between potential kernels and infinitely divisible probabilities on a locally compact Abelian group.

3. *Special classes of probability measures:* As a consequence of the recent development of the central limit problem infinitely divisible probability measures received a particular interest. The problem of embedding infinitely divisible measures in continuous convolution semigroups has been solved only partially. Very little is known about the continuous embedding of measures on algebraic groups. In K.R. PARTHASARATHY's paper it is shown that on SL(2,ℝ) and SL(2,ℂ) every infinitely divisible probability which generates the group is diadically embeddable. Another class of probability measures arising in central limit theory is the class of stable measures, whose generalization to general groups depends on the automorphism group of the underlying group. P. BALDI establishes for measures on the motion group a compactness lemma and obtains a characterization of stable measures as limits of normed powers. There is no result yet within this framework on domains of attraction of stable measures. For Gaussian measures on metric Abelian groups T. BYCZKOWSKI presents results on the integrability of seminorms and on zero-one laws. The framework of a metric Abelian group is chosen in order to apply the theorems obtained also to function spaces. Full measures as studied by V. LOSERT and K. SCHMIDT have their origin in ergodic theory. Their relations to cocycles, ergodic flows and eigenfunctions are described. Moreover the authors include results on the existence of full measures and of such subgroups which do not carry full measures. A. JANSSEN's work on the measurability of certain sets of measures on a topological space has implications to the theory of admissable translates.

4. *Random walks* on groups and homogeneous spaces play an eminent rôle in the potential theory on these structures. In recent years special emphasis has been given to noncommutative renewal theory and to the development of local limit theorems. The paper of G. FORST concerns renewal measures on the half line. A new proof is given for

Daley's result containing necessary and sufficient conditions for a given renewal measure to have the property that each of its positive multiples is again a renewal measure. The aim of L. ELIE's paper pertaining to renewal theory is a characterization of all almost compact locally compact groups having property (P). It turns out that this property (P) is the key tool in establishing renewal results for spread out transient measures generating the underlying group. In a second contribution L. Elie studies the cone of μ-invariant positive measures on the affine group and exhibits its extremal rays. The theory developed yields the study of the Martin boundary of the corresponding random walk on the group and new information on its positive harmonic functions. In the paper of H. HENNION a dichotomy theorem for random walks on homogeneous spaces is proved. Moreover it is conjectured that for rigid Lie groups a homogeneous space is recurrent if and only if it is of polynomial growth of degree ≤ 2. The object of the research done by L. HANTSCH and W. VON WALDENFELS is a concrete problem from atomic physics. Here the averaging of atomic decay rates by random impacts leads to a particular random walk on the general linear group. The limiting behavior of this random walk can be estimated. A. RAUGI and P. CREPEL have contributed to central limit theory for random walks on certain Lie groups. The first author presents a profound result on the convergence of normed powers of a measure to an absolutely continuous measure lying on a diffusion semigroup. The class of groups involved consists of semidirect products of rigid solvable Lie groups with compact groups. In contrast to this rather technical analysis we see Crépel's survey article on the methods and results from central limit theory for independent and dependent random variables. The well established methods of Fourier transforms, moments, those based on Trotter's and Skorohod's theorems and on the various kinds of dependence are exemplified on the motion and the Heisenberg groups.

5. *Group representations and probability:* In connection with the theory of ergodic automorphisms of compact groups the problem of characterizing all locally compact groups whose irreducible, unitary continuous representations are of uniformly bounded degree appears to be relevant. C.C. Moore showed that the groups with this property are exactly those admitting an Abelian normal subgroup of finite index. In his paper E. SCHLICHTING presents a new and direct proof of the theorem by reducing the problem to compact groups. Cocycles and coboundaries on the motion group are central in the contribution of

B.-J. FALKOWSKI. The author exhibits all infinitely divisible positive definite functions on the group which correspond to the cocycles of irreducible representations. Weakly stationary processes on a locally compact Abelian group are exactly the Fourier transforms of certain vector-valued measures on the group with values in some Hilbert space. P. RESSEL studies in his paper compact sets of such measures and derives a Lévy type continuity theorem.

Aside from the open problems stated in some of the papers in this volume several questions have been brought up at the conference which might stimulate future research. There is important work still to be done in the arithmetic of the semigroup of probability measures on a locally compact group. Fundamental classical results in this direction have been collected in the monographs [2] of R. Cuppens, [7] and [8] of Yu.V. Linnik and of Yu.V. Linnik and I.V. Ostrovskiǐ resp., and in the article [9] of L.Z. Livšic, I.V. Ostrovskiǐ and G.P. Čistjakov. Another field of interest is the theory of continuous convolution hemigroups on a locally compact group and its connection with evolution families of contraction operators and with additive stochastic processes. Pioneering work has been done in the papers [11] and [12] by D.W. Stroock and S.R.S. Varadhan, and [3] by P. Feinsilver. A wide range of open problems can be looked at in noncommutative potential theory as it has been exposed at the conference by F. Hirsch. And, after all, the application of the theory to problems outside mathematics will certainly gain priority in the course of years. The example presented by R.M. Dudley on the application of the central limit theorem on $GL(2,\mathbb{R})$ to high energy scattering gave a realistic insight into the discrepancy between theory and practice.

Herbert Heyer

References

[1] C. Berg, G. Forst: Potential theory on locally compact Abelian groups.
Berlin-Heidelberg-New York: Springer 1975

[2] R. Cuppens: Decomposition of multivariate probability.
New York-San Francisco-London: Academic Press 1975

[3] P. Feinsilver: Processes with independent increments on a Lie group.
Trans. Amer. Math. Soc. 242, 73-121 (1978)

[4] Y. Guivarc'h, M. Keane, B. Roynette: Marches aléatoires sur les groupes de Lie.
Lecture Notes in Math. Vol. 624. Berlin-Heidelberg-New York: Springer 1977

[5] W. Hazod: Stetige Halbgruppen von Wahrscheinlichkeitsmaßen und erzeugende Distributionen.
Lecture Notes in Math. Vol. 595 Berlin-Heidelberg-New York: Springer 1977

[6] H. Heyer: Probability measures on locally compact groups.
Berlin-Heidelberg-New York: Springer 1977

[7] Yu. V. Linnik: Décomposition des lois de probabilités.
Paris: Gauthier-Villars 1962

[8] Yu. V. Linnik, I. V. Ostrovskiǐ: Decomposition of random variables and vectors.
English translation: Amer. Math. Soc. 48 (1977)

[9] L. Z. Livšic, I. V. Ostrovskiǐ, G. P. Čistjakov: The arithmetic of probability laws.
Probability theory. Mathematical Statistics. Theoretical Cybernetics Vol. 21, pp 5-42
Moskow: Akad. Nauk SSSR Vsesojuz. Inst. Naučn. i. Techn. Informacii 1975 [Russian]

[10] A. Mukherjea, N. A. Tserpes: Measures on topological semigroups. Lecture Notes in Math. Vol. 547. Berlin-Heidelberg-New York: Springer 1976

[11] D. W. Stroock, S.R.S. Varadhan: Diffusion processes with continuous coefficients I, II. Comm. Pure Appl. Math. 12, 345-400, 479-530 (1969)

[12] D. W. Stroock, S.R.S. Varadhan: Limit theorems for random walks on Lie groups. Sankhya Ser. A 35, 277-294 (1973)

LIST OF PARTICIPANTS

P. Baldi, Paris
C. Berg, Kopenhagen
M.S. Bingham, Hull
P. Bougerol, Paris
T. Byczkowski, Wrocław
P. Crépel, Rennes
E. Dettweiler, Tübingen
Th. Drisch, Dortmund
R.M. Dudley, Cambridge
L. Elie, Paris
B.-J. Falkowski, München
P.J. Fernandez, Rio de Janeiro
G. Forst, Kopenhagen
P. Gerl, Salzburg
W. Guth, Konstanz
W. Hazod, Dortmund
H. Hennion, Rennes
H. Heyer, Tübingen
F. Hirsch, Cachan
A. Janssen, Dortmund
E. Kaniuth, Paderborn
A. Mukherjea, Tampa
P. Plaumann, Erlangen
A. Raugi, Rennes
P. Ressel, Hamburg
H. Rindler, Wien
J.-P. Roth, Mulhouse
B. Roynette, Nancy
E. Schlichting, München
K. Schmidt, Coventry
D.W. Strook, Boulder
C. Sunyach, Paris
W. von Waldenfels, Heidelberg

CONTENTS

LOIS STABLES SUR LES DEPLACEMENTS DE R^d

P.Baldi

Istituto di Matematica "L.Tonelli"

Pisa (Italie)

Cet article veut être une contribution à la généralisation aux groupes de la notion de loi stable. On va d'abord donner une définition de loi stable qui est une transposition à quelque chose près de celle classique (Def.2), on explicitera ensuite quelles sont les lois stables et on montrera que les lois stables sont les limites en loi de sommes normalisées de variables aléatoires i.i.d.

Dans toute la suite $G_d = R^d \times_\sigma SO(d)$ désignera le groupe des déplacements de R^d, $d \geq 3$, à savoir le groupe dont les éléments sont de la forme (x,u), $x \in R^d$, $u \in SO(d)$ muni de l'opération

$$(x,u)\cdot(x',u') = (x+u\cdot x',uu')$$

On appellera H le sous-groupe des éléments de la forme $(0,u)$, e sera l'élément neutre de $SO(d)$, m la mesure de Haar de $SO(d)$. Si ν est une loi de probabilité sur G_d, $\bar{\nu}$ sera sa distribution marginale sur $SO(d)$.

Proposition 1. Les automorphismes de G_d sont de la forme

$$A(x,u) = (y-vuv^{-1}y+cvx,vuv^{-1}) = (y,v)\cdot(cx,u)\ (y,v)^{-1}$$

où $c \in R$ et $(y,v) \in G_d$.

Démonstration. Soit A un automorphisme de G_d et $G' = R^d \times \{e\}$ le sous-groupe des translations; alors $A(G') \subset G'$. Soit en effet

$$\pi : G_d \to G_d/G' \simeq SO(d)$$

$\pi(A(G'))$ est un sous-groupe de G_d/G' abélien et distingué, donc discret comme $SO(d)$ est semisimple et donc réduit à l'élément neutre comme il est connexe; donc $A(G') \subset G'$.

Deuxièmement soit K l'image de H. Si $g=(0,a) \in H$, soit $A(g)=(\sigma(a),\tau(a))$. On a $\tau(aa')=\tau(a)\tau(a')$ et τ est donc un automorphisme de $SO(d)$. Comme $SO(d)$ est semi-simple il existe donc (voir [4] par ex.) $v \in SO(d)$ tel que $\tau(a)=vav^{-1}$. Ensuite

$$\sigma(aa')=\sigma(a)+\tau(a)\cdot\sigma(a')$$

et si on pose

$$y = \int\sigma(a')\ dm(a') = \int\sigma(aa')\ dm(a')$$

on obtient

$$y=\sigma(a)+vav^{-1}y$$

$$\sigma(a)=y-vav^{-1}y$$

Il existe donc $w\in Gl(d)$, $v\in SO(d)$, $y\in R^d$ tels que

$$A(x,u) = (y-vuv^{-1}y+wx, vuv^{-1})$$

Il suffit maintenant d'écrire la relation $A(gg')=A(g)A(g')$ et en déduire que $w=cv$, $c\in R$.

Dans la suite si A est un automorphisme de G_d on écrira $A=(y,v,c)$, avec les notations de la proposition 1. Si $A'=(y',v',c')$ on a facilement

$$A\circ A' = (y+cvy', vv', cc')$$

Définition 2. On dira qu'une loi de probabilité μ sur G_d est stable si pour tout $n\in N$ il existe un automorphisme A_n tel que

$$\mu^n = A_n\mu$$

On supposera de plus que $\bar{\mu}$ est adaptée, à savoir que son support engendre SO(d).

Proposition 3. Si μ est stable, $\bar{\mu}$ est la mesure de Haar m de SO(d).

Démonstration. Par la proposition 1 il existe $v_n\in SO(d)$ tel que

$$\bar{\mu} = \varepsilon_{v_n} * \bar{\mu}^n * \varepsilon_{v_n^{-1}}$$

Comme le support de $\bar{\mu}$ engendre SO(d), $\bar{\mu}$ n'est pas portée par une classe laterale d'un sous-groupe propre fermé distingué K de SO(d): sinon on aurait en effet, par passage au quotient, que SO(d)/K est engendré par un point et donc abélien, ce qui est absurde car SO(d) est semisimple. On peut donc appliquer un résultat classique ([1]) et en déduire que $\bar{\mu}^n \to m$; si v est un point d'adherence de $\{v_n\}$ on a finalement $\bar{\mu} = \varepsilon_v * m * \varepsilon_{v^{-1}} = m$.

Proposition 4. Si dans la définition 2 $A_n=(y_n,v_n,c_n)$, alors $c_2>1$, où bien μ est la mesure de Haar d'un sous-groupe compact de G_d.

Démonstration. Supposons $c_2<1$. On aurait alors

$$A_{2^n} = A_2\circ A_2\circ\ldots\circ A_2 = (y_2+c_2v_2y_2+\ldots+c_2^{n-1}v_2^{n-1}y_2,\ v_2^n,\ c_2^n) = (z_n, w_n, c_2^n)$$

Comme $\{z_n\}$ est bornée dans R^d, si z et w sont des valeurs d'adherence de $\{z_n\}$ et $\{w_n\}$ respectivement et si on pose $A=(z,w,0)$ on aurait que μ^n a comme valeur de adhérence la mesure $A\mu$, ce qui est absurde car $\mu^n \to 0$, à moins que elle ne soit portée par un sous-groupe compact ([2]).

Le cas $c_2=1$ est le plus difficile. Les représentations unitaires irreductibles sont données, par exemple, par la famille d'opérateurs T_g^ρ , $g\in G_d$, $\rho\in R^+$ définis sur l'espace de Hilbert $L^2(S_{d-1})$, où S_{d-1} est la surface de la sphère unitaire de R^d munie de la mesure de Lebesgue (d-1)-dimensionnelle et définis par

$$T^{\rho}_{g}\phi(\xi) = e^{i\rho(\xi,x)}\phi(u^{-1}\xi)$$

où $g=(x,u)$, $\xi\in S_{d-1}$, $\phi\in L^2(S_{d-1})$ (voir par ex. [7], chap.XI). La transformée de Fourier de μ sera l'opérateur défini par

$$\hat{\mu}(\rho) = \int T^{\rho}_{g}\, d\mu(g)$$

Si $c_2=1$ la définition 2 entraine que pour tout $n\in N$ il existe $g_n\in G_d$ tel que

$\varepsilon_{g_n} * \mu * \varepsilon_{g_n^{-1}} = \mu^n$, soit pour tout $\rho\in R^+$

(1) $$\hat{\mu}(\rho)^n = T^{\rho}_{g_n}\, \hat{\mu}(\rho)\, T^{\rho *}_{g_n}$$

Pour $\rho=0$ il est facile de vérifier, comme $\tilde{\mu}=m$, que si $\phi_1\equiv 1$, $\hat{\mu}(0)\phi_1=\phi_1$ et $\hat{\mu}(0)\phi=0$ si ϕ est orthogonale à ϕ_1. Pour simplifier la notation on posera $T^{\rho}_{g_2} = T$. Evidemment le spectre de $\hat{\mu}(0)$, $sp\hat{\mu}(0)$, est l'ensemble $\{0,1\}$.

Comme $\hat{\mu}(\rho)$ est une famille uniformement continue dans la topologie uniforme des opérateurs, on a que pour ρ assez petit $sp\hat{\mu}(\rho)$ est contenu dans la réunion de deux boules de rayon ε, de centre 0 et 1 respectivement; mais comme

$$\{\lambda^2,\ \lambda\in sp\hat{\mu}(\rho)\} = sp(\hat{\mu}(\rho)^2) = sp(T\,\hat{\mu}(\rho)\,T^{-1}) = sp\hat{\mu}(\rho)$$

on a que pour ρ assez petit $sp\hat{\mu}(\rho) = \{0,1\}$ et par connexité donc que cela est vrai pour tout $\rho\in R^+$. De plus ([5], th.3.16 pag.212) on a que pour tout $\rho\in R^+$ le sousespace propre associé à la valeur propre 1 est de dimension 1, comme cela est vrai pour $\rho=0$.

Soient maintenant Γ_0 et Γ_1 deux petites courbes fermées contenant 0 et 1 respectivement à leur intérieur; posons ([8])

$$\Pi_0 = (2\pi i)^{-1}\int_{\Gamma_0}(zI-\hat{\mu}(\rho))^{-1}dz$$

$$\Pi_1 = (2\pi i)^{-1}\int_{\Gamma_1}(zI-\hat{\mu}(\rho))^{-1}dz$$

Π_0 et Π_1 sont deux projecteurs qui commutent avec $\hat{\mu}(\rho)^2$ et $\hat{\mu}(\rho)$ et tels que $\Pi_0+\Pi_1=I$. Grâce aux propriétés de ce calcul fonctionnel et à (1) on a

$$\Pi_0 = (2\pi i)^{-1}\int_{\Gamma_0}(zI-\hat{\mu}(\rho)^2)^{-1}dz = T^{-1}\Pi_0 T$$

donc Π_0 commute avec T et cela est également vrai pour Π_1. On a facilement

$$(\Pi_0\hat{\mu}(\rho)\Pi_0) = T\Pi_0\hat{\mu}(\rho)\Pi_0 T^{-1}$$

$$(\Pi_0\hat{\mu}(\rho)\Pi_0)^{2^n} = T^n\Pi_0\hat{\mu}(\rho)\Pi_0 T^{-n}$$

et cela entraine que $\Pi_0\hat{\mu}(\rho)\Pi_0= 0$, car la norme du terme à gauche tend vers 0, comme le spectre est réduit à $\{0\}$, alors que la norme du terme à droite reste égale à la norme de $\Pi_0\hat{\mu}(\rho)\Pi_0$. On a donc $\hat{\mu}(\rho)=\Pi_1\hat{\mu}(\rho)\Pi_1$ et du fait que Π_1 est un projecteur unidimensionnel on a finalement $\hat{\mu}(\rho)^2=\hat{\mu}(\rho)$ pour tout $\rho\in R^+$. Donc

$\mu^2=\mu$ et μ est la mesure de Haar d'un sous-groupe compact de G_d (voir [3]).

Proposition 5. Soit μ une loi stable sur G_d;si μ n'est pas la mesure de Haar d'un sous-groupe compact il existe alors un automorphisme B_μ de la forme $(x,e,1)$ (un automorphisme intérieur donc) tel que si $\tilde{\mu} = B_\mu \mu$, $\tilde{\mu}$ est stable et

$$\tilde{A}_2 \tilde{\mu} = \tilde{\mu}^2$$

où $\tilde{A}_2$ est de la forme $(0,v_2,c_2)$.

Démonstration. On a facilement que $\tilde{A}_2 = B_\mu A_2 B_\mu^{-1}$ et si $A_2=(y_2,v_2,c_2)$

$$B_\mu A_2 B_\mu^{-1} = (x-c_2v_2x+y_2, v_2, c_2)$$

Il suffit donc que x soit tel que $x-c_2v_2x+y_2=0$, ce qui est toujours possible car $c_2>1$ et donc $I-c_2v_2$ est inversible.

Pour la définition et la structure des lois stables dans R^d on peut se reporter à [6].

Proposition 6. Si μ est stable, μ est le produit direct d' une loi dans R^d qui est stable et invariante par rotations par la mesure de Haar de $SO(d)$, ou bien l'image d'une telle loi par un automorphisme intérieur par un élément de la forme (x,e).

Démonstration. Si μ n'est pas la mesure de Haar d'un sous-groupe compact de G_d, grâce à la proposition précédente, quitte à remplacer μ par $\tilde{\mu}$,on peut supposer que A_{2^n} est de la forme $(0,v_2^n,c_2^n)$ et il suffit de montrer que μ est le produit direct d'une loi stable sur R^d invariante par rotations par la mesure de Haar normalisée de $SO(d)$. Soient (X,U), (X_i,U_i) des variables aléatoires i.i.d. de loi μ , f une fonction continue à support compact dans R^d, g une fonction continue dans $SO(d)$. On sait déjà que la loi de U est la mesure de Haar normalisée de $SO(d)$. Montrons que X et U sont indépendentes, soit que

$$E[f(X)g(U)] = E[f(X)]\, E[g(U)]$$

Soit $S_n=X_1+U_1X_2+\ldots+U_1U_2\ldots U_{n-1}X_n$, $V_n=U_1\ldots U_n$. Comme μ est stable on a

$$E[f(X)g(U)] = E[f(c_2^{-n}v_2^{-n}S_{2^n})g(v_2^{-n}V_{2^n}v_2^n)] = \lim_n E[f(c_2^{-n}v_2^{-n}S_{2^n-1})g(v_2^{-n}V_{2^n}v_2^n)]$$

$$\lim_n E[E[f(c_2^{-n}v_2^{-n}S_{2^n-1})g(v_2^{-n}V_{2^n}v_2^n) \mid (X_1,U_1),\ldots,(X_{2^n-1},U_{2^n-1})]]$$

$$= \lim_n E[f(c_2^{-n}v_2^{-n}S_{2^n-1})] \int g(u)\, m(du) = E[f(X)]\, E[g(U)]$$

où on a exploité l'indépendance des variables aléatoires et l'invariance par translations de m. Il ne reste plus qu'à déterminer la loi de X. Si $\check{\mu}$ est la loi image de μ par l'application $g \to g^{-1}$, $\check{\mu}$ est stable et sa loi est celle de

$(-U^{-1}X,U^{-1})$; comme U et X sont indépendantes et la loi de U est la mesure de Haar de SO(d), $-U^{-1}X$ est invariante par rotations. En revenant à μ on a donc que X est invariante par rotations. On a finalement

$$X_1+X_2+\ldots+X_n \simeq X_1+U_1X_2+\ldots+U_1\ldots U_{n-1}X_n \simeq y_n-v_nUv_n^{-1}y_n+c_nv_nX$$

Or le terme à gauche est invariant par rotations ainsi que c_nX et $v_nUv_n^{-1}y_n$ et donc y_n l'est aussi (en tant que variable aléatoire) ce qui entraine finalement $y_n=0$ et

$$X_1+X_2+\ldots+X_n \simeq c_nX$$

soit que X est une variable aléatoire stable dans R^d.

Proposition 7. (Lemme de compacité) Soient λ, μ, λ_n des lois de probabilité sur G_d, A_n des automorphismes de G_d tels que

$$\lambda_n \to \lambda$$

$$A_n\lambda_n \to \mu$$

Supposons de plus que $\bar{\lambda}=\bar{\mu}=m$. Alors si $A_n=(y_n,v_n,c_n)$ $\{|y_n|\}$ est une suite bornée. Si de plus le support de λ n'est pas contenu dans SO(d) $\{c_n\}$ est une suite bornée.

Démonstration. Comme les deux suites sont tendues, pour tout $\varepsilon>0$ il existe un compact K tel que

$$\lambda_n(K)\geq 1-\frac{\varepsilon}{2} \qquad A_n\lambda_n(K)\geq 1-\frac{\varepsilon}{2}$$

On pourra supposer $K = S_r \times SO(d)$, où S_r est la sphère de R^d de rayon r et de centre l'origine. On aura donc

$$\lambda_n(K\cap A_n^{-1}K)\geq 1-\varepsilon$$

pour tout $n\in N$. Posons $z_n^u = y_n+v_nuv_n^{-1}y_n$ et

$$C_n^u = (K\cap A_n^{-1}K)\cap R^d\times\{u\} = \{(x,u);\ x\in S_r,\ z_n^u + c_nv_nx\in S_r\}$$

On remarque que:

a) C_n^u est contenu dans une sphère de rayon r/c_n et centre z_n^u

b) C_n^u est vide si $|z_n^u|\gg(c_n+1)r$. En effet comme $z_n^u+c_nv_nx\in S_r$

$$|z_n^u|\leq c_n|x|+r\leq(c_n+1)r$$

Montrons que $\{|y_n|c_n^{-1}\}$ est une suite bornée. Sinon soit $\{a_n\}$ une sous-suite divergente et posons

$$M_n = \{u\in SO(d);\ |z_n^u|\geq a_n^{-1/2}|y_n|\}$$

Soit $\lambda_n=\int\lambda_n^u\bar{\lambda}_n(du)$ une désintégration de λ_n ; pour n assez grand et suivant la sous-suite considérée on a

$$\lambda_n(K\cap A_n^{-1}K) = \int \lambda_n^u(C_n^u)\,\bar{\lambda}_n(du) \leq \lambda_n(R^d\times M_n^C) = \bar{\lambda}_n(M_n^C)$$

Où on a utilisé le fait que C_n^u est vide si $u\in M_n$, car si $u\notin M_n$

$$|z_n^u| \geq a_n^{-1/2}|y_n| = (c_n|y_n|^{-1})^{1/2}|y_n| = a_n^{1/2}\,c_n$$

et donc pour n assez grand $|z_n^u| > (c_n+1)r$, ce qui par la remarque b) entraine que C_n^u est vide.

Evidemment comme $a_n\to +\infty, m(M_n^C)\to 0$.

<u>Lemme</u>. Il existe une suite $\{w_n\}$ dans SO(d) telle que $w_n M_n w_n^{-1} \subset M_{n+1}$.

<u>Démonstration</u>. On peut écrire

$$M_n = \{u\in SO(d);\ |uv_n^{-1}\tilde{y}_n - v_n^{-1}\tilde{y}_n| \geq a_n^{-1/2}\}$$

où $\tilde{y}_n|y_n| = y_n$. Quitte à se ramener à une sous-suite on peut supposer $\{a_n\}$ croissante. Soit w_n tel que

$$w_n v_n^{-1}\tilde{y}_n = v_{n+1}^{-1}\tilde{y}_{n+1}$$

Alors

$$|w_n u w_n^{-1} v_{n+1}^{-1}\tilde{y}_{n+1} - v_{n+1}^{-1}\tilde{y}_{n+1}| = |u w_n^{-1} v_{n+1}^{-1}\tilde{y}_{n+1} - w_n^{-1} v_{n+1}^{-1}\tilde{y}_{n+1}| =$$

$$= |uv_n^{-1}\tilde{y}_n - v_n^{-1}\tilde{y}_n| \geq a_n^{-1/2} \geq a_{n+1}^{-1/2}$$

et donc si $u\in M_n$, $w_n u w_n^{-1}\in M_{n+1}$.

Donc dans la notation du lemme si on pose $s_k = w_k w_{k-1}\cdots w_1$ et

$$N_n = s_{n-1}^{-1} M_n s_{n-1}$$

$\{N_n\}$ est une suite décroissante d'ensembles. Posons encore $\bar{\lambda}_n' = \varepsilon_{s_{n-1}} * \bar{\lambda}_n * \varepsilon_{s_{n-1}}$

On a que

$$\bar{\lambda}_n(M_n^C) = \bar{\lambda}_n'(N_n^C)$$

et

$$\bar{\lambda}_n' \to m$$

On a donc

$$m(M_k^C) \geq \limsup_n \bar{\lambda}_n'(N_k^C) \geq \limsup_n \bar{\lambda}_n'(N_n^C) = \limsup_n \bar{\lambda}_n(M_n^C)$$

et donc, comme $m(M_n^C)\to 0$, on a finalement

$$\lambda_n(K\cap A_n^{-1}K) \to 0$$

Et donc $\{|y_n|c_n^{-1}\}$ est bornée. On peut maintenant considérer la situation

$$\mu_n \to \mu$$
$$A_n^{-1}\mu_n \to \lambda$$

où on a posé $\mu_n = A_n\lambda_n$. Comme $A_n^{-1} = (-v_n^{-1}c_n^{-1}y_n, v_n^{-1}, c_n^{-1})$, si on applique le résultat précédent à A_n^{-1} on a que $\{|y_n|\}$est une suite bornée.

Allons maintenant montrer que si $\{c_n\}$ n'est pas bornée λ est portée par SO(d).

Soit (X,U) une variable aléatoire de loi λ et (X_n, U_n) des variables aléatoires indépendantes de loi λ_n.

$$(y_n - v_n U_n v_n^{-1} y_n + c_n v_n X_n,\ v_n U_n v_n^{-1})$$

est une variable aléatoire de loi $A_n\lambda_n$. Une fois de plus comme la suite est tendue , pour tout $\varepsilon>0$ il existe r tel que

$$1 - \varepsilon \leq A_n\lambda_n(K_r) = P(y_n - v_n U_n v_n^{-1} y_n + c_n v_n X_n \in S_r) = P(|y_n - v_n U_n v_n^{-1} + c_n v_n X_n| \leq r) \leq$$
$$\leq P(c_n|X_n| - 2M \leq r) = P(|X_n| \leq (r+2M)c_n^{-1})$$

où M est un majorant de $\{|y_n|\}$. Si $\{c_n\}$ n'était pas bornée, passant à la limite suivant une sous-suite divergente, on obtiendrait

$$P(|X| = 0) \geq 1 - \varepsilon$$

et comme ε est arbitraire on en déduirait que λ est portée par SO(d),

Théorème 8. Soit μ une loi de probabilité sur G_d. Alors μ est stable si et seulement si il existe une loi de probabilité ν telle que le support de $\bar{\nu}$ engendre SO(d) et une suite $\{A_n\}$ d'automorphismes de G_d telle que $A_n\nu^n \to \mu$.

Démonstration. Dans un sens l'implication est évidente. Viceversa soit μ telle que

$$A_n\nu^n \to \mu$$

où le support de $\bar{\nu}$ engendre SO(d), Il est clair que $\bar{\mu}$ est la mesure de Haar normalisée de SO(d). Si μ est la mesure de Haar d'un sous-groupe compact elle est évidemment stable.

Sinon soit $k \in N$, on a

$$A_{nk}\nu^{nk} = A_{nk}A_k^{-1}(\underbrace{A_k\nu^k, \ldots, A_k\nu^k}_{n \text{ fois}})$$

Or le terme entre parenthèses converge vers μ^k; on peut donc appliquer la proposition précédente: si

$$A_{nk}A_k^{-1} = (y_n, v_n, c_n)$$

$\{|y_n|\}$ et $\{c_n\}$sont bornées et si (y_0,v_0,c_0) est un point d'adhérence de (y_n,v_n,c_n) et A_0 est l'automorphisme associé, à la limite on a

$$\mu = A_0\mu^k$$

et donc μ est stable.

Remarque 1. Dans le cas du groupe des déplacements l'existence d'automorphismes qui "contractent", et qui jouent donc le rôle des homothéties dans le cas classique, nous a permis de donner une définition de loi stable qui est très semblable à celle de R^d. Malhereusement cela n'est pas le cas en général (dans un groupe semisimple, par exemple, tout automorphisme est un automorphisme intérieur) et il sera donc probablement nécessaire de donner une définition complètement differente. J'estime que dans le cas général on sera obbligés de remplacer les automorphismes par les procédés de normalisation utilisés dans les théorèmes central limite correspondants.

Remarque 2. Le complément naturel de ce travail serait, bien sûr, la détermination du domain d'attraction des lois stables, à savoir, si μ est stable, quelles sont les lois de probabilité sur G_d telles qu'il existe une suite de automorphismes $\{A_n\}$ telle que

$$A_n\nu^n \to \mu$$

Ce problème paraît assez difficile; en effet une des techniques principales dans la démonstration du théorème central limite sur le groupe des déplacements ([9], [10], [11]) consiste à se ramener à un théorème central limite dans R^d pour des variables aléatoires dépendantes. En ce moment je ne suis à connaissance d'aucun résultat de ce type, même dans R, pour les lois stables.

BIBLIOGRAPHIE

[1] H.S.Collins - Convergence of convolution iterates of mesures - Duke Math.J. 29, 2, (1962), pag 259.

[2] Y.Derriennic - Loi "zero ou deux" pour les processus de Markov. Applications aux marches aléatoires - Ann.Inst.H.Poincaré, vol.XII, n°2, (1976), pag.111.

[3] U.Grenander-Probabilities on algebraic structures - John Wiley, 1968.

[4] G.Hochschild - The structure of Lie groups - Holden-Day, 1965.

[5] T.Kato - Perturbation theory for linear operators - Springer Verlag, 1966.

[6] E.L.Rvačeva - On domain of attraction of multidimensional stable distributions - Selected translations in mathematical Statistics and Probability, 2, pag.183.

[7] N.Ja.Vilenkin - Special functions and the theory of group representations -

Translations of Mathematical Monographs, Providence, 1968

[8] K.Yosida - Functional Analysis - Springer Verlag, 1965

[9] L.G.Gorostiza - The central limit theorem for random motions of d-dimensional euclidean space - Annals of Probability, 1, 4, (1973), pag.603.

[10] V.N.Tutubalin - The central limit theorem for random motions of a Euclidean Space - Select.Transl.in Math.Stat.and Proba., 2, (1973).

[11] B.Roynette - Théorème central limite pour le groupe des déplacements de R^d - Ann.Inst.H.Poincaré, sect.B, vol.X, 4, (1974).

[12] B.V.Gnedenko, A.N.Kolmogorov - Limit distributions for sums of independent random variables - Addison Wesley, 1954.

Hunt convolution kernels which are continuous singular with respect to Haar measure

Christian Berg

0. Introduction.

Hunt convolution kernels play an important role in potential theory on locally compact abelian groups. The classical examples of Hunt convolution kernels, the most important example being the Newtonian kernel, are all very "nice" kernels, in the sense that we have simple and explicit expressions for them. It might therefore be of interest to examine how "bad" Hunt convolution kernels can be in a measure theoretic sense. We shall see that they can be "bad" in the sense that they can be continuous singular with respect to Haar measure, and at the same time have a support which is an essential part of the group. We shall give three examples, two on $\mathbb{R}$ and one on T^{∞}.

Before we get to the examples we fix the terminology, and after the examples we state an open problem.

1. Terminology.

Let G denote a locally compact abelian group with a countable base for the topology, and assume further that G is non-discrete.

A convolution semigroup on G is a family $(\mu_t)_{t>0}$ of positive measures on G satisfying

(i) $\mu_t * \mu_s = \mu_{t+s}$ for $t,s > 0$,

(ii) $\mu_t(G) \leq 1$ for $t > 0$,

(iii) $\lim_{t \to 0} \mu_t = \varepsilon_0$ vaguely.

Here ε_0 denotes the Dirac measure at the zero element of the group. It is easy to see that the assumptions above imply that the mapping $t \to \mu_t$ is vaguely continuous.

A positive measure κ on G is called a <u>Hunt convolution kernel</u>, if there exists a convolution semigroup $(\mu_t)_{t>0}$ on G such that

(1) $$\kappa = \int_0^{\infty} \mu_t \, dt \, .$$

From (1) follows that

(2) $$\operatorname{supp}(\kappa) = c\ell\left(\bigcup_{t>0} \operatorname{supp}(\mu_t) \right) ,$$

and it is known that supp(κ) is a subsemigroup of G, cf. [3].

A positive measure μ on G is called continuous if $\mu(\{x\}) = 0$ for all $x \in G$, and it is called singular, if there exists a Borel nullset A (with respect to Haar measure) on which μ is concentrated, i.e. $\mu(CA) = 0$.

A positive measure μ is called continuous singular, if it is continuous and singular.

Let μ be a positive measure such that $\mu(G) < \infty$. Then the Fourier-Stieltjes transform of μ is a function $\hat{\mu}$ on the dual group $\hat{G}$ of G defined as

$$\hat{\mu}(y) = \int_G \overline{(x,y)}\, d\mu(x) \qquad \text{for} \quad y \in \hat{G}.$$

If μ is absolutely continuous with respect to Haar measure, then $\hat{\mu}$ vanishes at infinity on $\hat{G}$, but the converse is not true. If $\hat{\mu}$ vanishes at infinity on $\hat{G}$ we can only conclude that μ is continuous.

We will use the following elementary result:

Lemma 1. *Let μ be a positive measure such that $\mu(G) < \infty$. If*

$$\limsup_{y \to \infty} |\hat{\mu}(y)| = \mu(G)$$

then μ is singular.

Proof. Let $\mu = \mu_a + \mu_s$ be the decomposition of μ as sum of the absolutely continuous part μ_a of μ and the singular part μ_s of μ. Then

$$\mu(G) = \limsup_{y \to \infty} |\hat{\mu}(y)| \le \limsup_{y \to \infty} |\hat{\mu}_a(y)| + \limsup_{y \to \infty} |\hat{\mu}_s(y)|$$

$$= \limsup_{y \to \infty} |\hat{\mu}_s(y)| \le \mu_s(G),$$

and it follows that $\mu_a(G) = 0$ so $\mu = \mu_s$, hence μ is singular. $\square$

There is a one-to-one correspondence between convolution semigroups $(\mu_t)_{t>0}$ on G and continuous negative definite functions ψ on $\hat{G}$, established by the formula

$$\hat{\mu}_t(y) = e^{-t\psi(y)} \qquad \text{for} \quad t > 0 \text{ and } y \in \hat{G},$$

cf. [3].

A symmetric convolution semigroup $(\mu_t)_{t>0}$ is called <u>Gaussian</u> if the corresponding negative definite function ψ is a quadratic form on $\hat{G}$, (plus a non-negative constant), i.e. satisfies

$$\psi(y_1+y_2) + \psi(y_1-y_2) = 2\psi(y_1) + 2\psi(y_2) , \quad y_1, y_2 \in \hat{G} , \tag{3}$$

cf. [3]. A Hunt convolution kernel κ for which the corresponding convolution semigroup is Gaussian is called of <u>local type</u>.

2. Example 1.

<u>Theorem 2. There exists a symmetric Hunt convolution kernel</u> κ <u>on</u> $\mathbb{R}$ <u>with</u> $\kappa(\mathbb{R}) = 1$ <u>such that</u>

(a) $\operatorname{supp}(\kappa) = \mathbb{R}$,

(b) κ <u>is continuous singular</u>.

For any number $a > 0$ the function $\psi_a(y) = 1 - \cos(ay)$ is a continuous negative definite function, and the corresponding convolution semigroup $(\mu_t^a)_{t>0}$ is given as

$$\mu_t^a = e^{-t} \sum_{k=0}^{\infty} \left[\frac{1}{2}\left(\varepsilon_{-a}+\varepsilon_a\right)\right]^{*k} \frac{t^k}{k!} , \tag{4}$$

which shows that $\operatorname{supp}(\mu_t^a) = a\mathbb{Z}$ for all $t > 0$.

<u>Lemma 3. Let</u> $a_1, a_2, \ldots$ <u>be a sequence of positive numbers such that</u> $\Sigma a_n^2 < \infty$. <u>Then the infinite series</u>

$$\psi(y) = \sum_{n=1}^{\infty} 1 - \cos(a_n y) , \quad y \in \mathbb{R} \tag{5}$$

<u>converges uniformly on compact subsets of</u> $\mathbb{R}$ <u>and defines a continuous negative definite function</u> ψ. <u>The corresponding convolution semigroup</u> $(\mu_t)_{t>0}$ <u>consists of continuous measures and</u> $\operatorname{supp}(\mu_t) = \mathbb{R}$ <u>for all</u> $t > 0$.

<u>Proof</u>. The first statement is clear so (5) defines a continuous negative definite function ψ. The Lévy measure of ψ in the Lévy-Khinchin formula for ψ is the discrete measure

$$\nu = \sum_{n=1}^{\infty} \frac{1}{2}\left(\varepsilon_{-a_n}+\varepsilon_{a_n}\right) ,$$

which is of infinite mass. By the theorem of Hartman and Wintner (cf. [6], [4] or [7]) it follows that each μ_t is a continuous measure. That $\mathrm{supp}(\mu_t) = \mathbb{R}$ for all $t > 0$ can be seen in several ways. It is known from [2] that $\mathrm{supp}(\mu_t)$ is a closed subgroup of $\mathbb{R}$ independent of t, so it is either of the form $a\mathbb{Z}$ or $\mathbb{R}$. Since μ_t is continuous we conclude that $\mathrm{supp}(\mu_t) = \mathbb{R}$. Another method is to remark that μ_t is equal to the infinite convolution

$$\mu_t = \mathop{*}_{n=1}^{\infty} \mu_t^{a_n} ,$$

and hence

$$\mathrm{supp}\mu_t = \lim_{n\to\infty}\left(\mathrm{supp}\left(\mu_t^{a_1}\right)+\cdots+\mathrm{supp}\left(\mu_t^{a_n}\right)\right) =$$

$$\lim_{n\to\infty}(a_1\mathbb{Z}+\cdots+a_n\mathbb{Z}) = \mathbb{R}.$$

For this way of reasoning we refer to the fundamental paper by Jessen and Wintner [8]. $\square$

<u>Remark</u>. By the Jessen-Wintner purity law (cf. [8]) we know that each μ_t in Lemma 3 is either continuous singular or absolutely continuous. There is not known any necessary and sufficient condition on (a_n) ensuring that μ_t is continuous singular.

We will now choose a special sequence (a_n) for which it can be seen that μ_t is singular. We will use Lemma 1.

Put $a_n = \frac{1}{n!}$, $n \in \mathbb{N}$. We will then see that

(6) $$\lim_{N\to\infty} \psi(2\pi N!) = 0 .$$

We find for $N \in \mathbb{N}$

$$\psi\left(2\pi N!\right) = \sum_{n=1}^{\infty} 1 - \cos\left(2\pi\frac{N!}{n!}\right) =$$

$$\left(1-\cos\frac{2\pi}{N+1}\right)\Big) + \left(1-\cos\left(\frac{2\pi}{(N+1)(N+2)}\right)\right) + \cdots .$$

Using the inequality $1 - \cos x \le \frac{1}{2}x^2$ we find

$$\psi(2\pi N!) \leq 2\pi^2\left(\frac{1}{(N+1)^2} + \frac{1}{((N+1)(N+2))^2} + \cdots\right)$$

$$\leq 2\pi^2 \sum_{k=1}^{\infty} (N+1)^{-2k} = 2\pi^2 \frac{1}{N^2+2N} ,$$

and (6) follows.

In particular

$$1 \geq \limsup_{|y|\to\infty} |\hat{\mu}_t(y)| \geq \lim_{N\to\infty} e^{-t\psi(2\pi N!)} = 1 ,$$

so μ_t is singular by Lemma 1.

Defining

(7) $$\kappa = \int_0^\infty e^{-t}\mu_t dt ,$$

κ is a probability measure with Fourier-Stieltjes transform

$$\hat{\kappa}(y) = \frac{1}{1+\psi(y)} , \quad y \in \mathbb{R} ,$$

and it follows by (6) and Lemma 1 that κ is singular. Since every μ_t is continuous, κ is continuous, and κ is a Hunt convolution kernel because $(e^{-t}\mu_t)_{t>0}$ is a convolution semigroup. The measure κ defined in (7) satisfies all the conditions of Theorem 2.

Remark. Recently G. Brown [5] has constructed a symmetric convolution semigroup $(\mu_t)_{t>0}$ on $\mathbb{R}$ of continuous singular measures for which $\text{supp}(\mu_t) = \mathbb{R}$ for all $t > 0$, and such that $\hat{\mu}_t$ vanishes at infinity for all $t > 0$. If κ is defined like in (7) from this semigroup then $\hat{\kappa}$ also vanishes at infinity, but I do not know whether κ is singular. The problem is that the nullset A_t, on which μ_t is concentrated, might vary with t, cf. the open problem of section 5.

3. Example 2.

Theorem 4. *There exists a Hunt convolution kernel* κ *on* $\mathbb{R}$ *with* $\kappa(\mathbb{R}) = \infty$ *such that*

(a) $\text{supp}(\kappa) = [0,\infty[$,

(b) κ *is continuous and singular*.

The construction needed for Theorem 4 is very similar to the construction in example 1, so we will leave out the details. Since the convolution semigroup we want to construct must be supported by $[0,\infty[$, it is convenient to use the Laplace transformation.

For $a > 0$ we consider the Poisson semigroup $(\pi_t^a)_{t>0}$ given as

$$\pi_t^a = e^{-t} \sum_{n=0}^{\infty} \frac{t^n}{n!} \varepsilon_{na} ,$$

and the corresponding Bernstein function (cf. [3]) is

$$f_a(s) = 1 - e^{-as} , \quad s \geq 0 .$$

Lemma 5. Let $a_1, a_2, \cdots$ be a sequence of positive numbers such that $\Sigma a_n < \infty$. Then the infinite series

(8) $$f(s) = \sum_{n=1}^{\infty} 1 - e^{-a_n s} , \quad s \geq 0$$

converges uniformly on compact subsets of $[0,\infty[$ and defines a Bernstein function f. The corresponding convolution semigroup $(\mu_t)_{t>0}$ on $\mathbb{R}$ consists of continuous measures and $\operatorname{supp}(\mu_t) = [0,\infty[$ for all $t > 0$.

The measure μ_t is the infinite convolution

$$\mu_t = \mathop{*}_{n=1}^{\infty} \pi_t^{a_n} .$$

The corresponding negative definite function ψ is related to the Bernstein function f by $\psi(y) = f(iy)$. Choosing again $a_n = \frac{1}{n!}$ it is easy to see that

$$\lim_{N\to\infty} \psi(2\pi N!) = 0 ,$$

so it follows in this special case that every μ_t is singular.

The Hunt convolution kernel

(9) $$\kappa = \int_0^\infty \mu_t dt$$

exists (cf. [3]) and $\kappa(\mathbb{R}) = \infty$, $\operatorname{supp}(\kappa) = [0,\infty[$. The continuity of κ is clear because every μ_t is continuous. Defining

(10) $$\kappa_1 = \int_0^\infty e^{-t} \mu_t dt$$

we see as in example 1 that the probability measure κ_1 is singular. There exists consequently a Borel nullset A such that $\kappa_1(CA) = 0$, and by (10) follows that $\mu_t(CA) = 0$ for Lebesgue almost all $t > 0$. By (9) we then get $\kappa(CA) = 0$, and it follows that κ is singular. This completes the construction needed for Theorem 4.

4. Example 3.

By $\mathbb{T}^\infty$ we denote a countable product of circle groups. The dual group $\mathbb{Z}^{(\infty)}$ consists of all sequences of integers which become zero eventually.

Theorem 6. *There exists a Hunt convolution kernel* κ *on* $\mathbb{T}^\infty$ *which is of local type and such that*

(a) $\operatorname{supp}(\kappa) = \mathbb{T}^\infty$,

(b) κ *is continuous singular*.

The following will be based on the paper [1], in which a class of Gaussian convolution semigroups on $\mathbb{T}^\infty$ is considered.

The Gaussian (Brownian) convolution semigroup on $\mathbb{T}$ is given by the density function

$$g_t(\theta) = 1 + 2\sum_{n=1}^{\infty} e^{-tn^2}\cos(n\theta), \quad t > 0, \quad \theta \in \mathbb{R}.$$

To every sequence $\mathcal{A} = (a_1, a_2, \cdots)$ of positive numbers we define a Gaussian convolution semigroup $(\mu_t^{\mathcal{A}})_{t>0}$ on $\mathbb{T}^\infty$ as the product measures

$$\mu_t^{\mathcal{A}} = \bigotimes_{k=1}^{\infty} g_{ta_k}, \quad t > 0.$$

The corresponding negative definite function ψ is the quadratic form $\psi: \mathbb{Z}^{(\infty)} \to \mathbb{R}$ given as

$$\psi(\underline{n}) = \sum_{k=1}^{\infty} a_k n_k^2 \quad \text{for} \quad \underline{n} = (n_1, n_2, \cdots) \in \mathbb{Z}^{(\infty)}.$$

For every $t > 0$ and every sequence $\mathcal{A}$ we have that $\mu_t^{\mathcal{A}}$ is a continuous measure and $\operatorname{supp}(\mu_t^{\mathcal{A}}) = \mathbb{T}^\infty$, cf. [1], where it is also proved that

$$\mu_t^{\mathcal{A}} \text{ is singular if and only if } \sum_{k=1}^{\infty} e^{-2ta_k} = \infty.$$

The measure

(11) $$\kappa^{\mathcal{A}} = \int_0^\infty e^{-t}\mu_t^{\mathcal{A}}\,dt$$

is a Hunt convolution kernel on T^∞ of local type such that $\operatorname{supp}(\kappa^{\mathcal{A}}) = T^\infty$, and $\kappa^{\mathcal{A}}$ is continuous.

If $\Sigma e^{-2ta_k} = \infty$ for all $t > 0$, then all the measures $\mu_t^{\mathcal{A}}$ are singular and one would expect $\kappa^{\mathcal{A}}$ to be singular. We have not been able to prove this, again because the nullset A_t, which carries $\mu_t^{\mathcal{A}}$, might vary with t in a way which is difficult to control.

However, if $\mathcal{A}$ is a bounded sequence, we can prove that $\kappa^{\mathcal{A}}$ is singular. We will do it here only in the special case where $\mathcal{A} = (1,1,\cdots)$. In this case we write μ_t and κ in stead of $\mu_t^{\mathcal{A}}$ and $\kappa^{\mathcal{A}}$.

<u>Proposition 7</u>. <u>The Hunt convolution kernel</u> κ <u>given by</u> (11) <u>is singular in the case where</u> $\mathcal{A}$ <u>is the sequence</u> $1,1,\cdots$.

The normalized Haar measure on T is denoted $d\theta$, and the normalized Haar measure on T^∞ is denoted m. For $k \in \mathbb{N}$ and $t > 0$ we define a subset $A_{k,t}$ of T^∞ as

(12) $$A_{k,t} = \{\underline{\theta} \in T^\infty \mid \prod_{j=1}^{k} g_t(\theta_j) > 1\} .$$

Denoting by π_k the projection of T^∞ onto T^k given as $\pi_k(\underline{\theta}) = (\theta_1,\cdots,\theta_k)$ we find

$$m(A_{k,t}) = \int_{\pi_k(A_{k,t})} d\theta_1\cdots d\theta_k < \int_{\pi_k(A_{k,t})} \left(\prod_{j=1}^{k} g_t(\theta_j)\right)^{\frac{1}{2}} d\theta_1\cdots d\theta_k \le \rho(t)^k,$$

where

(13) $$\rho(t) = \int_T \sqrt{g_t(\theta)}\,d\theta ,$$

cf. [1]. It is known from [1] that $0 < \rho(t) < 1$.

We now define

(14) $$A_t = \limsup_{k\to\infty} A_{k,t} := \bigcap_{n=1}^{\infty} \bigcup_{k=n}^{\infty} A_{k,t} , \quad t > 0 ,$$

and we get

$$m(A_t) = \lim_{n\to\infty} m\left(\bigcup_{k=n}^{\infty} A_{k,t}\right) \le \lim_{n\to\infty} \sum_{k=n}^{\infty} \rho(t)^k = 0 ,$$

so every A_t is a Borel nullset of T^∞.

Let $t > 0$ be fixed. Since

$$\frac{g_s(\theta)}{g_t(\theta)} \to 1 \quad \text{for} \quad s \to t, \quad \text{uniformly for} \quad \theta \in T ,$$

there exists an open neighbourhood U_t of t in $]0,\infty[$ such that

$$\frac{g_s(\theta)}{g_t(\theta)} \leq \rho(t)^{-\frac{1}{2}} \quad \text{for} \quad \theta \in T , \quad s \in U_t .$$

For $s \in U_t$ we find

$$\mu_s(CA_{k,t}) = \int_{T^k \setminus \pi_k(A_{k,t})} g_s(\theta_1)\cdots g_s(\theta_k)d\theta_1\cdots d\theta_k$$

$$= \int_{T^k \setminus \pi_k(A_{k,t})} \frac{g_s(\theta_1)}{g_t(\theta_1)}\cdots\frac{g_s(\theta_k)}{g_t(\theta_k)} g_t(\theta_1)\cdots g_t(\theta_k)d\theta_1\cdots d\theta_k$$

$$\leq \int_{T^k \setminus \pi_k(A_{k,t})} \rho(t)^{-k/2} \left(g_t(\theta_1)\cdots g_t(\theta_k)\right)^{\frac{1}{2}} d\theta_1\cdots d\theta_k \leq \rho(t)^{k/2} .$$

It follows that

$$\mu_s(CA_t) = \mu_s\left(\bigcup_{n=1}^{\infty} \bigcap_{k=n}^{\infty} CA_{k,t}\right) \leq \sum_{n=1}^{\infty} \mu_s\left(\bigcap_{k=n}^{\infty} CA_{k,t}\right) = 0 ,$$

because

$$\mu_s\left(\bigcap_{k=n}^{\infty} CA_{k,t}\right) \leq \inf_{k\geq n} \mu_s\left(CA_{k,t}\right) \leq \inf_{k\geq n} \rho(t)^{k/2} = 0 .$$

The system $(U_t)_{t>0}$ of open neighbourhoods in $]0,\infty[$ covers $]0,\infty[$, so there exists a sequence (t_n) of positive numbers such that

$$\bigcup_{n=1}^{\infty} U_{t_n} =]0,\infty[.$$

The set

$$A = \bigcup_{n=1}^{\infty} A_{t_n} \tag{15}$$

is a Borel nullset of T^∞, and for every $s > 0$ there exists $n \in \mathbb{N}$ such that $s \in U_{t_n}$, and hence

$$\mu_s(CA) \leq \mu_s(CA_{t_n}) = 0 .$$

This means that A is a Borel nullset which carries all the measures μ_s, $s > 0$, and by (11) it is then clear that also κ is carried by A, hence κ is singular. ▯

Remark. The above proof is closely related to the method used by Kakutani in the fundamental paper [9].

With the proof of Proposition 7 we have also established the result in Theorem 6.

Remark. K. Schmidt has drawn my attention to the fact that Proposition 7 is a special case of the following (unpublished) theorem, which can be proved using results on unique ergodicity due to Krieger and Schmidt (cf. [10]):

Let X be a compact (abelian) group and $(\mu_t)_{t>0}$ a convolution semigroup of probability measures on X such that μ_t is absolutely continuous with respect to Haar measure on X for all $t > 0$. For each $t > 0$ we denote by μ_t^∞ the product measure of countably many copies of μ_t on the product group G^∞, which is the product of countably many copies of G. Then $(\mu_t^\infty)_{t>0}$ is a convolution semigroup on G^∞, and each μ_t^∞ is singular with respect to Haar measure on G^∞ by the theorem of Kakutani [9]. Furthermore, and now the unique ergodicity is used, the Hunt convolution kernel

$$\kappa = \int_0^\infty e^{-t}\mu_t^\infty dt$$

on G^∞ is singular.

As mentioned above it can be proved that $\kappa^{\mathcal{A}}$, given by (11), is singular if $\mathcal{A} = (a_1, a_2, \cdots)$ is a bounded sequence. The idea is the following: If $\inf\{a_n \mid n \in \mathbb{N}\} = 0$, then one can conclude that $\kappa^{\mathcal{A}}$ is singular by means of Lemma 1 just as in example 1, because there exists a sequence $\underline{n}_k \in \mathbb{Z}^{(\infty)}$ such that $\lim_{k\to\infty} \psi(\underline{n}_k) = 0$. If $\inf_{n\in\mathbb{N}} a_n > 0$ there exist two numbers $0 < \alpha < \beta$ such that $\alpha \leq a_n \leq \beta$

for all $n \in \mathbb{N}$. Then using that the function $\rho:]0,\infty[\to]0,1[$, defined in (13), is increasing, a construction similar to that of Proposition 7 gives the result.

5. An open problem.

Motivated by the preceding examples it would be interesting to solve the following problem:

Let $(\mu_t)_{t>0}$ be a symmetric convolution semigroup consisting of singular measures, and define the Hunt convolution kernel κ by

$$\kappa = \int_0^\infty e^{-t}\mu_t dt.$$

Is it then true that κ is singular?

It amounts to find a Borel nullset A such that μ_t is concentrated on A for Lebesgue almost all $t > 0$. I have not been able to settle the question. Notice that the construction in example 4 gave much more: There existed a nullset A of T^∞ such that μ_t is concentrated on A for all $t > 0$.

References.

[1] Berg, C.: Potential theory on the infinite dimensional torus. Inventiones math. 32, 49-100 (1976).

[2] Berg, C.: On the support of the measures in a symmetric convolution semigroup. Math. Z. 148, 141 - 146 (1976).

[3] Berg, C., Forst, G.: Potential theory on locally compact abelian groups. Erg. der Math. Bd. 87, Berlin-Heidelberg-New York: Springer 1975.

[4] Blum, J.R., Rosenblatt, M.: On the structure of infinitely divisible distributions. Pac. J. Math. 9 , 1-7 (1959).

[5] Brown, G.: Singular infinitely divisible distributions whose characteristic functions vanish at infinity. Math. Proc. Camb. Phil. Soc. 82, 277-287 (1977).

[6] Hartman, P., Wintner, A.: On the infinitesimal generators of integral convolutions. Amer. J. Math. 64, 273-298 (1942).

[7] Huff, B.W.: On the continuity of infinitely divisible distributions. Sankhya 34, 443-446 (1972).

[8] Jessen, B., Wintner, A.: Distribution functions and the Riemann zeta function. Trans. Amer. Math. Soc. 38, 48-88 (1935).

[9] Kakutani, S.: Equivalence of infinite product measures. Ann. Math. 49, 214-224 (1948).

[10] Schmidt, K.: Unique ergodicity for quasi-invariant measures. Preprint february 1978 from Math. Inst. University of Warwick, Coventry.

Christian Berg
Matematisk Institut
Universitetsparken 5
2100 Copenhagen Ø
Denmark

Infinitely divisible probability measures and potential kernels

Christian Berg and Gunnar Forst

Introduction.

Let G be a locally compact abelian group. A potential kernel on G is a positive measure on G of the form

$$\kappa = \int_0^\infty \mu_t dt , \tag{1}$$

where $(\mu_t)_{t>0}$ is a vaguely continuous convolution semigroup of sub-probabilities on G which is transient in the sense that the integral in (1) converges vaguely and defines a Radon measure.

One of the simplest examples of a potential kernel is the Heaviside kernel Y on $\mathbb{R}$,

$$Y = 1_{]0,\infty[}(t)dt = \int_0^\infty \varepsilon_t dt ,$$

where ε_t is the Dirac measure at t; Y is thus the restriction of Lebesgue measure to the positive half-line. The measure Y has infinite mass and it is infinitely divisible in the sense that for $n \in \mathbb{N}$ there exists a positive measure ν_n on $\mathbb{R}$ such that $Y = \nu_n * \cdots * \nu_n$ (n convolution factors). In fact, for $\alpha \in]0,1]$ the positive measure

$$Y^\alpha = \frac{1}{\Gamma(\alpha)} t^{\alpha-1} 1_{]0,\infty[}(t)dt$$

is the α-th fractional power (with respect to convolution) of Y, i.e. we have $Y^1 = Y$ and

$$Y^\alpha * Y^\beta = Y^{\alpha+\beta} \quad \text{for} \quad \alpha,\beta,\alpha+\beta \in]0,1]. \tag{2}$$

Each of the measures Y^α, for $\alpha \in]0,1]$, is a potential kernel on $\mathbb{R}$ supported by $\mathbb{R}_+$.

Now one can prove the following: If κ is a potential kernel on G with semigroup $(\mu_t)_{t>0}$ and τ is a potential kernel on $\mathbb{R}$ supported by $\mathbb{R}_+$, then the measure κ_τ which is well-defined by the vague integral

$$\kappa_\tau = \int_0^\infty \mu_t d\tau(t) \tag{3}$$

is again a potential kernel on G.

For $\tau = \gamma$ equation (3) reduces to (1). This is the method of <u>subordination</u>, cf. Bochner [2], see also (8) below.

In particular, for $\alpha \in]0,1]$ the measure

$$\kappa^{\alpha} = \int_0^{\infty} \mu_t d\gamma^{\alpha}(t),$$

the "<u>α-th fractional power of</u> κ", is a potential kernel on G, and (using (2)) the family $(\kappa^{\alpha})_{\alpha \in]0,1]}$ has the property that $\kappa = \kappa^1$ and

$$\kappa^{\alpha} * \kappa^{\beta} = \kappa^{\alpha+\beta} \quad \text{for} \quad \alpha, \beta, \alpha+\beta \in]0,1].$$

For details of the above mentioned facts, we refer to [1].

It follows that any potential kernel κ on G is <u>infinitely divisible</u>, in the sense that for each $n \in \mathbb{N}$ there exists an n-th convolution root of κ which is a positive measure (namely $\kappa^{1/n}$), and moreover that κ has the special property that every n-th root is again a potential kernel. If κ is a probability measure then κ is <u>infinitely divisible in the usual sense</u>.

The above considerations indicate that it might be profitable to examine relationships between potential kernels and infinitely divisible probability measures. The above fact, that the measure κ_{τ} given by (3) is a potential kernel, is quite analogous to the well-known fact, cf. Feller [4], that if $(\mu_t)_{t>0}$ is a vaguely continuous convolution semigroup of probability measures on G and τ is an infinitely divisible probability measure on $\mathbb{R}$ supported by $\mathbb{R}_+$, then the probability measure on G defined by

$$\int_0^{\infty} \mu_t d\tau(t)$$

is an infinitely divisible probability measure on G. Furthermore, it seems that most of the known explicit conditions on a probability measure μ, which ensure that μ is infinitely divisible, also imply that μ is a potential kernel.

The purpose of the following is to review some results with special emphasis on similarities between potential kernels and infinitely divisible probabilities and also to interpret results of potential theory in the spirit of probability theory.

In the first two sections we introduce different classes of measures on $\mathbb{R}$ supported by respectively $[0,\infty[$ and $\mathbb{Z}_+$, and section 2 ends with a rather striking relationship between the corresponding classes on $[0,\infty[$ and $\mathbb{Z}_+$.

In section 3 we use the results from section 1 to associate various convex sets of potential kernels with a given potential kernel. This leads to two concrete classes of potential kernels on $\mathbb{R}^3$ and $\mathbb{R}$ respectively.

In section 4 we mention some recent concrete examples of potential kernels in $\mathbb{R}^n$, and finally we give a short summary of Thorin's work on generalized Γ-convolutions.

1. Measures on $\mathbb{R}_+$.

Let P denote the set of potential kernels on $\mathbb{R}_+$, i.e. measures τ of the form (1) on $G = \mathbb{R}$ with $\operatorname{supp}\tau \subseteq \mathbb{R}_+$, and let S (resp. H) be the set of non-zero measures on $\mathbb{R}$ which can be written

$$\tau = a\varepsilon_0 + k(t)dt , \tag{4}$$

where $a \geq 0$ (and ε_0 the Dirac measure at 0) and $k:]0,\infty[\to [0,\infty[$ is a completely monotone (resp. decreasing and logarithmic convex) function such that (4) defines a Radon measure, i.e.

$$\int_0^1 k(t)dt < \infty .$$

Then we have

$$S \subseteq H \subseteq P .$$

Here $S \subseteq H$ is clear by the Cauchy-Schwarz inequality, and $H \subseteq P$ has been known in potential theory for some time. For a proof see Ito [13], Hirsch [10] or [1].

The set P is a cone, but P is not convex, while S and H are convex cones, and S,H and P are all vaguely closed in the set of non-zero measures on $\mathbb{R}_+$.

Denoting by P^1 (resp. S^1 and H^1) the set of probability measures in P (resp. S and H) and by U the set of infinitely divisible probability measures on $\mathbb{R}$ supported by $\mathbb{R}_+$, we have by the introduction

$$P^1 \subseteq U ,$$

and also

$$S^1 \subseteq H^1 \subseteq P^1 \subseteq U .$$

Here it should be pointed out that $0 \in \operatorname{supp}\tau$ for $\tau \in P$ while this need not be the case for $\tau \in U$. The inclusion

$$\{\tau \in S^1 \mid \tau(\{0\}) = 0\} \subseteq U$$

is known as the Goldie-Steutel theorem, cf. Goldie [8] and Steutel [23].

One of the main tools in the discussion of the above classes of measures is the Laplace transformation. A C^∞-function $f\colon]0,\infty[\to [0,\infty[$ is called a Bernstein function if Df is completely monotone, and we denote by B the set of Bernstein functions and by B_0 the set $B_0 = \{f \in B \mid \lim_{s\to 0} f(s) = 0\}$. Then we have, cf. Feller [4] and [1], for a positive measure κ on $\mathbb{R}_+$ (denoting the Laplace transform of κ by $L\kappa$) that

$$\kappa \in U \Leftrightarrow \exists f \in B_0 \colon L\kappa = \exp(-f) ,$$

$$\kappa \in P \Leftrightarrow \exists f \in B\setminus\{0\} \colon L\kappa = \frac{1}{f} .$$

The set H is more complicated to characterize by Laplace transformation, cf. (7) below, while the set S can be characterized in terms of Stieltjes transforms, i.e. functions h of the form

$$h(s) = a + \int_0^\infty \frac{1}{s+t} d\mu(t) \quad \text{for} \quad s > 0 , \tag{5}$$

where $a \geq 0$ and μ is a non-negative measure on $[0,\infty[$ such that the integral in (5) converges for every $s > 0$. We have, cf. [1],

$$\kappa \in S \Leftrightarrow L\kappa \text{ is a non-zero Stieltjes transform.}$$

The Bernstein functions admit the following integral representation: f is a Bernstein function if and only if f can be written

$$f(s) = a + bs + \int_0^\infty (1-e^{-xs}) d\nu(x) \quad \text{for} \quad s > 0 , \tag{6}$$

where (a,b,ν) is a uniquely determined triple of numbers $a,b \geq 0$ and a positive measure ν on $]0,\infty[$ such that

$$\int_0^{\infty} \frac{x}{1+x} d\nu(x) < \infty .$$

The representation (6) of Bernstein functions is a special case of the <u>Lévy-Khinchin representation</u>, which in this simple case is an easy consequence of Bernstein's representation of completely monotone functions, and seems to go back to Schoenberg [22]. See also [1].

It follows in particular from (6) that every (<u>standard</u>) p-<u>function</u> in the sense of Kingman [18] is density for an element of P. Conversely $\kappa \in P$ has a continuous density $p: [0,\infty[\to]0,\infty[$ with respect to Lebesgue measure if and only if, in the representing triple (a,b,ν) for the Bernstein function f such that $L\kappa = \frac{1}{f}$, we have $b > 0$, and in the affirmative case, this density is proportional with a (standard) p-function, cf. Kingman [18]. See also [7].

Recently van Harn [9] considered a class F_∞ of infinitely divisible probability measures on $\mathbb{R}_+$ and most (if not all) of the properties of F_∞ noted in [9] are simple consequences of the fact that

$$\{\mu \in F_\infty | 0 \in \text{supp}(\mu)\} = P^1 ,$$

see e.g. Theorem 5.3 in [9].

2. Measures on $\mathbb{Z}_+$.

Let us start with a description of some classes of measures on $\mathbb{Z}_+ = \{0,1,2,\ldots\}$ which are analogous to the classes considered on $\mathbb{R}_+$. We will identify a positive measure μ on $\mathbb{Z}_+$ with the sequence $\underline{a} = (a_n)_{n\geq 0}$ such that

$$\mu = \sum_{n=0}^{\infty} a_n \varepsilon_n$$

(with ε_n point-mass at n).

Let

$$U_d = \{\underline{a} | \ \underline{a} \in U \text{ and } a_0 > 0\}$$

$$P_d = \{\underline{a} | \ \underline{a} \in P\}.$$

Note that if $\underline{a} \in P$ then $a_0 > 0$. Furthermore let H_d (resp. S_d) be the set of non-zero measures $\underline{a}$ on $\mathbb{Z}_+$ such that $\underline{a}$ is <u>decrea-</u>

sing and logarithmic convex, i.e. $a_{n-1} \geq a_n$ and $a_n^2 \leq a_{n-1}a_{n+1}$ for $n = 1,2,\ldots$ (resp. completely monotone). The sequences $\underline{a}$ in H_d with $a_0 = 1$ are called Kaluza sequences in Kingman [18].

Finally let P_d^1 (resp. S_d^1 and H_d^1) be the subset of P_d (resp. S_d and H_d) consisting of probability measures.

It is easy to see that $\underline{a}$ is a renewal sequence, cf. Kingman [18], if and only if $\underline{a} \in P_d$ and $a_0 = 1$, cf. [6], or in other words a sequence $\underline{a}$ belongs to P_d if and only if it is a positive multiple of a renewal sequence.

As above we have

$$P_d^1 \subseteq U_d \quad \text{and} \quad S_d \subseteq H_d \subseteq P_d .$$

Here $P_d^1 \subseteq U_d$ is a special case of $P^1 \subseteq U$, and $S_d \subseteq H_d$ follows by the Cauchy-Schwarz inequality and Hausdorff's representation of completely monotone sequences as moment sequences. Finally $H_d \subseteq P_d$ goes back to Kaluza [16], using the above mentioned relation between P_d and the set of renewal sequences, see also Lamperti [21].

Another curious fact is the following: Let μ be a positive measure on $\mathbb{R}_+$ for which the Laplace transform $L\mu(s)$ exists for $s > 0$, and define for $s > 0$ the sequence $\underline{a}(s)$ by

$$\underline{a}(s)_n = (-1)^n s^n \frac{1}{n!} D^n(L\mu)(s) \quad \text{for} \quad n \geq 0.$$

Then we have:

$$\begin{aligned} \mu \in U &\Leftrightarrow \forall s > 0 : \underline{a}(s) \in U_d \\ \mu \in P &\Rightarrow \forall s > 0 : \underline{a}(s) \in P_d \\ \mu \in H &\Leftrightarrow \forall s > 0 : \underline{a}(s) \in H_d \\ \mu \in S &\Leftrightarrow \forall s > 0 : \underline{a}(s) \in S_d . \end{aligned} \tag{7}$$

Here the first equivalence is essentially a combination of a result of Goldie [8] and a result of Katti [17]; the third is in Hirsch [10] while the rest is from [6].

3. Convex sets of potential kernels.

Let $\kappa = \int_0^\infty \mu_t dt$ be a potential kernel on G. For $\tau \in P$ we have by (3) defined a potential kernel κ_τ on G called the potential kernel subordinated κ by means of τ. If $\tau \in P$ has the form $\tau = \int_0^\infty \eta_t dt$ the family $(\mu_t^\tau)_{t>0}$ defined by

$$\mu_t^\tau = \int_0^\infty \mu_s d\eta_t(s) \quad , \quad t > 0 \tag{8}$$

is a vaguely continuous convolution semigroup of subprobabilities on G and one finds

$$\kappa_\tau = \int_0^\infty \mu_t^\tau dt \ .$$

The semigroup $(\eta_t)_{t>0}$ supported by $[0,\infty[$ is often called a <u>subordinator</u>.

We define three sets of potential kernels associated with κ, namely

$$P(\kappa) = \{\kappa_\tau \mid \tau \in P\},$$
$$H(\kappa) = \{\kappa_\tau \mid \tau \in H\},$$
$$S(\kappa) = \{\kappa_\tau \mid \tau \in S\}.$$

Then clearly $S(\kappa) \subseteq H(\kappa) \subseteq P(\kappa)$, and $S(\kappa)$ and $H(\kappa)$ are convex cones while $P(\kappa)$ is a cone but in general not convex. The set $S(\kappa)$ is particularly important in potential theory (cf. Itô [14]) and is often called the <u>Stieltjes cone associated with</u> κ. One can prove that $S(\kappa) \cup \{0\}$ is vaguely closed.

The sets $H_1(\kappa)$ (resp. $S_1(\kappa)$) of probability measures in $H(\kappa)$ (resp. $S(\kappa)$) are convex sets of infinitely divisible probabilities associated with the potential kernel κ.

Using the measures

$$\tau_p = e^{-pt} 1_{]0,\infty[}(t)dt \quad , \quad p > 0,$$

from S we get the <u>resolvent family</u> $(\kappa_p)_{p>0}$ <u>for</u> κ (or for $(\mu_t)_{t>0}$):

$$\kappa_p := \kappa_{\tau_p} = \int_0^\infty e^{-pt}\mu_t dt \quad , \quad p > 0.$$

It is convenient to put $\kappa_0 = \kappa$, and notice that $\lim_{p\to 0} \kappa_p = \kappa$. We shall see that $S(\kappa)$ is the convex cone generated by the family $(\kappa_p)_{p\geq 0}$ and ε_0. In fact, a measure $\tau \in S$ has the form

$$\tau = a\varepsilon_0 + L\sigma(t)dt \ ,$$

where $a \geq 0$ and σ is a positive measure on $[0,\infty[$ such that

$$\int_1^\infty \frac{d\sigma(t)}{t} < \infty ,$$

hence

$$\kappa_\tau = a\varepsilon_0 + \int_0^\infty (\mu_t \int_0^\infty e^{-pt} d\sigma(p)) dt = a\varepsilon_0 + \int_0^\infty \kappa_p d\sigma(p) ,$$

so it follows that

$$S(\kappa) = \{a\varepsilon_0 + \int_0^\infty \kappa_p d\sigma(p)\} \tag{9}$$

with a and σ as above.

<u>Example</u>. Let $G = \mathbb{R}^n$, and consider the <u>Brownian convolution semi-group</u> $(\mu_t^{(n)})_{t>0}$ on $\mathbb{R}^n$ of normal distributions, i.e.

$$\mu_t^{(n)} = (4\pi t)^{-\frac{n}{2}} \exp(-\frac{\|x\|^2}{4t}) dx , \quad x \in \mathbb{R}^n, \quad t > 0.$$

The vague integral $\int_0^\infty \mu_t^{(n)} dt$ converges only for $n \geq 3$ and its value is the <u>Newtonian kernel</u> $N^{(n)}$ given as

$$N^{(n)} = \frac{1}{(n-2)\alpha_n} \frac{1}{\|x\|^{n-2}} dx ,$$

where α_n denotes the (n-1)-dimensional surface area of the unit sphere.

For $n = 3$ we get in particular

$$N^{(3)} = \frac{1}{4\pi} \frac{1}{\|x\|} dx ,$$

and the resolvent family $(N_p^{(3)})_{p>0}$ is given by

$$N_p^{(3)} = \frac{1}{4\pi\|x\|} e^{-\sqrt{p}\|x\|} dx , \quad p > 0.$$

It follows from (9) that the Stieltjes cone $S(N^{(3)})$ can be described as the set of non-zero measures κ of the form

$$\kappa = a\varepsilon_0 + \frac{f(\|x\|)}{\|x\|} dx , \tag{10}$$

where $a \geq 0$ and $f :]0,\infty[\to [0,\infty[$ is a completely monotone function satisfying

$$\int_0^1 f(t)\,t\,dt < \infty.$$

In particular, any probability measure of the form (10) is infinitely divisible.

For $n \geq 4$ the expressions for the resolvent family $(N_p^{(n)})_{p>0}$ for the Newtonian kernel $N^{(n)}$ are rather complicated. We mention that $S(N^{(n)})$ nevertheless has been described in the following beautiful way by Itô [14]:

A positive non-zero measure κ on $\mathbb{R}^n$ belongs to $S(N^{(n)})$ if and only if the following three conditions are satisfied:

(i) κ is rotation invariant,

(ii) κ tends to zero at infinity, i.e. $\lim_{\|x\|\to\infty} \kappa(B(x,1)) = 0$,

where $B(x,1)$ is the open ball with center x and radius 1.

(iii) $\Delta^k \kappa \geq 0$ in $\mathbb{R}^n \setminus \{0\}$ in the sense of distributions for all $k \geq 1$, where Δ^k is the k-fold iteration of the Laplace operator Δ.

The formalism of subordination extends to some non-transient semigroups. Let $(\mu_t)_{t>0}$ be a vaguely continuous convolution semigroup of probabilities on G for which the vague integral $\int_0^\infty \mu_t dt$ does not (necessarily) converge.

The integral

$$\int_0^\infty \mu_t d\tau(t) \tag{11}$$

may however converge for some $\tau \in P$ (but not for $\tau = Y$). This is the case e.g. for all $\tau \in P$ of finite total mass. Whenever the integral (11) converges its value is a potential kernel. In fact, (8) defines a convolution semigroup $(\mu_t^\tau)_{t>0}$, and it is clear that

$$\int_0^\infty \mu_t d\tau(t) = \int_0^\infty \mu_t^\tau dt .$$

This remark can be applied to the Brownian convolution semigroup in the case $n = 1$ to give a simple proof of the following result:

Let $f :]0,\infty[\to]0,\infty[$ be a completely monotone function satisfying

$$\int_0^1 f(t)dt < \infty \quad \text{and} \quad \lim_{t\to\infty} f(t) = 0.$$

Then the symmetric measure $\rho = f(|x|)dx$ on $\mathbb{R}$ is a potential kernel, and in particular infinitely divisible if $\rho(\mathbb{R}) = 1$.

In fact, for $n = 1$ the resolvent family $(N_p^{(1)})_{p>0}$ for the Brownian convolution semigroup $(\mu_t^{(1)})_{t>0}$ is given as

$$N_p^{(1)} = \frac{1}{2\sqrt{p}}\, e^{-|x|\sqrt{p}}dx \quad , \quad p > 0.$$

There exists a positive measure σ_1 on $]0,\infty[$ such that

$$f(s) = \int_0^\infty e^{-st}d\sigma_1(t) \quad \text{and} \quad \int_1^\infty \frac{d\sigma_1(s)}{s} < \infty .$$

By $\tilde{\sigma}_1$ we denote the image measure of σ_1 under the mapping $x \to x^2$, and for $s > 0$ we then find

$$f(s) = \int_0^\infty e^{-s\sqrt{t}}d\tilde{\sigma}_1(t) = \int_0^\infty \frac{1}{2\sqrt{t}}\, e^{-s\sqrt{t}}d\sigma(t) \quad ,$$

where $d\sigma(t) = 2\sqrt{t}d\tilde{\sigma}_1(t)$. It follows that

$$\int_1^\infty \frac{1}{s}\, d\sigma(s) = 2\int_1^\infty \frac{1}{s}\, d\sigma_1(s) < \infty$$

and

$$\rho = \int_0^\infty N_p^{(1)} d\sigma(p) = \int_0^\infty \mu_t^{(1)} L\sigma(t)dt \ ,$$

hence ρ is of the form (11).

The above result is actually a special case of the following statement which is essentially due to Kishi [19], and may be considered as an extension of the Goldie-Steutel Theorem.

Let $f,g :]0,\infty[\to [0,\infty[$ be completely monotone functions, not both zero, and satisfying

$$\int_0^1 f(t)dt < \infty \ , \quad \int_0^1 g(t)dt < \infty \quad \text{and} \quad \lim_{t\to\infty} f(t)g(t) = 0.$$

Then the measure

$$\kappa = f(t)1_{]0,\infty[}(t)dt + g(-t)1_{]-\infty,0[}(t)dt$$

on $\mathbb{R}$ is a potential kernel, and in particular infinitely divisible if $\kappa(\mathbb{R}) = 1$.

In fact, Kishi proves in [19] that κ satisfies the complete maximum principle, and to conclude that κ is a potential kernel one may proceed in two different ways: One can establish that κ has the dominated convergence property of Kishi which by a theorem of Kishi, cf. [20], gives that κ is a potential kernel. Or even easier, one can remark that κ has the renewal property, which by a theorem of Hirsch and Taylor, cf. [11] implies that κ is a potential kernel.

In all the examples of infinitely divisible probability distributions in this section one has the possibility of "putting" some mass at the origin. In general it is not true that the convex combination $\alpha\varepsilon_0 + (1-\alpha)\mu$, where $0 < \alpha < 1$, is infinitely divisible whenever μ is. It is however true, when μ is a potential kernel, due to the following result established by Deny around 1960; for a proof see e.g. [1] p. 100:

Let κ be a potential kernel on G and $\alpha,\beta > 0$. Then $\alpha\varepsilon_0 + \beta\kappa$ is a potential kernel.

4. Various further results.

In the case of $\mathbb{R}^n$, $n \geq 2$, very little is known about general conditions ensuring either that a measure is a potential kernel or infinitely divisible.

We mention the paper [12] by Horn and Steutel about infinitely divisible distributions in $\mathbb{R}^n$.

In recent papers by Itô [15] and Flandrin [5] one can find sufficient conditions on a rotation invariant measure of the form $\kappa = f(\|x\|)dx$, where $f :]0,\infty[\to [0,\infty[$ is a continuous function, ensuring that κ is a potential kernel. The conditions in both papers are complicated, and while a certain smoothness of f is required in Itô's paper, this is not the case in Flandrin's paper.

In this paper we have tried to throw some light on infinitely divisible distributions by means of the theory of potential kernels. Let us conclude by mentioning some recent results on infinitely divisible distributions which are not covered by this aspect.

In the paper [26] Thorin introduced a class of infinitely divisible distributions, which he called generalized Γ-convolutions, and he proved that the Pareto distribution belongs to this class (cf. [26]).

In a subsequent paper [27] the lognormal distribution was shown to be a generalized Γ-convolution, hence infinitely divisible.

The Γ-distribution $\Gamma_{a,p}$, where $a,p > 0$, is given as

$$\Gamma_{a,p} = \frac{1}{\Gamma(p)} a^{-p} t^{p-1} \exp\left(-\frac{t}{a}\right) 1_{]0,\infty[}(t)\,dt.$$

Its Laplace transform is

$$L(\Gamma_{a,p})(s) = e^{-p \log(1+as)} , \quad s > 0 ,$$

and since $p \log(1+as)$ is a Bernstein function we find that $\Gamma_{a,p} \in U$. Moreover $\Gamma_{a,p} \in P^1$ for $p \in]0,1]$ as is easily seen.

A probability distribution μ on $[0,\infty[$ is called a <u>generalized Γ-convolution</u> if it is weak limit of distributions of the form

$$\Gamma_{a_1,p_1} * \cdots * \Gamma_{a_n,p_n} * \varepsilon_b ,$$

with $a_1,\dots,a_n,p_1,\dots,p_n > 0$ and $b \geq 0$.

It follows that μ is a generalized Γ-convolution if and only if

$$L\mu(s) = e^{-f(s)} , \quad s > 0 ,$$

with f belonging to the convex cone

$$T = \left\{ bs + \int_0^\infty \log(1+sx)\,d\sigma(x) \right\} ,$$

where $b \geq 0$ and σ is a positive measure on $]0,\infty[$ for which the integral converges for all $s > 0$, or equivalently

$$\int_0^\infty \log(1+x)\,d\sigma(x) < \infty .$$

Note that T is a subcone of B_0, in fact it is not difficult to see that

$$T = \{ f \in B_0 \mid Df \in S \} .$$

Using the technique developped by Thorin in [27], L. Bondesson has given sufficient conditions on densities of the form $Cx^\beta h(x)$ on $]0,\infty[$, ensuring that they define infinitely divisible distributions, cf. [3]. Here $\beta > -1$ and h is completely monotone, but further technical conditions on h are needed.

References.

[1] Berg, C. and Forst, G.: Potential Theory on Locally Compact Abelian Groups. Berlin: Springer 1975.

[2] Bochner, S.: Harmonic analysis and the theory of probability. Berkeley and Los Angeles: University of California Press 1955.

[3] Bondesson, L.: A general result on infinite divisibility. Preprint. University of Lund. Sweden 1977.

[4] Feller, W.: An Introduction to Probability Theory and Its Applications. Vol. II, 2 ed. New York: Wiley 1970.

[5] Flandrin, E.: Détermination explicite de noyaux de Hunt sur $\mathbb{R}^n$. C.R. Acad. Sc. Paris 284, 1451 - 1453. (1977). (For more details see: Thèse: Title as above, Fac. des. Sciences d'Orsay Univ. de Paris XI, 1977).

[6] Forst, G.: A Characterization of Potential Kernels on the Positive Half-Line. Z. Wahrscheinlichkeitstheorie verw. Gebiete 41, 335 - 340 (1978).

[7] Forst, G.: Multiples of renewal functions. Remark on a result of D.J. Daley. This volume.

[8] Goldie, C.: A class of infinitely divisible random variables. Proc. Camb. Phil. Soc. 63, 1141 - 1143. (1967).

[9] van Harn, K.: Degrees of infinite divisibility, and a relation with p-functions. Preprint. Eindhoven University of Technology. The Netherlands. 1977.

[10] Hirsch, F.: Familles d'opérateurs potentiels. Ann. Inst. Fourier 25^{3-4}, 263 - 288 (1975).

[11] Hirsch, F. and Taylor, J.C.: Renouvellement et existence de résolvantes. Séminaire de Théorie du Potentiel. Paris. To appear in: Lecture Notes in Mathematics. Berlin: Springer 1978.

[12] Horn, R.A. and Steutel, F.W.: On multivariate infinitely divisible distributions. Stochastic Processes Appl. 6, 139 - 151 (1978).

[13] Itô, M.: Sur une famille sous-ordonnée au noyau de convolution de Hunt donné. Nagoya Math. J. 51, 45 - 56 (1973).

[14] Itô, M.: Sur les cônes convexes de Riesz et les noyaux de convolution complètement sous-harmoniques. Nagoya Math. J. 55, 111 - 144 (1974).

[15] Itô, M.: Les noyaux de Frostman-Kunugui et les noyaux de Dirichlet. Ann. Inst. Fourier 27^3, 45 - 95 (1977).

[16] Kaluza, T.: Über die Koeffizienten reziproker Potenzreihen. Math. Z. 28, 161 - 170 (1928).

[17] Katti, S.K.: Infinite divisibility of integer valued random variables. Ann. Math. Stat. 38, 1306 - 1308 (1967).

[18] Kingman, J.F.C.: Regenerative Phenomena. London: Wiley 1972.

[19] Kishi, M.: An example of a positive non-symmetric kernel satisfying the complete maximum principle. Nagoya Math. J. 48, 189 - 196 (1972).

[20] Kishi, M.: Some remarks on the existence of a resolvent. Ann. Inst. Fourier, 25^{3-4}, 345 - 352 (1975).

[21] Lamperti, J.: On the coefficients of reciprocal power-series. Am. Math. Monthly 65, 90 - 94 (1958).

[22] Schoenberg, I.J.: Metric spaces and completely monotone functions. Ann. Math. 39 , 811 - 841 (1938).

[23] Steutel, F.W.: Note on the infinite divisibility of exponential mixtures. Ann. Math. Stat. 38, 303 - 305 (1967).

[24] Steutel, F.W.: Preservation of infinite divisibility under mixing and related topics. Math. Centre Tracts 33. Amsterdam: Math. Centre: 1970.

[25] Steutel, F.W.: Some recent results in infinite divisibility. Stochastic Processes Appl. 1, 125 - 143 (1973).

[26] Thorin, O.: On the Infinite Divisibility of the Pareto Distribution. Scand. Actuarial J. 1977, 31 -40.

[27] Thorin, O.: On the Infinite Divisibility of the Lognormal Distribution. Scand. Actuarial J. 1977, 121 - 148.

Christian Berg
Gunnar Forst
Matematisk Institut
Universitetsparken 5
2100 København Ø
Denmark

UNE MAJORATION UNIVERSELLE DES FONCTIONS DE CONCENTRATION SUR LES GROUPES LOCALEMENT COMPACTS NON COMPACTS

par

Philippe BOUGEROL

Y. Derriennic [3] et A. Mukherjea [7] ont montré indépendamment que la suite des puissances de convolution d'une probabilité sur un groupe localement compact converge vaguement vers zéro dès que le support de cette probabilité n'est pas porté par un sous-groupe compact. Dans cet article on précise ce résultat en montrant que, pour une large classe de groupes localement compacts non compacts G, si μ est une probabilité suffisamment régulière sur G et K un compact de ce groupe, il existe une constante c telle que la fonction de contentration définie sur N par $F_\mu(n,K) = \sup_{x,y\in G} \mu^{*n}(xKy)$ est majorée par $c\, n^{-1/2}$.

1 - RESULTAT PRINCIPAL

Rappelons d'abord quelques définitions :

Si G est un groupe localement compact à base dénombrable (noté LCD) et μ une probabilité sur G, on dit que

. μ est adaptée si son support n'est contenu dans aucun sous-groupe fermé propre de G.

. μ est apériodique si μ est adaptée et si son support n'est contenu dans aucune classe latérale d'un sous-groupe fermé distingué propre de G.

. μ est étalée si elle possède une puissance de convolution non étrangère à une mesure de Haar.

Les deux propositions suivantes sont démontrées dans [1], [2].

PROPOSITION 1 - Si G est un groupe LCD, G n'est pas moyennable si et seulement si pour toute probabilité μ adaptée sur G il existe un réel ρ strictement compris entre 0 et 1 tel que, si K est un compact de G, il existe un réel positif α vérifiant :

$$\forall n \in \mathbb{N} \quad , \quad \sup_{x,y\in G} \mu^{*n}(xKy) \leq \alpha\, \rho^n.$$

PROPOSITION 2 - Soit G un groupe LCD à génération compacte, extension compacte d'un groupe abélien de rang d. Si μ est une probabilité apériodique étalée sur G et K un compact de ce groupe il existe un réel positif α tel que :

$$\forall n \in \mathbb{N}\ ,\quad \sup_{x,y\in G} \mu^{*n}(xKy) \leq \alpha\, n^{-d/2}$$

Enonçons les résultats principaux.

THEOREME - Soit G un groupe LCD à génération compacte, non compact, supposé, ou non unimodulaire, ou non moyennable, ou encore extension compacte d'un groupe résoluble. Si μ est une probabilité étalée apériodique sur G et K un compact de ce groupe il existe un réel positif α tel que

$$\forall n \in \mathbb{N} \qquad F_\mu(n,K) = \sup_{x,y\in G} \mu^{*n}(xKy) \leq \frac{\alpha}{\sqrt{n}}$$

COROLLAIRE - Si G est un groupe LCD non compact, extension compacte de sa composante connexe de l'élément neutre ou sous-groupe fermé d'un groupe de Lie connexe, si μ est une probabilité sur G étalée apériodique et K un compact de G, il existe un réel positif α tel que

$$\forall n \in \mathbb{N}\quad ,\quad \sup_{x,y\in G} \mu^{*n}(xKy) \leq \frac{\alpha}{\sqrt{n}}$$

Démonstration : Un groupe du type envisagé dans le corollaire étant soit non moyennable, soit extension compacte d'un groupe résoluble [4], [5], ce corollaire est un cas particulier du théorème que nous allons démontrer :

Sous les hypothèses du théorème, si H est un sous-groupe distingué fermé de G et π la surjection canonique de G sur G/H il est clair que

$$\sup_{x,y\in G} \mu^{*n}(xKy) \leq \sup_{u,v\in G/H} \pi(\mu)^{*n}\ (u\ \pi(K)v).$$

Puisque l'image $\pi(\mu)$ de μ par π est établie apériodique dès que μ l'est [2], il suffit donc, pour montrer le théorème, de trouver un quotient de G sur lequel il soit vrai.

Si G possède un sous groupe distingué fermé résoluble R tel que G/R soit compact, soit $\mathcal{D}^0 R = R$, $\mathcal{D}^1 R = \overline{[R,R]}, \dots, \mathcal{D}^n R = \overline{[\mathcal{D}^{n-1}R, \mathcal{D}^{n-1}R]}$, la suite des groupes dérivés. Si m_0 est le plus petit entier tel que $\mathcal{D}^m R/\mathcal{D}^{m+1}R$ soit non compact, $R/\mathcal{D}^{m_0}R$ est compact et donc $G/\mathcal{D}^{m_0}R$ l'est aussi. Alors le quotient $G/\mathcal{D}^{m_0+1}R$ de G est extension compacte du groupe abélien non compact $\mathcal{D}^{m_0}R/\mathcal{D}^{m_0+1}R$ et, d'après la proposition 2, le théorème y est vrai.

Si G est un groupe non unimodulaire, de module Δ, G/KerΔ est un sous groupe à génération compacte, non compact, de $\mathbb{R}$ où le théorème est vrai.

Le cas où G est non moyennable est traité dans la proposition 1.

2 - REMARQUES

A. Remarque 1 - On doit pouvoir supprimer l'hypothèse d'étalement de μ dans le théorème (il suffit d'ailleurs de le faire dans la proposition 2). Par contre, si le théorème est vrai sur un groupe abélien dès que μ n'est pas portée par une classe latérale d'un sous-groupe compact, l'hypothèse d'apériodicité est en général nécessaire, comme le montre l'exemple suivant :

Soit G le groupe $\mathbb{R} \times_\sigma \mathbb{Z}$ où le produit est donné par $(x,m)\,(y,n) = (x+2^m y, m+n)$. Si ν est une probabilité sur $\mathbb{R}$ possédant une densité à support compact et si ε_{-1} est la masse de dirac au point -1 de $\mathbb{Z}$, $\mu = \nu \otimes \varepsilon_{-1}$ est une probabilité sur G étalée adaptée (mais pas apériodique). Si (A_n, B_n) sont des variables aléatoires sur un espace $(\Omega, \mathcal{A}, P)$ indépendantes, de loi μ, et si I étant un intervalle de $\mathbb{R}$, $K = I \times \{0\} \subset G$, on a

$$\sup_{x \in G} \mu^{*n}(Kx) = \sup_{\substack{a \in \mathbb{R} \\ b \in \mathbb{Z}}} P\Big(\Big(\sum_{i=1}^{n} \frac{A_i}{2^{i-1}}, -n\Big) \in (I+a) \times \{b\}\Big)$$

$$= \sup_{a \in \mathbb{R}} P\Big(\sum_{i=1}^{n} \frac{A_i}{2^{i-1}} \in I + a\Big)$$

La série $\sum_{i=1}^{n} \frac{A_i}{2^{i-1}}$ convergeant presque surement, il est clair que

$$\lim_{x \to +\infty} \sup_{x \in G} \mu^{*n}(Kx) \neq 0$$

B. Remarque 2 - Notons d'abord que les seuls groupes (LCD) connexes portant une probabilité μ étalée telle que, pour un ouvert relativement compact V, il existe une constante α vérifiant $\mu^{*n}(V) \sim \alpha\, n^{-1/2}$, sont les groupes G possèdant un sous-groupe distingué compact K tel que G/K soit isomorphe à $\mathbb{R}$. En effet, de tels groupes doivent être récurrents donc posséder un sous groupe compact K tel que G/K soit isomorphe à $\mathbb{R}$, $\mathbb{R}^2$ ou au groupe des déplacements de $\mathbb{R}^2$ [6]. La proposition 2 donne donc le résultat.

La proposition suivante indique dans quelle mesure ce résultat s'étend aux fonctions de concentration :

PROPOSITION 3 - Soit G un groupe LCD connexe unimodulaire ainsi que ses quotients, μ une probabilité etalée sur G, K un compact de ce groupe. Si G ne possède pas de sous-groupe distingué compact H tel que G/H soit isomorphe à $\mathbb{R}$ il existe un réel positif α tel que

$$\forall n \in \mathbb{N} \quad , \quad \sup_{x,y\in G} \mu^{*n}(xKy) \leq \frac{\alpha}{n}$$

Par contre, sur les groupes $G = \mathbb{R}^d \times_\sigma \mathbb{R}$, où le produit est défini par $((x_1,\ldots,x_d),\alpha)\ ((y_1,\ldots,y_d),\beta) = ((x_1 + e^\alpha y_1,\ldots,x_d + e^\alpha y_d),\alpha+\beta)$, on peut trouver des probabilités μ étalées telles que, si V est un ouvert relativement compact de G, il existe un réel α tel que $\sup_{x\in G} \mu^{*n}(Vx)$ soit équivalent à $\frac{\alpha}{\sqrt{n}}$.

<u>Démonstration</u> : Pour montrer la première partie on peut supposer que G est un groupe de Lie connexe moyennable grace à la proposition 1. G est alors de type R et s'il n'est pas isomorphe à $\mathbb{R}$ à un sous-groupe compact près il admet comme quotient $\mathbb{R}^2$ ou le groupe des déplacements de $\mathbb{R}^2$. [6]. La proposition 2 montre alors l'inégalité.

La seconde partie de la proposition est une conséquence de [2].

C. <u>Remarque 3</u> - Le théorème étudie $\sup_{x,y\in G} (\varepsilon_x * \mu^{*n} * \varepsilon_y)(f)$ quand f est une fonction bornée à support compact. En renforçant les hypothèses de régularité sur μ on peut obtenir la convergence vers zéro pour une classe plus large de fonctions. Montrons par exemple :

<u>PROPOSITION 4</u> - Soit G un groupe LCD unimodulaire non compact, de mesure de Haar λ, satisfaisant aux hypothèses du théorème ou du corollaire. Si μ est une probabilité apériodique sur G possédant une densité dans $L^2(G,\lambda)$, pour toute fonction f de $L^2(G,\lambda)$, $\sup_{x,y\in G} \int f(t)\, d(\varepsilon_x * \mu^{*n} * \varepsilon_y)(t)$ tend vers zéro.

<u>Démonstration</u> : Si ψ_n est la densité de μ^{*n} par rapport à λ et si $f, g \in L^2$

$$\left|\int f(t)\, d(\varepsilon_x * \mu^{*n} * \varepsilon_y)(t) - \int g(t)\, d(\varepsilon_x * \mu^{*n} * \varepsilon_y)(t)\right| \leq$$

$$\int |f(t)-g(t)|\ \psi_n(x^{-1}t\, y^{-1})\, d\lambda(t) \leq \|f - g\|_{L^2} \|\psi_n(x^{-1}.y^{-1})\|_{L^2}$$

$$\leq \|f - g\|_{L^2} \|\psi_1\|_{L^2}$$

Ayant la convergence vers zéro de $\sup_{x,y\in G} \int f(t)\, d(\varepsilon_x * \mu^{*n} * \varepsilon_y)(t)$ pour f continue à support compact, l'inégalité précédente permet de conclure par densité.

D. <u>Remarque 4</u> - On peut se poser le même type de problème dans le cadre des espaces homogènes. Plus précisément soit G un groupe LCD, H un sous-groupe fermé de G, π la surjection canonique de G sur G/H. Si μ est une probabilité sur G soit $P(.,.)$ le noyau de transition de la marche aléatoire de loi μ sur G/H défini par, si A est un borélien de G/H , $P(\bar{x},A) = \int 1_A \circ \pi(yx)\, d\mu(y)$ si $x \in G$ et $\bar{x} = \pi(x)$ auquel cas $P^n(\bar{x},A) = \int 1_A \circ \pi(yx)\, d\mu^{*n}(y)$. A-t-on, si A est compact, $\sup_{x\in G/H} P^n(\bar{x},A)$ tend vers zéro quand n tend vers l'infini ? Notons seulement ici qu'il ne suffit pas que G/H soit non compact. Si en effet G est le groupe affine de $\mathbb{R}$, H le sous-groupe des homothéties et μ une probabilité à support compact sur G telle que $\int \operatorname{Log} a(g)\, d\mu(g) < 0$, où a est la projection de G sur H, alors $\varliminf_{n\to+\infty} \sup_{x\in G/H} P^n(\bar{x},A) \neq 0$ si A est ouvert.

BIBLIOGRAPHIE

[1] BOUGEROL P. "Fonctions de concentration sur les extensions compactes de groupes abéliens" C.R.A.S. t. 283 p. 527-529 (1976).

[2] BOUGEROL P. "Fonctions de concentration sur certains groupes localement compacts' A paraître.

[3] DERRIENNIC Y. "Lois zéro ou deux pour les processus de Markov. Applications aux marches aléatoires" Ann. Inst. Poincaré Section B 12 1976(n°2) p.111-129 (1976).

[4] GREENLEAF F.P. "Invariant means on topological groups" Van Nostrand (1969).

[5] GUIVARC'H Y. "Croissance polynomiale et période des fonctions harmoniques" Bull. Soc. Math. France, 101, p. 333-379 (1973).

[6] GUIVARC'H Y., KEANE M., ROYNETTE B. "Marches aléatoires sur les groupes de Lie" Lecture Notes in Math. n° 624 Springer Verlag.

[7] MUKHERJEA A. "Limit theorems for probability measures on non compact groups and semi-groups". Z. Wahrscheinlichkeitstheorie verw. Gebiete (33) p.273-284 (1976)

Philippe BOUGEROL
U.E.R. de Mathématiques
Université PARIS VII
2, Place Jussieu,
75005 PARIS

GAUSSIAN MEASURES ON METRIC ABELIAN GROUPS

T. Byczkowski

Some well-known results of the theory of Gaussian measures on Banach spaces (more generally, on locally convex spaces) which seemed to be connected with the convex structure of these spaces are, in fact, of more general nature.

We present two of such results for Gaussian measures on metric abelian groups: The integrability of seminorms and the zero-one law.

These results are well-known when G is a Banach space (see e.g. [6] and [1]) and have many applications, e.g., in the theory of Gaussian stochastic processes.

Since there are metric linear spaces (or more generally metric groups) very natural from the point of view of the probability theory, which are neither Banach spaces nor locally convex spaces, these results can be of some interest in this more general situation.

1. Preliminaries.

DEFINITION 1. Let G be an abelian group and let $\mathcal{B}$ be a σ-field of subsets of G. $(G,\mathcal{B})$ is called a measurable group (m.g.) if the mapping

$$(x,y) \longrightarrow x-y$$

is measurable with respect to $\mathcal{B}$ and the product σ-algebra $\mathcal{B}\times\mathcal{B}$.

The typical example of a m.g. is a metric separable abelian group with its Borel σ-field. The space $D[0,1]$ of all left-continuous real functions defined on the unit interval, without discontinuities of the second kind, with the Borel σ-field (with respect to the Skorohod topology) is also a m.g., although it is not a topological group.

Roughly speaking, a m.g. is a group on which the convolution $\mu*\nu$ can be defined for arbitrary pair of probability measures μ, ν on $(G,\mathcal{B})$:

$$\mu*\nu\,(A) = \mu\times\nu(\{(x,y);\ x+y\in A\})$$

for every $A\in\mathcal{B}$, where $\mu\times\nu$ denotes the product of μ and ν.

A mapping X defined on a probability space $(\Omega, \mathfrak{S}, P)$ with values in a m.g. $(G, \mathfrak{B})$ is called a random element (r.e.) if it is measurable with respect to $\mathfrak{S}$ and $\mathfrak{B}$.

If X, Y are two r.e!s defined on a common probability space with values in a m.g., then $X \pm Y$ are also r.e!s. If X, Y are independent (in the usual sense), then the distribution of X+Y is equal to $\mu * \nu$, where μ, ν are the distributions of X, Y, respectively.

If G, $\mathfrak{B}$ is a m.g. we can consider Gaussian measures on G.

DEFINITION 2. Let $(G, \mathfrak{B})$ be a m.g.. A probability measure μ is called Gaussian if for every pair of independent r.e!s X_1, X_2 having the distribution μ, the r.e!s

$$X_1 + X_2 \quad \text{and} \quad X_1 - X_2$$

are independent.

Let $\psi: G \times G \longrightarrow G \times G$ be defined by the following formula

(1) $$\psi(x, y) = (x+y, x-y).$$

The above definition can be stated equivalently:

μ is Gaussian iff there are probability measures ν_1, ν_2 such that

(2) $$\psi(\mu \times \mu) = \nu_1 \times \nu_2,$$

that is

$$\mu \times \mu(\psi^{-1}(A)) = \nu_1 \times \nu_2(A) \quad \text{for every } A \in \mathfrak{B} \times \mathfrak{B}.$$

The measures ν_1, ν_2 are uniquely determined by μ:

$$\nu_1(E) = \nu_1 \times \nu_2(E \times G) = \mu \times \mu(\psi^{-1}(E \times G)) = \mu \times \mu(\{(x,y); x+y \in E\})$$
$$= \mu * \mu(E)$$

for every $E \in \mathfrak{B}$. Hence

(3) $$\nu_1 = \mu * \mu \quad \text{and} \quad \nu_2 = \mu * \bar{\mu},$$

where $\bar{\mu}(B) = \mu(-B)$ for $B \in \mathfrak{B}$.

Definition 2 has been used by Frechet in [7] in case separable Banach spaces and by Corvin in [5] in case LCA groups.

2. Elementary properties of Gaussian measures.

PROPOSITION. (i). If μ_1, μ_2 are Gaussian measures on a m.g. $(G, \mathfrak{B})$ then $\mu_1 * \mu_2$, $\bar{\mu}_2$ are Gaussian.

(ii). If $\pi: (G_1, \mathfrak{B}_1) \longrightarrow (G_2, \mathfrak{B}_2)$ is a measurable homomorphism between two m.g.s and if μ is a Gaussian measure on $(G_1, \mathfrak{B}_1)$, then $\pi(\mu)$

is a Gaussian measure on $(G_2, \mathcal{B}_2)$.
(iii). If $(G,\mathcal{B})$ is a metric separable abelian group with its Borel σ-algebra $\mathcal{B}$, then the family of all Gaussian measures on $(G,\mathcal{B})$ is closed in the weak topology.
(iv). Let $(G,\mathcal{B})$ be a metric separable abelian group (with its Borel σ-algebra $\mathcal{B}$) such that $x \longrightarrow 2x$ is a homeomorphism. Let μ be a Gaussian measure on $(G,\mathcal{B})$. The support of μ is a coset of a closed subgroup H of G. Moreover, H is closed under division by 2 (that is, $2x \in H \Longrightarrow x \in H$). Hence, if G is, in addition, a real vector space, then the support of μ is a coset of a closed vector subspace of G.

Proof. Observe that Definition 2 can be stated equivalently:

μ is Gaussian iff there are probability measures ν_1, ν_2 such that

(4) $$\iint f(x+y)\,g(x-y)\,\mu(dx)\mu(dy) = \iint f(x)g(y)\,\nu_1(dx)\nu_2(dy)$$

for every real-valued, bounded and $\mathcal{B}$-measurable functions f, g, from G into R.

The conclusion of (i) is now a consequence of [4] as well as the standard properties of convolution.

Next, (ii) follows easily from the application of the change of variables formula.

Further, if $(G,\mathcal{B})$ is a separable metric group, then μ is a Gaussian measure on $(G,\mathcal{B})$ iff there are probability measures ν_1, ν_2 such that (4) holds, for every continuous bounded f, g. From this fact and from the definition of the weak topology we obtain (iii).

Finally, we prove (iv). Denote $\theta(x) = 2x$. Let $C(\mu)$ denote the support of μ. Since θ is a homeomorphism, we have

(5) $$C(\theta(\mu)) = \theta(C(\mu)) = 2\,C(\mu).$$

From (2) we obtain

$$\psi^2(\mu\times\mu) = \psi(\nu_1\times\nu_2).$$

Since $\psi^2(x,y) = (2x,2y)$, we have

$$\psi(\nu_1\times\nu_2) = \theta(\mu)\times\theta(\mu).$$

By this formula we obtain

(6) $$\theta(\mu) = \nu_1 * \nu_2.$$

Using (3) and the standard formula on the support of $\mu*\mu$ (or $\mu*\mu$ we obtain

(7) $$x,y \in C(\mu) \Longrightarrow x+y \in C(\nu_1) \text{ and } x-y \in C(\nu_2),$$

(8) $x \in C(\nu_1)$ and $y \in C(\nu_2)$ $\qquad$ $x+y \in C(\nu_1 * \nu_2)$ and $x-y \in C(\nu_1 * \nu_2)$.

By (5) and (6) we obtain

(9) $$C(\nu_1 * \nu_2) = 2\,C(\mu).$$

Hence, if we assume that μ is symmetric, then $\nu_1 = \nu_2$, so by (7) -(9) we obtain

$$x, y \in C(\mu) \Longrightarrow 2x-2y \in 2\,C(\mu),$$

which means that $C(\mu)$ is a subgroup of G. In particular, $0 \in C(\mu)$. Hence $x \in C(\mu)$ implies $x \in 2\,C(\mu)$, that is, $C(\mu)$ is closed under division by 2.

In the general case observe that $\nu_2 = \mu * \bar{\mu}$ is a symmetric Gaussian measure; hence, by the first part of the proof, we obtain that $H = C(\nu_2)$ has all the desired properties. If $x_0 \in C(\mu)$, then

$$C(\mu) - x_0 \subseteq C(\mu) + C(\mu) \subseteq C(\nu_2) = H.$$

Using this inclusion as well as (8) and (9) we obtain

$$2x_0 + 2\,H = C(\nu_1) + C(\nu_2) \subseteq C(\nu_1 * \nu_2) = 2\,C(\mu) \subseteq 2x_0 + 2\,H.$$

Thus, we have

$$C(\mu) = x_0 + H,$$

which completes the proof of (iv).

3. <u>The integrability of seminorms.</u> Let p be a function defined on a m.g. $(G, \mathcal{B})$ with values in R^+. p will be called a seminorm if it is subadditive, that is if

$$p(x+y) \leqslant p(x)+p(y)$$

for every $x, y \in G$.

By a modification of Fernique's method (see [6] and [10]) we show that for every symmetric Gaussian measure there exists an $\varepsilon > 0$ such that $\exp(\varepsilon p(x))$ is integrable, under the assumption that p is $\mathcal{B}$-measurable.

THEOREM 1. Let μ be a symmetric Gaussian measure on $(G, \mathcal{B})$ and let p be a $\mathcal{B}$-measurable seminorm satisfying $p(x) \leqslant p(2x)$. Then there exists an $\varepsilon > 0$ such that

$$\int \exp(\varepsilon\, p(x))\,\mu(dx) < \infty.$$

<u>Proof.</u> Since μ is symmetric, we obtain $\nu_1 = \nu_2 = \mu * \mu$ in the formula (2). Let us denote $\nu = \mu * \mu$. Let X_1, X_2 (Y_1, Y_2) be independent r.e's defined on a probability space $(\Omega, \mathfrak{S}, P)$ with the dis-

tributions μ (ν, respectively).

Observe that

$$P\{2X_i \in A\} = \mu\{x;\ 2x \in A\} = \theta(\mu)(A) = \nu_1 * \nu_2(A) = P\{Y_1+Y_2 \in A\}$$
$$= P\{Y_1-Y_2 \in A\}, \quad i = 1, 2.$$

Thus $2X_i$ has the same distribution as Y_1+Y_2 (and Y_1-Y_2), $i = 1, 2$.

Next

$$(10)\qquad \begin{aligned} &P\{p(2X_1) \leqslant s,\ p(-2X_1) \leqslant s\} P\{p(2X_1) > t\} \\ &= P\{p(Y_1-Y_2) \leqslant s,\ p(Y_2-Y_1) \leqslant s,\ p(Y_1+Y_2) > t\} \\ &\leqslant P\{p(Y_1)-p(Y_2) \leqslant s,\ p(Y_2)-p(Y_1) \leqslant s,\ p(Y_1)+p(Y_2) > t\} \\ &\leqslant P\{p(Y_1) > (t-s)/2,\ p(Y_2) > (t-s)/2\} = P\{p(Y_1) > (t-s)/2\}^2. \end{aligned}$$

Since $p(Y_i) \leqslant s/2$ and $p(-Y_i) \leqslant s/2$ $(i = 1, 2)$ implies $p(Y_1-Y_2) \leqslant p(Y_1) + p(-Y_2) \leqslant s$ and $p(Y_2-Y_1) \leqslant p(Y_2)+p(-Y_1) \leqslant s$, we have

$$(11)\qquad \begin{aligned} &P\{p(Y_1) \leqslant s/2,\ p(-Y_1) \leqslant s/2\}^2 \quad P\{p(Y_1-Y_2) \leqslant s,\ p(Y_2-Y_1) \leqslant s\} \\ &= P\{p(2X_1) \leqslant s,\ p(-2X_1) \leqslant s\}. \end{aligned}$$

From (10) we obtain

$$(12)\qquad \begin{aligned} &P\{p(Y_1-Y_2) \leqslant s,\ p(Y_2-Y_1) \leqslant s\} P\{p(Y_1+Y_2) > t\} \\ &\leqslant P\{p(Y_1) > (t-s)/2\}^2 \end{aligned}$$

for every symmetric independent Gaussian r.e!s Y_1, Y_2 with the same distributions.

Using the property $p(x) \leqslant p(2x)$ as well as (10), (11) and (12) (applied to X_1, X_2 instead of Y_1, Y_2) we obtain

$$(13)\qquad \begin{aligned} &P\{p(Y_1) \leqslant s/2,\ p(-Y_1) \leqslant s/2\}^6\ P\{p(X_1) > t\} \\ &P\{p(Y_1) \leqslant s/2,\ p(-Y_1) \leqslant s/2\}^2\ P\{p(2X_1) \leqslant s,\ p(-2X_1) \leqslant s\}\quad P\{p(2X_1) > t\} \\ &\leqslant P\{p(Y_1) \leqslant s/2,\ p(-Y_1) \leqslant s/2\}^2\ P\{p(Y_1) > (t-s)/2\}^2 \\ &\leqslant P\{p(X_1) > (t-2s)/4\}^4. \end{aligned}$$

Finally, let s be a fixed real number such that

$$P\{p(Y_1) \leqslant s/2,\ p(-Y_1) \leqslant s/2\}^2 > 1/2.$$

Denote $t_0 = 2s$, $t_{n+1} = 2s+4t_n$ and

$$x_n = \frac{P\{p(X_1) > t_n\}}{P\{p(Y_1) \leqslant s/2,\ p(-Y_1) \leqslant s/2\}}.$$

From (13) we have $x_{n+1} \leqslant x_n^4$. Hence

$$x_n \leqslant x_0^{2^{2n}} = \exp(2^{2n} \ln x_0).$$

The rest of the proof is almost the same as in [6] and is omitted.

COROLLARY 1. Let $(G,\mathcal{B})$ be a m.g.. Assume that G is a real vector space such that the mapping $\alpha \longrightarrow \alpha x$ is Borel measurable, for every $x \in G$. Let p be a $\mathcal{B}$-measurable seminorm on G. Then for every symmetric Gaussian measure μ on $(G,\mathcal{B})$ there exists an $\varepsilon > 0$ such that

$$\int \exp\ \varepsilon\, p(x))\,\mu(dx) < \infty .$$

Proof. For every x G the mapping

$$\alpha \longrightarrow p(\alpha x)$$

is subadditive and Borel measurable, hence bounded in any finite interval (see Theorem 7.4.1 in [9]). Thus, if we define

$$p'(x) = \sup\{\, p(x/2^n),\ n = 0,1,\dots\}$$

then p' is a $\mathcal{B}$-measurable seminorm having the property:

$$p' \geqslant p \quad \text{and} \quad p'(2x) \geqslant p'(x).$$

Now, the conclusion follows from Theorem 1.

Next, observe that in the definition of Gaussian measure we have only used the fact that the mapping ψ defined by (1) is measurable.

However, the formula (2) makes sense if ψ is only measurable with respect to the completion of $\mathcal{B}\times\mathcal{B}$ under $\mu\times\mu$. The latter condition is satisfied if we assume that μ is tight, that is, if

$$\mu(A) = \sup\{\,\mu(K);\ K \text{ compact} \subseteq A\}$$

for every Borel subset A of G (see [13]).

In the sequel, whenever a topological structure on G appears, we will assume that $\mathcal{B}$ is the Borel σ-field of G and that all considered probability measures are defined on $\mathcal{B}$ and are tight.

Hence, we can consider (tight) Gaussian measures on Hausdorff groups. It is easy to see that the statements (iii) and (iv) of Proposition as well as the proof of Theorem 1 remain valid in this slight different situation.

Thus, we have the following

COROLLARY 2. Let G be a Hausdorff topological real vector space. Let p be a Borel measurable seminorm on G. Then for every symmetric Gaussian measure μ on G there exists an $\varepsilon > 0$ such that

$$\int \exp(\varepsilon\, p(x))\,\mu(dx) < \infty .$$

4. The zero-one law. Throughout this section G denotes a Hausdorff group and $\mathcal{B}$ its Borel σ-field. As mentioned in the final part of the previous section, we assume that all considered probability measures are defined on $\mathcal{B}$ and are tight.

A Borel measurable mapping φ: $G \longrightarrow G$ will be called bi-measurable if $\varphi(A) \in \mathcal{B}$ whenever $A \in \mathcal{B}$. If μ is a probability measure on G, then the completion of $\mathcal{B}$ with respect to μ will be denoted by $\mathcal{B}^{\mu}$.

DEFINITION 3. Let μ_1, μ_2 be two probability measures on G. μ_1, μ_2 are called associated probability measures if there exist probability measures ν_1, ν_2 such that

$$\psi(\mu_1 \times \mu_2) = \nu_1 \times \nu_2. \tag{14}$$

LEMMA 1. Let $x \longrightarrow 2x$ be one-to-one and bi-measurable. If μ_1, μ_2 are associated and symmetric, then $\mu_1 = \mu_2$. If μ_1 and μ_2 are associated and are translations of some symmetric measures, then μ_1 is a translation of μ_2.

Proof. Assume that μ_1, μ_2 are symmetric probability measures satisfying (14). Then, we have

$$\psi^2(\mu_1 \times \mu_2) = \psi(\nu_1 \times \nu_2).$$

Since $\nu_1 = \mu_1 * \mu_2$ and $\nu_2 = \mu_1 * \bar{\mu}_2 = \mu_1 * \mu_2$, we have

$$\nu_1 = \nu_2 = \bar{\nu}_1 = \bar{\nu}_2.$$

By (14) we obtain

$$\mu_1\{x;\ 2x \in A\} = \nu_1 * \nu_2(A),$$

$$\mu_2\{x;\ 2x \in A\} = \nu_1 * \bar{\nu}_2(A);$$

hence

$$\mu_1(A) = \nu_1 * \nu_2(2A) = \nu_1 * \bar{\nu}_2(2A) = \mu_2(A).$$

Next, if $\mu_1 = \mu_1' * x_1$ and $\mu_2 = \mu_2' * x_2$, where μ_1', μ_2' are symmetric, then by (14) and the change of variables formula it follows that

$$\psi(\mu_1' \times \mu_2') = \nu_1' \times \nu_2',$$

where $\nu_1' = \nu_1 * (-x_1 - x_2)$ and $\nu_2' = \nu_2 * (-x_1 + x_2)$. Thus, μ_1' and μ_2' are associated and symmetric; hence

$$\mu_1' = \mu_2',$$

so that

$$\mu_2 = \mu_1 * (x_2 - x_1),$$

which completes the proof.

LEMMA 2. Let G be a countable abelian group endowed with the discrete topology. Let μ be a Gaussian measure on G. If $\mu(\{0\}) > 0$, then μ is the normed Haar measure concentrated on a finite subgroup I of G.

The arguments needed to prove this lemma are quite standard and can be found, e.g., in [11], p. 101, Remark 2; for a detailed proof, see [3], Lemma 2.

Now, we are in a position to state the zero-one law.

THEOREM 2. Let G be a Hausdorff abelian group such that $x \longrightarrow 2x$ is one-to-one and bi-measurable. Assume that every Gaussian measure on G without idempotent factors is a translation of a symmetric measure. Let μ be a Gaussian measure without idempotent factors. Then for every $\mathcal{B}^{\mu}$-measurable subgroup F of G we have

$$\mu(F) = 0 \quad \text{or} \quad \mu(F) = 1.$$

Proof. The proof of the theorem consists of several steps.

1. First of all, observe that it suffices to prove our theorem for σ-compact subgroups of G. Indeed, if F is a $\mathcal{B}^{\mu}$-measurable subgroup of G such that $\mu(F) > 0$, then there exists a compact subset $K \subseteq F$ such that $\mu(K) > 0$. The subgroup D generated by K is σ-compact (hence Borel), $D \subseteq F$ and $\mu(D) > 0$. Thus, if we prove that $\mu(D) = 1$, then also $\mu(F) = 1$. Hence, throughout the remaining part of the proof, we can assume without loss of generality that F is a Borel subgroup of G such that $\mu(F) > 0$.

2. Let E be the subgroup of G generated by the set

$$\{x \in G;\ \mu(F+x) > 0\}.$$

We show that $\mu(E) = 1$. First of all, observe that E consists of countably many cosets of F and therefore $E \in \mathcal{B}$. By the definition of E we have

$$\mu(E+x) = 0 \quad \text{if} \quad x \notin E.$$

Hence

$$0 < \mu(F) \leqslant \mu(E) \leqslant \{x;\ 2x \in E\} = \nu_1 * \nu_2(E) = \int \mu(E-x)\bar{\mu}^{*3}(dx)$$
$$\mu(E)\ \mu^{*3}(E) = \ldots = ((\mu(E))^4,$$

which gives $\mu(E) = 1$.

3. From step 2 we infer that μ restricted to E is Gaussian. Let π be the canonical homomorphism of E onto E/F with the discrete

topology). Since E/F is countable, π is Borel measurable. Now, the statement (ii) of Proposition yields that $\pi(\mu)$ is a Gaussian measure on E/F (with the discrete topology). By Lemma 2, $\pi(\mu)$ is the normed Haar measure concentrated on E/F and E/F is finite. Hence E/F is isomorphic to the direct sum of cyclic groups. Assume that card $E/F > 1$. Then

$$E/F \cong Z(k_1) \oplus \dots \oplus Z(k_n),$$

where $Z(k_i)$ denotes the cyclic group of order k_i, $k_i > 1$, $i = 1,\dots,n$. Let

$$F_1 = \pi^{-1}(Z(k_2) \oplus \dots \oplus Z(k_n)).$$

Then F_1 is a Borel subgroup of G, $E/F_1 = Z(k_1)$ and $\pi_1(\mu)$ is the Haar measure on $Z(k_1)$, where π_1 is the canonical homomorphism of E onto E/F_1. Moreover,

$$0 < \mu(F_1) < 1. \tag{15}$$

Thus, in order to prove our theorem it suffices to show that (15) leads to a contradiction. Hence, throughout the remainder we assume that $E/F = Z(k)$, $k > 1$, and that $\pi(\mu)$ is the normed Haar measure on $Z(k)$.

4. By step 3, we have

$$\mu(F+x) = \mu(F) = 1/k$$

for every $x \in E$. Let $u \in E \setminus F$ be such that $iu \notin F$ for $0 < i < k$. Let us denote

$$\mu_i'(A) = k\,\mu(A \cap (F+iu)),\ i = 0, 1,\dots,k;\ A \in \mathcal{B}.$$

μ_i' is a probability measure on E concentrated on $F+iu$. Define

$$\mu_i = (-iu) * \mu_i'.$$

μ_i is a probability measure concentrated on F. It can be checked that μ_i, μ_j are associated Gaussian measures on F without idempotent factors for $i, j = 0, 1,\dots, k$, and that

$$\mu_{i-n} * \mu_n = \mu_i * \mu_0 \quad \text{for } 0 \le n \le i \le k. \tag{16}$$

This verification, although elementary, is tedious and is left to the reader.

5. By step 4, μ_i, μ_j are associated Gaussian measures without idempotent factors, $i, j = 0, 1,\dots k$. From Lemma 1 it follows that

$$\mu_i = x_i * \mu_0$$

for some x_i's, $x_i \in F$, $i = 1, \dots, k$. Moreover, since μ_i's have no no idempotent factors, from (16) we obtain (see [13])

$$x_{l-n} + x_n = x_l, \text{ for } 0 < n < l < k.$$

Hence $x_i = ix_1$. Since $\mu_k = (-ku) * \mu_0$, we have

$$k(u+x_1) * \mu_0 = \mu_0.$$

Since, by step 4, μ_0 has no idempotent factors, we obtain

$$k(u+x_1) = 0.$$

Let $u' = u+x_1$. Then $u' \in E \setminus F$, u' generates a cyclic group of order k and

$$\{iu \; ; \; i = 0, 1, \ldots, k-1\} \cap F = 0.$$

Hence

$$E \cong Z(k) \oplus F.$$

Moreover,

$$\mu_i' = iu' * \mu_0,$$

hence

$$\mu = \lambda_Z * \mu_0,$$

where λ_Z is the normed Haar measure concentrated on Z , where

$$Z = \{iu' \; ; \; i = 0, \ldots, k-1\}.$$

This contradicts the assumption that μ has no idempotent factors and completes the proof of the theorem.

The assumption that every Gaussian measure μ on G without idempotent factors is a translation of a symmetric measure is known to be satisfied in many situations: if G is an LCA group such that the mapping $x \longrightarrow 2x$ is an automorphism (see [8]), or if G is a complete separable locally convex vector space (see [4]). This assumption is also satisfied if G is a separable Orlicz space (not necessarily locally convex), see [2]. It is also satisfied if we consider the product G^T with the product σ-field $\mathcal{B}^T$, for arbitrary set T (observe that $(G^T, \mathcal{B}^T)$ is a m.g.). The following is also true

THEOREM 3. Let G be a metric group having no nonzero elements of order two. Let μ be a Gaussian measure on G without idempotent factors. Then μ is a translation of a symmetric Gaussian measure.

The proof of this theorem as well as a more detailed version of the proof of Theorem 2 will appear elsewhere.

Combining Theorems 2 and 3 and using the Kuratowski Isomorphism Theorem we obtain

THEOREM 4. Let G be a metric group having no nonzero elements of order two. Let μ be a Gaussian measure on G without idempotent factors. Then for every $\mathcal{B}^{\mu}$-measurable subgroup F of G we have

$$\mu(F) = 0 \quad \text{or} \quad \mu(F) = 1.$$

It is also possible to obtain a version of the zero-one law for Gaussian measures on arbitrary metric groups. That version of the zero-one law depends on the embedding property of Gaussian measures and will not be discussed here.

In the end of this section we give some examples of metric groups (or measurable groups) which indicate possible applications of the preceding results.

Examples.

(i). Let $(T, \mathcal{F}, m)$ be a finite measure space and let G be an LCA group satisfying the second axiom of countability. By $L_G^0 = L_G^0(T, \mathcal{F}, m)$ we will denote the space of all measurable functions x defined on T, with values in G, with the topology induced by the seminorm

$$|x| = \int \frac{\|x(t)\|}{1 + \|x(t)\|}\, m(dt) \quad ,$$

where $\|\cdot\|$ is a fixed (complete) seminorm on G. Assume that T is a separable metric space and $\mathcal{F}$ its Borel σ-algebra. Then L_G^0 is a complete separable metric group. As it has been proved in [3], there is a correspondence between Gaussian measures on L_G^0 and measurable Gaussian stochastic processes with values in G. If G is without nonzero elements of order two, so is L_G^0. Hence , the zero-one law for Gaussian measures on L_G^0 can be formulated in terms of Gaussian measurable stochastic processes with values in G. We can also consider, more generally, Orlicz spaces L_G^{ϕ}.

(ii). Let T be a set and let $(G^T, \mathcal{B}^T)$ be the product of G with the product σ-field. It is easy to see that $(G^T, \mathcal{B}^T)$, with the coordinatewise addition, is a m.g.. As previously, every Gaussian measure on $(G^T, \mathcal{B}^T)$ can be regarded as a Gaussian stochastic process with values in G and the final remark of (i) applies to this situation.

(iii). Let $C_G = C_G[0,1]$ be the space of all continuous functions defined on the unit interval with values in G with the uniform convergence topology. C_G is a complete separable metric group. Moreover, the Borel σ-field of C_G is equal to the relative σ-field induced from $(G^T, \mathcal{B}^T)$. Hence, we have again a one-to-one correspondence between Gaussian measures on C_G and Gaussian stochastic processes with continuous sample paths (with values in G).

(iv). Let G be a complete separable metric (real) vector space. We show that there exists a Gaussian measure μ on G such that $C(\mu) = G$. Let $\|\cdot\|$ be a seminorm on G. We can assume, without loss of generality that $\|\cdot\|$ is non-decreasing, that is, $\|tx\| \leqslant \|sx\|$ if $|t| \leqslant |s|$, $t, s \in R$ (see [12]). Let x_n be a sequence which is dense in the unit sphere of G. Put $y_n = \alpha_n x_n$, where α_n is such that $\|y_n\| \leqslant 1/2^n$. Let (X_n) be a sequence of independent real Gaussian random variables with mean zero and variance 1. Let $(\lambda_n) \in l_2$, $\lambda_n \neq 0$, $n = 1, 2, \ldots$. Then $(\lambda_n X_n)$ defines a Gaussian random element with values in the sequence space l_2 (see [14], 2.2.5). In particular, $\lim \lambda_n X_n = 0$ with probability one. Therefore, the following series

$$\sum_{n=1}^{\infty} \lambda_n X_n y_n$$

is convergent, in the seminorm of G, with probability one (hence defines a r.e., say Y, with values in G). Let Y_n be the n-th partial sum of this series and let μ_n, μ be the distributions of Y_n, Y, respectively. Since Y_n converges to Y with probability one, μ_n converges weakly to μ. Since μ_n is (symmetric) Gaussian, by part (iii) of Proposition, μ is (symmetric) Gaussian. Moreover,

$$C(\mu) \supseteq C(\mu_n) \supseteq \{\lambda_i y_i\}_{i=1}^{\infty}.$$

Since $C(\mu)$ is, by part (iv) of Proposition, a closed linear subspace, we obtain

$$C(\mu) \supseteq \overline{\text{lin}}\,\{y_i\} = \overline{\text{lin}}\,\{x_i\} = G.$$

References

[1]. C.R. Baker, Zero-one laws for Gaussian measures on Banach spaces, Trans. Amer. Math. Soc., 186 (1973), 291-308.

[2]. T. Byczkowski, Norm convergent expansion for L_ϕ-valued Gaussian randomelements, Studia Math., 64 (to appear).

[3]. ———, Some results concerning Gaussian measures on metric linear spaces, Springer Lecture Notes in Math., vol. 656 (1978), 1-16 (to appear).

[4]. S.D. Chatterji, Sur l'integrabilité de Pettis, Math. Zeitschrift, 136 (1974), 53-58.

[5]. L. Corwin, Generalized Gaussian measure and a functional equation, I, J. Funct. Analysis, 5 (1970), 481-505.

[6]. X. Fernique, Intégrabilité des vecteurs gaussiens, C. R. Paris, 270 (1970), 1698-1699.

[7]. M. Fréchet, Généralisation de la loi de probabilité de Laplace, Ann. Inst. H. Poincaré, 12 (1951).

[8]. H. Heyer and C. Rall, Gaußsche Wahrscheinlichkeitsmaße auf Corwinschen Gruppen, Math. Zeitschrift, 128 (1972), 343-361.

[9]. E. Hille and R.S. Phillips, Functional analysis and semi-groups, Amer. Math. Soc. Colloq. Publ., vol. 31, Providence 1957.

[10]. T. Inglot and A. Weron, On Gaussian random elements in some

non-Banach spaces, Bull. Acad. Pol. Sci., 22 (1974), 1039-1043.
[11]. K.R. Parthasarathy, Probability measures on metric spaces, New York 1967.
[12]. S. Rolewicz, Metric linear spaces, Warszawa 1972.
[13]. A. Tortrat, Lois de probabilité sur un espace topologique complètement régulier et produits infinis à termes indépendent dans un group topologique, Ann. Inst. H. Poincaré 1 (1966), 217-237.
[14]. N.N. Vakhania, Probability distributions on linear spaces in Russian , Tbilisi 1971.

Wrocław Technical University
50-370 Wrocław, Poland

GRENZWERTSÄTZE FÜR ABHÄNGIGE ZUFALLSVARIABLE UND IRRFAHRTEN AUF GRUPPEN

Pierre CRÉPEL

1. Einleitung

Sei G eine lokalkompakte Gruppe, (ξ_n) eine Folge von unabhängigen gleichverteilten Zufallsvariablen mit Werten in G (μ ist die gemeinsame Verteilung) . $Z_n = \xi_1 \dots \xi_n$ (Verteilung : μ^{*n}) ist eine μ-Rechtsirrfahrt auf G . Sei m ein Rechtshaarmass auf G .

Wir nehmen immer an , dass keine abgeschlossene echte Untergruppe existiert, die den Träger von μ enthält.

Definitionen

* μ heisst strikt aperiodisch ("strap"), falls keine abgeschlossene normale echte Untergruppe H und kein $g \in G$ existieren, sodass der Träger von μ in gH liegt.

* μ heisst ausgebreitet, wenn ein $n_0 \in N$ existiert, sodass: $\mu^{*n_0} \not\perp m$

Es ist bekannt, dass, wenn G kompakt ist, μ^{*n} vag (nach m) konvergiert, genau dann, wenn μ strap ist;

und dass, wenn G nicht kompakt ist, μ^{*n} vag nach 0 konvergiert.

Problem

Wie kann man im nicht kompakten Falle Z_n normalisieren, um einen nichtdegenerierten Grenzwert zu bekommen? Oder, anders gesagt, wie kann man eine Folge von Funktionen $\psi_n : G \to E$ finden, wo E reich genug ist (z.B. $E=G$ oder $E=\mathbb{R}^d$ mit korrektem d), sodass $\psi_n(Z_n)$ in Verteilung konvergiert; sei ν_n die Verteilung von $\psi_n(Z_n)$: wenn $\nu_n \xrightarrow{vag} \nu$ ($\neq$ Diracmass), was ist ν ?

Beispiel

* Wenn $G=(\mathbb{R},+)$ und $\sigma^2=\int x^2 d\mu(x) < \infty$, $\int x d\mu(x) = 0$, so konvergiert $\frac{Z_n}{\sigma\sqrt{n}}$ nach $N(0,1)$

* Wenn $G=(\mathbb{R},+)$ und keine weiteren Bedingungen verlangt werden, so kann $\frac{Z_n}{a_n} - b_n$ (wo a_n und b_n Folgen von reellen Zahlen sind) nur nach einer stabilen Verteilung konvergieren; dann hat μ notwendigerweise eine gewisse bekannte Form.

Bemerkung

In diesem Artikel werden wir mehr über die Methoden als über die Ergebnisse berichten.

2. Wie kann man eine Konvergenz in Verteilung beweisen?

Es seien (ν_n) eine Folge von W-Massen auf G und ν ein W-Mass auf G .

$C_b(G)$: Banachraum der stetigen beschränkten Funktionen auf G.

$C_o(G) = \{f \in C_b(G) : f$ verschwindet in $\infty\}$

$C_c(G) = \{f \in C_c(G) :$ der Träger von f ist kompakt$\}$

$\nu_n \xrightarrow{vag} \nu$ bedeutet: $\forall f \in C_b(G)$ $\int f d\nu_n \longrightarrow \int f d\nu$.

[oder $C_o(G)$,

oder $C_c(G)$]

a) Schwerindustrie

Um eine solche Konvergenz zu zeigen, beweist man meistens, dass ν_n in der schwachen Topologie relativ Kompakt ist und ν der einzig mögliche Häufungspunkt ist.

Dafür ist es oft hinreichend zu zeigen, dass

$$\int f d\nu_n \longrightarrow \int f d\nu$$

für f in einer gewissen kleineren Familie F von Funktionen, für die es nicht so schwierig ist, diese Konvergenz zu bekommen.

Beispiele : $G=\mathbb{R}$

(1) - $F = \{f_t : x \to e^{itx} \quad / t \in \mathbb{R}\}$

(2) - $F = \{f_\lambda : x \to e^{-\lambda x} \quad / \lambda \in \mathbb{R}^+\}$

(3) - $F = \{f_k : x \to x^k \quad / k \in \mathbb{N}\}$

(4) - $F = C_c^\infty(\mathbb{R}) = \{f \in C_c(\mathbb{R}) :$ f unendlich oft differenzierbar $\}$

oder $C_c^2(\mathbb{R}) = \{f \in C_c(\mathbb{R}) :$ f 2-mal stetig differenzierbar $\}$

Im Falle (1) gebraucht man die Fouriertransformierte.
Im Falle (2) gebraucht man die Laplacetransformierte.
Im Falle (3) gebraucht man die Methode der Momente: [B'],S.181-2

[Satz:
Wenn $\int x^k \, d\nu_n$ für jedes $k \in \mathbb{N}$ (existiert! und) konvergiert (sei $\alpha_k = \lim_n \int x^k \, d\nu_n$) und $\overline{\lim_k} \dfrac{|\alpha_k|^{1/k}}{k} < \infty$, dann:

$$\exists ! \nu \text{ , so dass } \int x^k \, d\nu = \alpha_k$$

$$\nu_n \overset{vag}{\to} \nu \text{]}$$

Im allgemeinen muss man, wenn man diese Methode benutzt, Trunkaturen machen.

Im Falle (4) ist die Benutzung von infinitesimalen Generatoren [F], S. 252 ...

b) Leichtindustrie

Eine andere Arbeitsweise ist Halbfabrikate zu benutzen, das heisst, Sätze anzuwenden, die aus der Schwerindustrie stammen.

Beispiele

- Satz von Trotter: [K]

Sei $T_t^{(n)}$ eine Folge von starkstetigen Halbgruppen von linearen

Kontraktionen auf einem Banachraum X ; seien A_n die infinitesimalen Generatoren mit Definitionsbereich $D(A_n)$. Wenn ein dichter Unterraum $F \subset X$ existiert, so dass

(i) $\forall f \in F \quad \lim\limits_n A_n f$ existiert $(= Af)$

(ii) $\exists \lambda_o > 0$, sodass $\overline{(\lambda_o I - A)(F)} = X$,

so ist A der infinitesimale Generator von einer starkstetigen Halbgruppe von Kontraktionen : T_t , und:

$$\|T_t^{(n)} f - T_t f\| \to 0 \quad \forall f \in X .$$

- Satz von Skorokhod: [S] , S. 13

  Dieser Satz gibt hinreichende Bedingungen für die Konvergenz einer Folge $X_t^{(n)}$ von stochastischen Prozessen.

- Klassische oder weniger klassische Grenzwertsätze für Partialsummen von unabhängigen oder abhängigen Zufallsvariablen, oder Maxima von Partialsummen...

3. Wie kann man diese Ideen für Grenzwertsätze auf Gruppen anwenden?

Sei $\begin{cases} \nu_n & \text{die Verteilung von } \psi_n(Z_n) \\ \nu & \text{die unbekannte Grenzverteilung} \end{cases}$

a) Zwei Beispiele

* Sei G_d die Gruppe der Bewegungen des euklidischen Raumes $\mathbb{R}^d$
$G_d \simeq So(d) \times \mathbb{R}^d$ mit dem Produkt:

$$(u_1, y_1)(u_2, y_2) = (u_1 u_2, y_1 + u_1 y_2) .$$

Es seien (U_n, Y_n) eine Folge von unabhängigen gleichverteilten,

Zufallsvariablen mit Werten in $G_d \quad So(d) \times \mathbb{R}^d$

sowie mit $E\|Y_1\|^2 < \infty$ und U_1 strap.

Dann hat man

$$(G_d)\quad Z_n = \prod_{i=1}^{n} (U_i, Y_i) = (U_1 \ldots U_n, Y_1 + U_1 Y_2 + \ldots + U_1 \ldots U_{n-1} Y_n)$$

Wir studieren hier

$$\psi_n(Z_n) = \left(U_1 \ldots U_n\ ,\ \frac{Y_1 + U_1 Y_2 + \ldots + U_1 \ldots U_{n-1} Y_n}{\sqrt{n}}\right) \in G$$

oder

$$\tilde{\psi}_n(Z_n) = \frac{Y_1 + U_1 Y_2 + \ldots + U_1 \ldots U_{n-1} Y_n}{\sqrt{n}} \in \mathbb{R}^d$$

<u>Satz 1</u>

> Unter diesen Bedingungen gilt
>
> $$\tilde{\psi}_n(Z_n) \xrightarrow{\text{in Verteilung}} N(o, \sigma^2 I),$$
>
> wo σ eine gewisse Konstante ist

*Sei H_1 die erste Heisenberggruppe, d.h. die Gruppe der Matrizen

$$\begin{pmatrix} 1 & u & w \\ 0 & 1 & v \\ 0 & 0 & 1 \end{pmatrix} \qquad u, v, w \in \mathbb{R},$$

d.h. $\mathbb{R}^3$ mit dem Produkt

$$(u_1, v_1, w_1)(u_2, v_2, w_2) = (u_1 + u_2, v_1 + v_2, w_1 + w_2 + u_1 v_2)$$

[Diese Gruppe ist isomorph zu $\mathbb{R}^3$ mit dem Produkt:

$$(H_1)\quad (u_1, v_1, w_1) \circ (u_2, v_2, w_2) = (u_1 + u_2, v_1 + v_2, w_1 + w_2 + \tfrac{1}{2}(u_1 v_2 - u_2 v_1)) \ ]$$

Es seien (U_n, V_n, W_n) eine Folge von unabhängigen gleichverteilten Zufallsvariablen mit Werten in H_1, und:

$$\operatorname{cov}(U_1, V_1) = I\ , \quad E|W_1| < \infty$$

$$EU_1 = EV_1 = 0$$

Dann bekommt man

$$Z_n = \prod_{i=1}^{n} (U_i, V_i, W_i)$$

$$= (U_1 + \ldots + U_n;\ V_1 + \ldots + V_n;\ W_1 + \ldots + W_n + \{V_1 + U_1 V_2 + \ldots + (U_1 + \ldots + U_{n-1}) V_n\})$$

Wir studieren hier

$\psi_n(Z_n) =$

$$(H_1\ni)\quad \left(\frac{V_1+\ldots+U_n}{n}, \frac{V_1+\ldots+V_n}{n}, \frac{W_1+\ldots+W_n}{n} + \frac{V_1+U_1V_2+\ldots+(U_1+\ldots+U_{n-1})V_n}{n}\right)$$

oder (um zu vereinfachen, falls $w_i = 0$) :

$$\tilde{\psi}_n(Z_n) = \frac{V_1+U_1V_2+\ldots+(U_1+\ldots+U_{n-1})V_n}{n}$$

[oder

$$\frac{1}{2n}\{[V_1+U_1V_2+\ldots+(U_1+\ldots+U_{n-1})V_n]-[U_1+V_1U_2+\ldots+(V_1+\ldots+V_{n-1})U_n]\},$$

wenn wir die zweite Form des Produktes in H_1 benutzen.]

Satz 2.

> Mit dieser Form und unter diesen Bedingungen gilt:
>
> $\psi_n(Z_n) \xrightarrow{\text{in Verteilung}} B_1$ (wo B_t die Brownsche Bewegung mit infinitesimalem Generator $\frac{\delta^2}{\delta u^2} + \frac{\delta^2}{\delta v^2}$ auf H_1 ist)
>
> insbesondere $\tilde{\psi}_n(Z_n) \xrightarrow{\text{in Verteilung}}$ Verteilung mit Dichte $\frac{1}{\operatorname{ch}\pi x}$

b) Benutzung der Schwerindustrie

Fourier

*Wenn $\psi_n(Z_n)$ in einer abelschen Gruppe liegt, so kann man die gewöhnliche Fouriertransformierte verwenden und beweisen, dass

$$E(e^{it\psi_n(Z_n)}) = \hat{\nu}_n(t)$$

konvergiert.

~Im Beispiele (G_d) kann man das auf folgende Weise tun: man betrachtet die $\tilde{\psi}_{nk}(Z_{nk})$:

$$\psi_{nk}(Z_{nk}) = \frac{1}{\sqrt{nk}} (Y_1+U_1Y_2+\ldots+U_1\ldots U_{nk-1}Y_{nk})$$

$$= \frac{1}{\sqrt{n}} \{[\frac{Y_1+U_1Y_2+\ldots+U_1\ldots+U_{k-1}Y_k}{\sqrt{k}}]+U_1\ldots U_k[\frac{Y_{k+1}+\ldots+U_{k+1}\ldots U_{2k-1}Y_{2k}}{\sqrt{k}}]$$

$$+\ldots+U_1\ldots U_{(n-1)k}[\frac{Y_{(n-1)k+1}+\ldots+U_{(n-1)k+1}\ldots U_{nk-1}Y_{nk}}{\sqrt{k}}]\}$$

- Die [ ] sind unabhängig;
- und die $(U_1\ldots U_{(i-1)k})$ sind beinahe wie das Haarmass verteilt, wenn U_1 strap ist.

Elementare Rechnungen und Betrachtungen über unabhängige Zufallsvariable in $\mathbb{R}^d$ ergeben, dass, wenn man n und k gross wählt, so ist die Fouriertransformierte von $\tilde{\psi}_{nk}(Z_{nk})$ fast die Funktion $e^{\frac{-\sigma^2|t|^2}{2}}$, wo σ eine gewisse Konstante ist [R']

~Mit einer analogen Idee kann man auch das Beispiel (H_1) behandeln [G'KR], [CR']

* Wenn $\psi_n(Z_n) \in G$, so kann man anstelle gewöhnlicher Fouriertransformierter unitäre irreduzible Darstellungen (u.i.D.) verwenden, d.h. zu zeigen:

$$T_{\nu_n} =: \int_G T_g \, d\nu_n(g) \longrightarrow \int_G T_g \, d\nu(g) =: T_\nu \quad \text{in der starken Topologie,}$$

für jede u.i.D. T_g von G.

Ebenso kann man das Beispiel (G_d) behandeln [T]

Momente

*Wenn $\psi_n(Z_n)$ zu $\mathbb{R}^d$ oder zu einem Raum gehört, in dem man Polynome definieren kann, so kann man hoffen, die Methode der Momente zu gebrauchen.

~Im Beispiele (H_1) ist das möglich:
um zu vereinfachen, nehmen wir an, dass $E|U_1|^k < \infty \quad \forall k$.
$E|V_1|^k < \infty$

(der allgemeine Fall kann auf diesen Fall reduziert werden.)

Das einzige Problem ist, den folgenden Grenzwert zu berechnen:

$$\alpha_k =: \lim_{n\to\infty} E[\tilde{\psi}_n(Z_n)]^k = \lim_n \frac{E\{\frac{1}{2}[V_1+\ldots+(U_1+\ldots+U_{n-1})V_n]-\ldots\}^k}{n^k}$$

Man entwickelt die Potenz k.
Wegen der Unabhängigkeitseigenschaften der Paare (U_i, V_i), und der Bedingung $EU_1 = EV_1 = 0$, $\mathrm{cov}(U_1, V_1) = I$, ist es nicht schwer zu finden, dass die α_k genau die Momente der Verteilung ν, mit Dichte $\frac{1}{\mathrm{ch}\pi x}$ und Fouriertransformierte $\frac{1}{\mathrm{ch}t}$ sind.

Diese Methode der Momente kann man in allgemeineren Situationen gebrauchen, es ist uns aber noch nicht gelungen, sie für alle nilpotenten Gruppen anzuwenden.

c) Benutzung der Leichtindustrie

*Trotter

* Die Idee könnte die folgende sein:
Sei $T_t^{(n)}$ die Kontraktionshalbgruppe von $C_b(G)$:

$$T_t^{(n)} f(x) = E[f(x \circ \psi_n(Z_{nt}))] \quad \text{oder etwas ähnliches}$$

$$(\forall f \in C_b(G), \ \forall x \in G).$$

Man zeigt, dass $T_t^{(n)}$ die Bedingungen des Satzes von Trotter erfüllen; daher gibt es eine Halbgruppe T_t, sodass

$$T_t^{(n)} f(x) \longrightarrow T_t f(x),$$

insbesondere

$$\int f d\nu_n = T_1^{(n)} f(e) \longrightarrow T_1 f(e) = \int f d\nu$$

Beispiel (H_1)
Es gibt einige Sachen zu tun:

- $T_t^{(n)}$ ist keine Halbgruppe; daher muss man die Definition für $T_t^{(n)}$ etwas abändern
- Um zu beweisen, dass A_n gegen den infinitesimalen Generator A der Brownschen Bewegung strebt, hat man eine Taylorformel zu benutzen [H], [CR], u.s.w.

*Skorokhod

Beispiel (H_1) ... : siche in diesem Buch : $[R_2]$

*Benutzung der zentralen Grenzwertsätze für abhängige Zufallsvariablen

α) Φ-mischende Folgen : Beispiel (G_d)

Seien $X_n = U_1 \ldots U_{n-1} Y_n$ $(\in \mathbb{R}^d)$, und F_n die σ-Algebra, die von $(U_1,Y_1),\ldots,(U_n,Y_n)$ generiert wird . Es ist nicht sehr schwierig zu zeigen, dass gilt :

Ist die Verteilung von U_1 ausgebreitet, so sind F_n und die von $(X_{m+n},X_{m+n+1},\ldots)$ generierte σ-Algebra fast unabhängig, für m gross; genauer bekommt man:

(X_n) ist ϕ-mischend mit einem exponentialen ϕ-Koeffizient.

z.B. [CLP] , [B]

So kann man die berühmten Grenzwertsätze für ϕ-mischende Folgen anwenden.

β) Martingale Beispiel (H_1)

Sei $X_n = (U_1+\ldots+U_{n-1})V_n$ $(\in \mathbb{R})$ und F_n die von $(U_1,V_1),\ldots,(U_n,V_n)$ generierte σ-Algebra.

Es ist klar, dass:

$$E\left(X_n \mid F_{n-1}\right) = E\left[(U_1+\ldots+U_{n-1})V_n \mid F_{n-1}\right] = (U_1+\ldots+U_{n-1})EV_n = 0 ,$$

d.h. $S_n = X_1+\ldots+X_n$ ist ein Martingal

γ) Markoffketten

Wenn man einen Grenzwertsatz auf halbeinfachen Gruppen (z.B. $Sl(2,\mathbb{R})$) zeigen will, so kann man beweisen, dass gewisse Variable , die in diesem Problem vorkommen, die Doeblinbedingungen für Markoffketten erfüllen $[R_1]$ u.s.w.

δ) "Mischungale" ("mixingale" in [ML])

Betrachten wir die folgende Bedingung:

Sei (X_n) eine Folge von L^2 Zv , (F_n) eine Folge von σ-Algebren, so dass X_n F_n- messbar ist, (X_n,F_n) heisst Mischungal, falls:

$$||E(X_n|F_{n-k})||_2 \leq ||X_n||_2 \; \Psi_k \quad \text{mit } \Psi_k \searrow 0.$$

Diese Bedingung kommt oft vor, z.B.:

- wenn (X_n) eine Folge von Martingalzunahmen ist (denn $E(X_n|F_{n-k}) = 0$)
- wenn (X_n) ϕ-mischend ist
- in vielen anderen Fällen

Beispiel (G_d) : Wenn die Verteilung von U_1 nicht ausgebreitet ist, so ist wahrscheinlich (X_n) nicht ϕ-mischend.
Man kann aber zeigen, dass (X_n) ein Mischungal mit exponentialem Koeffizienten ist: die Ursache ist, dass man für ϕ-Mischung $E[f(X_n)|F_{n-k}]$ mit Uniformität in f studieren muss, während man für Mischungale $E[f(X_n)|F_{n-k}]$ nur für eine gewisse Funktion $(f : x \longrightarrow x)$ studieren muss.

ε) Andere Grenzwertsätze für abhängige Zv wurden in der Literatur für Irrfahrten auf Gruppen angewendet, insbesondere in G .

d) Fragen

1/ Betrachten wir eine Abhängigkeitsbedingung (z.B. Martingal, ψ-Mischung, Mischungal ...).
Welche Irrfahrten auf Gruppen oder homogenen Räumen können diese speziellen Bedingungen erfüllen?

2/ Dann kommt eine andere natürliche Frage: kann man die zentralen Grenzwertsätze für solche abhängige Zv anwenden?
Für G_d gibt es kein besonderes Problem
Aber für H_1 bekommt man ein interessantes Phänomen:
-Dafür muss man folgende Ergebnisse zurückrufen: [P]

> Sei (X_n) eine Folge von unabhängigen nicht notwendig gleichverteilten reellen Zv . Sei $S_n = X_1 + \ldots + X_n$.
> So kann $\frac{S_n}{a_n} - b_n$ (wo a_n und b_n Folgen von reellen Zahlen sind) nur nach gewissen Verteilungen konvergieren; die Klasse dieser (unendlich teilbaren) Verteilungen heisst die Klasse L und wird durch eine besondere Form der Lévy-Hinčin Formel

charakterisiert (sie enthält die Klasse der stabilen Verteilungen)

Wenn die (X_n) zentriert und L^2 sind, mit $\frac{\max_{1\le k\le n} EX_k^2}{B_n} \to 0$

$(B_n = \sum_{i=1}^{n} EX_i^2 = ES_n^2)$, so:

$$\left(\frac{S_n}{\sqrt{B_n}} \xrightarrow{vag} N(0,1)\right) \iff \left(\text{d.h. } \frac{1}{B_n}\sum_{k=1}^{n}\int_{|X_k|>\varepsilon\sqrt{B_n}} X_k^2\, dP \to 0\right) \quad \text{[Lindebergbedingung]}$$

- Wenn (X_n) eine Folge von Martingalzunahmen ist, so gibt es zentrale Grenzwertsätze, die eine ähnliche Form wie (*) haben; man braucht aber immer eine zusätzliche Bedingung, z.B.:

$$\frac{1}{B_n}\overline{\lim_n}\sum_{i\neq j} EX_i^2 X_j^2 \le 1 \qquad [ML]$$

Ich habe kein Gegenbeispiel ohne eine solche Bedingung gesehen, unsere Zv hier in H_1 geben ein solches Gegenbeispiel:
Sei $X_n = (U_1+\ldots+U_{n-1})V_n$ (mit Momenten der Ordnung 4)

Es ist leicht zu zeigen, dass $\frac{1}{B_n^2}\sum_{i=1}^{n} EX_i^4 = 0(\frac{1}{n}) \to 0$,

daraus folgt ebenso leicht die Lindebergbedingung,

aber $\frac{S_n}{\sqrt{B_n}} \not\longrightarrow N(0,1)$

(in der Wirklichkeit strebt $\frac{S_n}{\sqrt{B_n}}$ gegen eine andere Verteilung!)

<u>Bemerkungen</u> : -es kann auch gezeigt werden, dass:

$$\frac{1}{B_n}\overline{\lim_n}\sum_{i\neq j} EX_i^2X_j^2 > 1$$

-diese Zv geben auch ein Gegenbeispiel für das Gesetz des iterierten Logarithmus, weil:

$$0 < \overline{\lim_n}\frac{S_n}{n \log\log n} < \infty \quad \text{f.s.}$$

also $\overline{\lim_n} \frac{S_n}{\sqrt{B_n \log \log B_n}} = \infty$ f.s. $(B_n \simeq n^2)$

- es kann auch bemerkt werden, dass die Grenzverteilung mit Dichte $\frac{1}{\mathrm{ch}\,\pi x}$ in der Klasse L liegt.

3/ Die folgende Frage stellt sich natürlich:

Wenn (X_n) $\begin{cases} \text{eine Folge von Martingalzunahmen} \\ \text{ein Mischungal} \\ \text{eine } \psi\text{-mischende Folge} \end{cases}$ ist,

welche Grenzverteilung kann man bekommen, wenn man S_n normalisiert?

Unter welchen Bedingungen über (X_n) kann man eine gewisse Grenzverteilung bekommen?

4/ Welche anderen Grenzwertsätze (lokale Theoreme, grosse Abweichungen, Fastsicher-Konvergenzsätze...) kann man für abhängige Zv bekommen und bei Irrfahrten auf Gruppen anwenden?

Literatur

[B] P. BOUGEROL: Thèse de 3e-cycle Universitè de Paris VII (1976)

[B'] L. BREIMAN: "Probability" - Addison Wesley (1968)

[CLP] P. CREPEL, E. LE PAGE: "Loi du logarithme itéré fonctionnelle pour des marches aléatoires sur le groupe des deplacements de R^d" C.R.A.S. 281 (1975) S. 475-8

[CR] P. CREPEL, A. RAUGI: "Théorème central limite sur les groupes nilpotents" C.R.A.S. 281 (1975), S. 605-8, und Ann. IHP : wird veröffentlicht.

[CR'] P. CREPEL, B. ROYNETTE: Une loi du logarithme itéré sur le groupe de Heisenberg" Zeit. für Wahr. 39 (1977) S. 217-229

[F] W. FELLER: "An introduction to probability theory and its applications" Band II, Wiley (1966)

[G] L.G. GOROSTIZA: "The central limit theorem for random motions of d-dimensional euclidean space" Ann. Prob. 1 (1973) S.603-12

[G'KR] Y.GUIVARC'H, M. KEANE, B. ROYNETTE: "Marches aléatoires sur les groupes de Lie" Lect. notes N. 624 Springer (1977)

[H] H. HENNION: "Théorème central limite et théorème central limite fonctionnel sur un group de Lie nilpotent" Séminaires Université de Rennes (1975)

[K] T.G.KURTZ : "Extensions of Trotter's operator semigroups approximation theorems" Journ.Funct. Anal. 3 (1969) S. 111-132

[ML] D. MC LEISH: "Invariance principles for dependent variables" Zeit.für Wahr. 32 (1975) S. 165-178

[P] V.V. PETROV: "Sums of independent random variables" Springer (1975)

[R1] A. RAUGI: "Fonctions harmoniques et théorèmes limites pour les marches aléatoires sur les groupes" Bull. SMF, Mémoire N 54(1977)

[R2] A. RAUGI: "Théorème central limite pour les groupes de type rigide" Lect. Notes(in diesem Band)

[R'] B. ROYNETTE: " Théorème central limite pour le groupe des déplacements de R^d" Ann. IHP 10 (1974) , S.391-8

[S] A.V.SKOROKHOD: "Studies in the theory of random processes" Addison-Wesley (1965)

[T] V.N. TUTUBALIN: "The central limit theorem for random motions of a euclidean space" Sel. transl.in math.stat. and proba. 12 (1973) , S. 47-57

Laboratoire de probabilités
Université de Rennes
35042 RENNES Cedex
FRANKREICH

Faltungshalbgruppen sind analytische Funktionen ihrer infinitesimalen Funktionale

Th. Drisch

Abstract: Let $(\mu_t : t \in \mathbb{R}_+)$ be a vaguely continuous convolution semigroup of positive bounded Radon-measures on a locally compact abelian group $\mathfrak{G}$ and let A denote the infinitesimal functional of the semigroup. Then the exponential series

$\sum \frac{t^j}{j!} A^{*j}$ converges to μ_t on a function space $\mathfrak{Z}$ not depending on the convolution semigroup, and $\mathfrak{Z}$ is dense in the space of continuous functions on $\mathfrak{G}$ vanishing at infinity.

Therefore the convolution semigroups are the restrictions of entire weakly analytic (distribution valued) functions to the positive real numbers.

Sommaire: Soit $(\mu_t : t \in \mathbb{R}_+)$ un semi-groupe, vaguement continu, de mesures de Radon positives bornées sur un groupe $\mathfrak{G}$ abélien localement compact, soit A le générateur infinitésimal du semi-groupe. Alors, la série exponentielle

$\sum \frac{t^j}{j!} A^{*j}$ converge vers μ_t par rapport à un espace $\mathfrak{Z}$ de fonctions, qui ne dépend pas du semi-groupe; $\mathfrak{Z}$ est dense dans l'espace de fonctions continues sur $\mathfrak{G}$ tendant vers 0 à l'infini.

Par conséquence, les semi-groupes sont les restrictions des fonctions (à valeurs dans l'espace de distributions) entières faiblement analytiques à la semi-droite positive réelle.

Einführung.

$(\mu_t : t \in \mathbb{R}_+)$ sei eine schwach stetige Faltungshalbgruppe positiver beschränkter Radon-Maße auf einer zusammenhängenden abelschen Lie-Gruppe, A bezeichne das infinitesimale Funktional der Halbgruppe, $\hat{A}$ sei die Fouriertransformierte von A. Dann gilt bekanntlich $\hat{\mu}_t = e^{t\hat{A}}$.

A ist eine Distribution mit der Eigenschaft, daß die n-fache Faltung A^{*n} existiert; es wird gezeigt, daß die Partialsummen

$$\sum_{j=0}^{k} \frac{t^j}{j!} \; A^{*j}$$

schwach auf einem - allen Faltungshalbgruppen gemeinsamen- Funktionenraum konvergieren, der dicht in den stetigen im Unendlichen verschwindenden Funktionen liegt, daß der Limes e^{tA} für $t \geq 0$ stetig bezüglich der Supremumsnorm ist, sich mithin zu einem beschränkten Radon-Maß fortsetzen läßt, das ebenfalls mit e^{tA} bezeichnet werde, und daß $\mu_t = e^{tA}$ gilt.

Es stellt sich heraus, daß die Darstellung von μ_t als Exponentialreihe eine schwache Taylorentwicklung in $t = 0$ ist, die in der ganzen komplexen Ebene konvergiert, d.h. die Faltungshalbgruppen sind die Restriktionen ganzer analytischer Funktionen auf die positive reelle Achse (im schwachen Sinne).

Im vierten (letzten) Paragraphen wird skizziert, wie sich die Ergebnisse auf beliebige lokalkompakte abelsche Gruppen verallgemeinern lassen.

1 Präliminarien über Faltung und Fouriertransformation integrierbarer Distributionen.

1.1 Die Testfunktionen.

$\mathbb{G}$ sei eine zusammenhängende abelsche Lie-Gruppe, d.h. $\mathbb{G} = \mathbb{R}^q \times \mathbb{T}^p$. Für $m = 0,1,2,\ldots$ bezeichne

$\mathcal{C}^m(\mathbb{R}^q, \mathbb{R}^p)$ den Vektorraum der m-mal stetig differenzierbaren Funktionen $f:\mathbb{R}^{q+p} \longrightarrow \mathbb{C}$, die in den letzten p Variablen die Periode $\omega:=(2\pi,\dots,2\pi)$ haben und die samt ihren m Ableitungen beschränkt sind. Es sei

1.1.1 $$\|f\|_m(x):=\sum_{j=0}^{m}\frac{1}{j!}\,\|D^j f(x)\|\,,\quad \|f\|_m := \sup_{x\in\mathbb{R}^{q+p}} \|f\|_m(x).$$

Dabei sei $D^0 f:=f$ vereinbart, und $\|D^j f(x)\|$ bezeichne die Operatornorm des j-fach multilinearen Funktionals $D^j f(x)$ (mit $j=1,\dots,m$):

$$\|D^j f(x)\| := \sup_{\|y_1\|,\dots,\|y_j\| \leqslant 1} |D^j f(x)(y_1,\dots,y_j)|\,,$$

$$\|D^0 f(x)\| := |f(x)|.$$

$\mathcal{C}^m(\mathbb{R}^q,\mathbb{R}^p)$ trage stets die Norm 1.1.1. $\mathcal{C}^m_\infty(\mathbb{R}^q,\mathbb{R}^p)$ sei der lineare Unterraum aller $f\in\mathcal{C}^m$ derart, daß $x \longrightarrow \|f\|_m(x)$ auf $\mathbb{R}^q\times[0,2\pi]^p$ im Unendlichen verschwindet, ebenfalls mit der Norm 1.1.1. ϑ bezeichne die natürliche Surjektion von $\mathbb{R}^{q+p}$ auf $\mathbb{R}^q \times (\mathbb{R}^p/(2\pi\mathbb{Z})^p) = \mathfrak{G}$. $\mathcal{C}^0(\mathfrak{G})$ bezeichne den Banach-Raum der stetigen beschränkten Funktionen $f:\mathfrak{G} \longrightarrow \mathbb{C}$ (in der Supremumsnorm). Zu jedem $f\in\mathcal{C}^0(\mathbb{R}^q,\mathbb{R}^p)$ existiert genau ein $f_\mathfrak{G}\in\mathcal{C}^0(\mathfrak{G})$ mit $f = f_\mathfrak{G}\circ\vartheta$, und die Abbildung $\Theta: f \longmapsto f_\mathfrak{G}$ ist ein isometrischer Isomorphismus. $\mathcal{C}^m(\mathfrak{G})$ bzw. $\mathcal{C}^m_\infty(\mathfrak{G})$ bezeichne den Vektorraum $\Theta(\mathcal{C}^m(\mathbb{R}^q,\mathbb{R}^p))$ bzw. $\Theta(\mathcal{C}^m_\infty(\mathbb{R}^q,\mathbb{R}^p))$ mit der durch Θ induzierten Norm. $\mathcal{C}^m_*(\mathfrak{G})$ bezeichne den linearen Unterraum aller $f\in\mathcal{C}^m(\mathfrak{G})$ mit kompaktem Träger, $\mathcal{C}^\infty(\mathfrak{G})$ sei der Durchschnitt aller $\mathcal{C}^m(\mathfrak{G})$, $\mathcal{C}^\infty_\infty(\mathfrak{G})$ bzw. $\mathcal{C}^\infty_*(\mathfrak{G})$ seien die entsprechenden linearen Unterräume.

Meist werden die Funktionenräume über $\mathbb{R}^q \times \mathbb{R}^p$ mit den entsprechenden Räumen über $\mathfrak{G}$ identifiziert; dann schreibe ich einfach $\mathcal{C}^m$, $\mathcal{C}^m_\infty$, $\mathcal{C}^m_*$ usw.

Bekanntlich gilt:

1.1.2 $\mathcal{C}^m$ ist eine komplexe Banach-Algebra.

1.1.3 $\mathcal{C}^m_\infty$ ist abgeschlossen in $\mathcal{C}^m$, also ebenfalls eine Banach-Algebra.

1.1.4 $\mathcal{C}^\infty_\infty$ liegt dicht in jedem $\mathcal{C}^m_\infty$.

1.2 Definition: Ein stetiges lineares Funktional auf $\mathcal{C}^m_\infty(\mathfrak{G})$ heißt integrierbare Distribution von der Ordnung $\leq m$ auf $\mathfrak{G}$.

Ich identifiziere meist $A \in (\mathcal{C}^m_\infty(\mathfrak{G}))'$ mit dem transponierten Funktional $\Theta^t(A) \in (\mathcal{C}^m(\mathbb{R}^q, \mathbb{R}^p))'$, definiert durch $\langle \Theta^t(A), f \rangle := \langle A, \Theta(f) \rangle$, und schreibe in diesem Fall $A \in (\mathcal{C}^m_\infty)'$. $\|A\|_m$ bezeichne die Operatornorm von $A \in (\mathcal{C}^m_\infty)'$.

1.3 Darstellungssatz für integrierbare Distributionen+): Sei $A \in (\mathcal{C}^m_\infty(\mathfrak{G}))'$. Dann existieren beschränkte Radon-Maße μ_l auf der Gruppe $\mathfrak{G}$, wobei l die q+p-fachen Multi-indizes $(l_1, \ldots l_{q+p})$ mit $|l| := l_1 + \ldots + l_{q+p} \leq m$ durchläuft, so daß

1.3.1 $\langle A, f \rangle = \sum_{|l| \leq m} \langle \mu_l, \partial^l f \rangle$ (genauer: $\langle A, f \rangle = \sum_{|l| \leq m} \langle \Theta^t(\mu_l), \partial^l (f \cdot \vartheta) \rangle$)

gilt. Dabei ist $\partial^l (f \cdot \vartheta) := \left(\frac{\partial}{\partial x_1}\right)^{l_1} \ldots \left(\frac{\partial}{\partial x_{q+p}}\right)^{l_{q+p}} (f \cdot \vartheta)$. Darüber hinaus lassen sich die μ_l so wählen, daß gilt:

1.3.2 $\|A\|_m = \sup_{|l| \leq m} l_1! \ldots l_{q+p}! \, \|\mu_l\|$.

Beweisskizze: r^{q+p}_m bezeichne die Anzahl der q+p-fachen Multi-indizes der Ordnung $\leq m$. Für $f \in \mathcal{C}^m_\infty(\mathfrak{G})$ sei $\xi_l(f) := \Theta(\partial^l (f \cdot \vartheta)) \in \mathcal{C}^0_\infty(\mathfrak{G})$. Man bettet $\mathcal{C}^m_\infty(\mathfrak{G})$ vermöge $\xi : f \longmapsto (\xi_l(f))_{|l| \leq m}$ in $(\mathcal{C}^0_\infty(\mathfrak{G}))^{r^{q+p}_m}$ ein. Das stetige lineare Funktional A auf dem Unterraum $\xi(\mathcal{C}^m_\infty(\mathfrak{G}))$ läßt sich nach Hahn-Banach ohne Vergrößerung der Norm zu einem stetigen linearen Funktional $\bar{A}$ auf $(\mathcal{C}^0_\infty(\mathfrak{G}))^{r^{q+p}_m}$ fortsetzen. Durch $\langle \mu_l, g \rangle := \langle \bar{A}, (0, \ldots, g, \ldots, 0) \rangle$ (g an der l-ten Stelle) sind beschränkte Radon-Maße auf $\mathcal{C}^0_\infty(\mathfrak{G})$ definiert, die das in 1.3.1 und 1.3.2 Verlangte leisten. □

1.4 Faltung integrierbarer Distributionen. Ist f eine Funktion

---

+) Man vergleiche mit [8], ch. III, § 8, th. XXXI.

$\mathbb{R}^{q+p} \longrightarrow \mathbb{C}$ und $x \in \mathbb{R}^{q+p}$, so bezeichne f_x die Funktion $y \longmapsto f(x-y)$ und $\overset{\circ}{f}$ die Funktion $y \longmapsto f(-y)$. Es sei $f \in \mathcal{C}^\infty_*$ und $B \in (\mathcal{C}^n_\infty)'$, $n=0,1,2,\ldots$; $B*f$ bezeichne die Funktion $x \longmapsto \langle B, f_x \rangle$. Mithilfe des Darstellungssatzes 1.3 beweist man ohne Schwierigkeiten:

1.4.1 $\quad B*f \in \mathcal{C}^\infty_\infty$.

1.4.2 Zu $m=0,1,2,\ldots$ existiert ein (von B und f unabhängiges) $c_m \in \mathbb{R}_+$ mit

$$\|B*f\|_m \leq c_m \|B\|_m \|f\|_{m+n}.$$

Seien $A \in (\mathcal{C}^m_\infty)'$, $B \in (\mathcal{C}^n_\infty)'$. Dann ist wegen 1.4.1 durch

1.4.3 $$\langle A*B, f \rangle := \langle A, (B*\overset{\circ}{f})^\circ \rangle =: \langle A_x, \langle B_y, f(x+y) \rangle\rangle$$

ein lineares Funktional auf $\mathcal{C}^\infty_*$ definiert, das nach 1.4.2 stetig bezüglich der Norm $f \longmapsto \|f\|_{m+n}$ ist und gemäß 1.1.4 auf genau eine Weise zu einem stetigen linearen Funktional auf $\mathcal{C}^{m+n}_\infty$ fortgesetzt werden kann. Diese Fortsetzung werde ebenfalls mit $A*B$ bezeichnet und <u>Faltung</u> von A und B genannt. Wegen 1.4.2 existiert ein (von A und B unabhängiges) $c_{m,n} \in \mathbb{R}_+$ mit

1.4.4 $$\|A*B\|_{m+n} \leq c_{m,n} \|A\|_m \|B\|_n.$$

1.5 Straffe Fortsetzung integrierbarer Distributionen.

Satz 1.3 zeigt:

1.5.1 $A \in (\mathcal{C}^m_\infty)'$ läßt sich zu einem stetigen linearen Funktional $\bar{A}$ auf $\mathcal{C}^m$ fortsetzen, das der folgenden <u>Straffheitsbedingung</u> genügt:

1.5.2 Zu jedem $\varepsilon > 0$ existiert eine kompakte Teilmenge K_ε von G, so daß für $f \in \mathcal{C}^m$ mit $\|f\|_m \leq 1$ und mit $f|_{K_\varepsilon} = 0$ gilt: $|\langle \bar{A}, f \rangle| \leq \varepsilon$.

Nun läßt sich zeigen:

1.5.3 Sei $f \in \mathcal{C}^m$. Dann existiert zu jeder kompakten Teilmenge K von G

und jedem $\varepsilon > 0$ ein $f_{K,\varepsilon} \in \mathcal{C}^m_\infty(G)$ mit

$$(f - f_{K,\varepsilon})\Big|_K = o \quad \text{und} \quad \Big| \|f\|_m - \|f_{K,\varepsilon}\|_m \Big| < \varepsilon.$$

Daher ist $\bar{A}$ durch A eindeutig bestimmt und werde im folgenden mit A identifiziert. Es sei vermerkt, daß $\|\bar{A}\|_m = \|A\|_m$ gilt. Die soeben beschriebene Fortsetzung von A zu $\bar{A}$ motiviert die folgende Definition:

1.6 Definition: Ein stetiges lineares Funktional auf $\mathcal{C}^m$ mit der Eigenschaft 1.5.2 heißt straffe Distribution von der Ordnung $\leq m$.

1.7 Fouriertransformation integrierbarer (bzw. straffer) Distributionen.

E_y bezeichne die Funktion $x \longmapsto e^{ixy}$ von $\mathbb{R}^{q+p}$ in $\mathbb{C}$ (für $x := (x_1, \dots, x_{q+p})$, $y := (y_1, \dots, y_{q+p})$ ist $xy := \sum_{j=1}^{q+p} x_j y_j$). Die Fouriertransformierte $\hat{A}$ von $A \in (\mathcal{C}^m_\infty)'$ ist die Funktion $y \longmapsto \langle A, E_y \rangle$ von $\hat{G} := \mathbb{R}^q \times \mathbb{Z}^p$ in $\mathbb{C}$, (wobei die soeben beschriebene straffe Fortsetzung von A auf $\mathcal{C}^m$ benutzt wird). Es gilt:

1.7.1 $\hat{A}$ ist stetig.

1.7.2 $\hat{A}$ ist beschränkt durch ein Polynom m-ten Grades; genauer:

$$|\hat{A}(y)| \leq \|A\|_m \sum_{j=o}^{m} \frac{\|y\|^j}{j!} .$$

Es seien noch die beiden folgenden Sätze angeführt, die sich mit Hilfe von 1.3 beweisen lassen. Dabei bezeichne $L^1_*(\hat{G})$ den Vektorraum der integrierbaren Funktionen $f : \hat{G} \longrightarrow \mathbb{C}$ mit kompaktem Träger. $\mathfrak{Z} := (L^1_*(\hat{G}))^\wedge$ besteht offenbar aus ganzen analytischen Funktionen, deren Restriktionen auf $\mathbb{R}^q \times [0, 2\pi]^p$ im Unendlichen verschwinden; speziell gilt $\mathfrak{Z} \subseteq \mathcal{C}_\infty$.

1.8 <u>Satz (Parseval):</u> Sei $A\in(\mathcal{C}^m_\infty)'$, $f\in L^1_\varkappa(\hat{G})$. Dann gilt:

1.8.1 $$\langle A,\hat{f}\rangle = \langle \hat{A},f\rangle := \int \hat{A}f := \sum_{y_2\in\mathbb{Z}^p} \int_{\mathbb{R}^q} \hat{A}(y_1,y_2)f(y_1,y_2)dy_1 .$$

(1.8.1 zeigt, daß die Definition von $\hat{A}$ mit der Fouriertransformierten temperierter Distributionen übereinstimmt.)

1.9 <u>Faltungssatz:</u> Sei $A\in(\mathcal{C}^m_\infty)'$, $B\in(\mathcal{C}^n_\infty)'$. Dann gilt: $(A*B)^\wedge=\hat{A}\,\hat{B}$.

2 Die Exponentialfunktion fast positiver Funktionale und der Logarithmus unbegrenzt teilbarer Maße.

2.1 <u>Die Testfunktionen:</u> $T(\mathbb{R}^q,\mathbb{R}^p)$ sei der Vektorraum aller $f\in\mathcal{C}^0(\mathbb{R}^q,\mathbb{R}^p)$, die im Nullpunkt eine Entwicklung 2.Ordnung gestatten, versehen mit der folgenden Norm:

2.1.1 $$\|f\|_T := |f(o)| + \|f'(o)\| + \tfrac{1}{2}\|f''(o)\| + \sup_{x\in\mathbb{R}^{q+p}} \|Df(x)\|$$

$$\text{mit } (Df)(x) := \begin{cases} \left[f(x)-f(o)-f'(o)(\zeta(x))-\tfrac{1}{2}f''(o)(\zeta(x),\zeta(x))\right]\dfrac{1}{\psi(x)} & \text{für } x\neq o \\ 0 & \text{für } x=o \end{cases}$$

wobei $\frac{1}{2}f''(o)$ die symmetrische Bilinearform in der Entwicklung 2. Ordnung von f bezeichnet und $\zeta:\mathbb{R}^{q+p}\longrightarrow\mathbb{R}^{q+p}$ bzw. $\psi:\mathbb{R}^{q+p}\longrightarrow\mathbb{R}$ Abbildungen mit folgenden Eigenschaften sind:

> $\zeta(o) = o$, $\|\zeta\|_0 \le 1$, ζ ist ω-periodisch in den letzten p Variablen, ζ ist auf einer Nullumgebung $\mathcal{C}^2$-diffeomorph;
>
> $\psi(o) = o$, $\|\psi\|_0 \le 1$, ψ ist ω-periodisch in den letzten p Variablen, ψ ist auf einer Nullumgebung $\mathcal{C}^2$-Funktion; zu jeder Umge-

bung U von $o \in \mathbb{R}^{q+p}$ existiert ein $c_U > o$ mit $\Psi \geq c_U$ auf $\mathbb{R}^q \times [-\pi,\pi]^p \setminus U$.

(Abbildungen ζ und Ψ bzw. $\Theta(\zeta)$ und $\Theta(\Psi)$ mit den genannten Eigenschaften heißen Koordinatenabbildung bzw. Majorantenfunktion. Typische Vertreter sind $\zeta(x) = \frac{x}{1+x^2}$, $\Psi(x) = \frac{x^2}{1+x^2}$ im Falle der reellen Geraden und $\zeta(x) = \sin x$, $\Psi(x) = \frac{1-\cos x}{2}$ im Falle des Torus; im mehrdimensionalen Fall bildet man z.B. direkte Produkte der genannten ζ-Funktionen bzw. Tensorprodukte der Ψ-Funktionen.)

Sei $T := \Theta(T(\mathbb{R}^q, \mathbb{R}^p))$ mit der durch Θ induzierten Norm. Man kann zeigen:

2.1.2 Bei verschiedener Wahl von ζ und Ψ ergeben sich äquivalente Normen.

2.1.3 $(T, \| \ \|_T)$ ist eine komplexe Banach-Algebra.

2.1.4 Der Vektorraum $\mathcal{C}^2_{loc}$ der Funktionen $f \in \mathcal{C}^o$, die in einer (von f abhängigen) Umgebung des Nullpunktes zweimal stetig differenzierbar sind, liegt dicht in T.

2.2 Definition: Ein lineares Funktional $A: T \longrightarrow \mathbb{C}$ heißt fast positiv, wenn gilt:

2.2.1 Für jedes $f \in T$ mit $f(e) = o$, $f \geq o$ ist $\langle A,f \rangle \geq o$.

A heißt straff, wenn gilt:

2.2.2 Zu jedem $\varepsilon > o$ existiert eine kompakte Teilmenge K_ε von $\mathbb{R}^q \times [o, 2\pi]^p$, so daß für alle $f \in T$ mit $\|f\|_o \leq 1$, $f|_{K_\varepsilon} = o$ gilt: $|\langle A,f \rangle| \leq \varepsilon$.

2.3 Bemerkung: Ein fast positives lineares Funktional auf T ist stetig. Ein stetiges straffes lineares Funktional auf T ist eine straffe Distribution der Ordnung ≤ 2 (mit Straffheitsbedingung der Ordnung 0).

Beweis: Ist A ein lineares Funktional auf T, so gilt für $f \in T(\mathbb{R}^q, \mathbb{R}^p)$:

$$\langle \Theta^t(A), f\rangle = \langle A, 1\rangle f(o) + \langle A, \Theta(f'(o)\xi)\rangle + \langle A, \Theta(\gamma D f)\rangle .$$

Ist A fast positiv, so folgt für $g \in \mathcal{C}^o(\mathbb{R}^q, \mathbb{R}^p)$:

$$|\langle A, \Theta(\gamma g)\rangle| \leqslant \langle A, \Theta\rangle \|g\|_o$$

(man beachte $\gamma g \in T(\mathbb{R}^q, \mathbb{R}^p)$); beides zusammen zeigt die Stetigkeit von $\Theta^t(A)$ bezüglich der Norm 2.1.1, womit die erste Behauptung bewiesen ist.

Als nächstes überzeugt man sich - etwa mit Hilfe des Mittelwertsatzes - von der Ungleichung

2.3.1 $$\|f\|_T \leqslant c \|f\|_2 \quad \text{für } f \in \mathcal{C}^2(\mathbb{R}^q, \mathbb{R}^p)$$

(c von f unabhängige Konstante), womit $A \in (\mathcal{C}^2)'$ gezeigt ist. Zum Nachweis von 2.2.2 $\Longrightarrow$ 1.5.2 (letztere Bedingung von der Ordnung 0) wählt man die kompakte Menge K_ε in der Bedingung 2.2.2 als Nullumgebung. □

Im Folgenden werden zwei Dichteaussagen benötigt, deren Beweise dem Leser überlassen bleiben; dabei sei $T_\infty := T \cap \mathcal{C}^o_\infty$ gesetzt:

2.4 $\mathcal{C}^\infty_*$ liegt dicht in T_∞.

2.4.1 $(\mathcal{C}^\infty_*(\hat{G}))^\wedge$ liegt dicht in T_∞.

Daraus folgt unter Beachtung von 2.3 und 1.5.3:

2.4.2 Ein $\| \|_T$-stetiges lineares Funktional auf $\mathcal{C}^\infty_*$ bzw. $(\mathcal{C}^\infty_*(\hat{G}))^\wedge$

läßt sich auf höchstens eine Weise zu einem straffen $\| \|_T$-stetigen linearen Funktional auf T fortsetzen.

(Man kann zeigen, daß eine derartige Fortsetzung tatsächlich stets möglich ist.) Der Eindeutigkeitssatz der Fouriertransformation temperierter Distributionen liefert aufgrund von 2.4.2 (unter Beachtung von 2.3 und der Bemerkung im Anschluß an 1.8) einen Eindeutigkeitssatz der Fouriertransformation fast positiver straffer linearer Funktionale

Aufgrund von 2.4.2 läßt sich ein fast positives (und daher $\| \|_T$- stetiges) straffes lineares Funktional aus seiner Restriktion auf $\mathcal{C}^2_\infty$ rekonstruieren. Daher seien die fast positiven straffen Funktionale auf T und die sich durch Restriktion auf $\mathcal{C}^2_\infty$ (bzw. $\mathcal{C}^2$) ergebenden integrierbaren (bzw. straffen) Distributionen von der Ordnung ≤ 2 identifiziert.

2.5 Die Exponentialfunktion fast positiver straffer Funktionale.

2.5.1 Lemma: Für jedes $f \in L^1_*(\hat{G})$ und jedes $z \in \mathbb{C}$ existiert

$$\langle A_z, \hat{f} \rangle := \lim_{k\to\infty} \sum_{j=0}^{k} \frac{z^j}{j!} \langle A^{*j}, \hat{f} \rangle = \lim_{k\to\infty} \langle (\varepsilon_0 + \frac{z}{k} A)^{*k}, \hat{f} \rangle .$$

Beweis: Die Sätze 1.8 und 1.9 zeigen:

$$\left| \sum_{j=0}^{k} \frac{z^j}{j!} \langle A^{*j}, \hat{f} \rangle \right| \leq \sum_{j=0}^{k} \frac{|z|^j}{j!} \langle |\hat{A}|^j, |f| \rangle \leq \int e^{|z||\hat{A}|} |f| < \infty$$

(man beachte $e^{z|\hat{A}|}|f| \in L^1_*(\hat{G})$ wegen 1.7.1), so daß der Satz von Lebesgue die Konvergenz der Exponentialreihe sichert und überdies zeigt:

2.5.2 $$\langle A_z, \hat{f} \rangle = \int e^{z\hat{A}} f .$$

Dieselbe Argumentation mit Hilfe von 1.8, 1.9, 1.7.1 und dem Satz von Lebesgue zeigt

$$\lim_{k\to\infty} \langle (\varepsilon_0 + \frac{z}{k} A)^{*k}, \hat{f} \rangle = \int e^{z\hat{A}} f ,$$

so daß sich mit 2.5.2 die zweite Gleichung des Lemmas ergibt. □

2.5.3 <u>Lemma:</u> Ist $t \in \mathbb{R}_+$ und A reell, so existiert genau eine Fortsetzung von A_t zu einem beschränkten Radon-Maß bzw. endlichen straffen Maß $\mu_t : \mathcal{C}^0_\infty \longrightarrow \mathbb{C}$ bzw. $\mathcal{C}^0 \longrightarrow \mathbb{C}$. +) μ_t ist positiv. Es gilt:

2.5.4 $$\hat{\mu}_t = e^{t\hat{A}}.$$

<u>Beweis:</u> T_∞ liegt dicht in $\mathcal{C}^0_\infty$, so daß aus 2.4.1 folgt, daß $\mathfrak{Z}$ dicht in $\mathcal{C}^0_\infty$ liegt, womit die Eindeutigkeit der Fortsetzung gezeigt ist.

(Da $\mathfrak{Z}$ eine separierende und selbstadjungierte Algebra ist, folgt die Dichte von $\mathfrak{Z}$ in $\mathcal{C}^0_\infty$ auch unmittelbar aus dem Satz von Stone-Weierstraß).

Da A reell ist, ist $\hat{A}$ <u>hermitesch</u> und fast <u>positiv definit</u>, so daß das Lemma von Schönberg (z.B. [4], lemme 1) liefert:

2.5.5 $e^{t\hat{A}}$ ist positiv definit für $t \in \mathbb{R}_+$.

Nach 2.5.5 existiert ein positives endliches straffes Maß μ_t auf $\mathfrak{G}$ mit $\hat{\mu}_t = e^{t\hat{A}}$ (Satz von Bochner), und mit Hilfe des Satzes von Fubini und von 2.5.2 erhält man für $f \in L^1_*(\hat{\mathfrak{G}})$:

$$\langle \mu_t, \hat{f} \rangle = \int \hat{\mu}_t f = \langle A_t, \hat{f} \rangle. \quad \square$$

<u>Bemerkung zur Wahl des Funktionenraumes</u> $\mathfrak{Z}$: Für die Fortsetzung μ_t von A_t gilt Formel 2.5.2 für alle $f \in L^1(\hat{\mathfrak{G}})$, jedoch konvergieren

---

+) Ist $\mathfrak{G}$ nicht kompakt, so gilt $\mathcal{C}^o_* \cap \mathfrak{Z} = \{o\}$, d.h. die A_t leben auf einem zum Definitionsraum der Radon-Maße komplementären Unterraum von $\mathcal{C}^0_\infty$.

die Partialsummen in 2.5.1 nicht einmal für alle $f \in L^1(\hat{G})$, deren Fouriertransformierte ganz-analytisch sind (siehe Beispiel 3.7.3).

Aufgrund von 2.5.1 und 2.5.4 definiert man (A reell, $t \geq 0$):

2.5.6 μ_t heißt Exponential von A und wird mit $\exp_* t\, A$ oder e^{tA} bezeichnet.

2.6 Satz: $(\exp_* t\, A)^\wedge = \exp t\, \hat{A}$.

Beweis: Sei $x = (x_1, x_2)$ mit $x_1 \in \mathbb{R}^q$, $x_2 \in \mathbb{F}^p$, sei $y \in \mathbb{R}^q \times \mathbb{Z}^p$, sei $\sigma > 0$.

Sei $g_{y,\sigma}(x) := \exp\left[ixy - \frac{\|x_1\|^2 \sigma^2}{2}\right]$, sei $h_{y,\sigma} : \mathbb{R}^q \times \mathbb{Z}^p \longrightarrow \mathbb{C}$ die Funktion mit $\hat{h}_{y,\sigma} = \Theta(g_{y,\sigma})$. Dann gilt gemäß 2.5.2:

$$(\exp_* t\, A)^\wedge(y) = \lim_{\sigma \downarrow 0} \langle \exp_* t\, A, \Theta(g_{y,\sigma}) \rangle = \lim_{\sigma \downarrow 0} \int e^{t\hat{A}} h_{y,\sigma} = e^{t\hat{A}(y)}. \square$$

2.7 Der Logarithmus unbegrenzt teilbarer Maße.

Eine Faltungshalbgruppe auf G sei eine Familie $(\mu_t : t \in \mathbb{R}_+)$ beschränkter positiver Radon-Maße auf G mit $\mu_s * \mu_t = \mu_{s+t}$ und mit $\lim_{t \to 0} \mu_t = \varepsilon_0$ in der vagen Topologie (ε_x bezeichne das Punktmaß in $x \in G$). Die reellen fast positiven straffen linearen Funktionale sind die infinitesimalen Funktionale der Faltungshalbgruppen auf G; genauer (man vergleiche mit [5], Satz 4.2, Zusatz 4.3):

Ist (μ_t) eine Faltungshalbgruppe auf G, so existiert

2.7.1 $$\langle A, f \rangle := \left.\frac{d^+}{dt}\right|_{t=0} \langle \mu_t, f \rangle := \lim_{t \downarrow 0} \left\langle \frac{\mu_t - \varepsilon_0}{t}, f \right\rangle$$

für alle $f \in T$, und durch 2.7.1 ist ein reelles fast positives straffes lineares Funktional A definiert, so daß gilt:

2.7.2 $\hat{\mu}_t = e^{t\hat{A}}$.

Ist umgekehrt ein derartiges Funktional A gegeben, so existiert eine (wegen 2.7.2 eindeutig bestimmte) Faltungshalbgruppe (μ_t) auf $\mathfrak{G}$, so daß 2.7.1 erfüllt ist.

Aus 2.6 und 2.7.2 folgt: $\mu_t = \exp_* t\, A$.

Dies motiviert folgende Begriffsbildung:

2.7.5 <u>Definition:</u> μ sei ein unbegrenzt teilbares beschränktes positives Radon-Maß auf $\mathfrak{G}$. Dann werde das infinitesimale Funktional einer Faltungshalbgruppe (μ_t) mit $\mu_1 = \mu$ <u>Logarithmus</u> von μ genannt und mit $\log \mu$ bezeichnet.*)

2.8 <u>Satz:</u> A sei ein fast positives straffes lineares Funktional auf $\mathfrak{T}$, μ ein unbeschränkt teilbares beschränktes positives Radon-Maß auf $\mathfrak{G}$. Dann gilt:

$$\langle \log \exp_* A, f\rangle = \langle A, f\rangle \quad \text{für } f\in T,$$
$$\langle \exp_* \log \mu, f\rangle = \langle \mu, f\rangle \quad \text{für } f\in \mathcal{C}^0.$$

<u>Beweis:</u> 2.6 und 2.7.2 zeigen

$$(\exp_* \log \mu)^\wedge = e^{(\log \mu)^\wedge} = \hat{\mu}.$$

Umgekehrt sei $(\mu_t : t\in \mathbb{R}_+)$ die Faltungshalbgruppe mit $\mu_1 = \exp_* A$. Dann zeigen 2.7.1 und 2.7.2:

$$(\log \exp_* A)^\wedge = \lim_{t\downarrow 0} \frac{(\mu_t - \varepsilon_0)^\wedge}{t} = \lim_{t\downarrow 0} \frac{e^{t\hat{A}} - 1}{t} = \hat{A}.$$

In beiden Fällen führt daher der <u>Eindeutigkeitssatz der Fouriertransformation</u> zum Erfolg (für fast positive Funktionale folgt dieser Satz aus dem entsprechenden Resultat für temperierte Distributionen (siehe Bemerkung im Anschluß an 2.4.2)). □

---

*) $\log \mu$ ist genau dann durch μ eindeutig bestimmt, wenn $p=0$ gilt.

3 Einbettung in Faltungsgruppen und schwache Analytizität.

3.1 Der Testfunktionenraum. Ist K eine kompakte Teilmenge von $\hat{G} = \mathbb{R}^q \times \mathbb{Z}^p$, so bezeichne $L^1_K(\hat{G})$ den Banach-Raum der integrierbaren Funktionen $\hat{G} \longrightarrow \mathbb{C}$, die auf K konzentriert sind (mit der L^1-Norm). $L^1_*(\hat{G})$ trage die induktive Topologie der $L^1_K(\hat{G})$, $\mathfrak{Z} = (L^1_*(\hat{G}))^\wedge$ trage die durch die Fouriertransformation von $L^1_*(\hat{G})$ induzierte Topologie. Man erhält ([7], VII, §1, Satz 3):

3.1.1 $\mathfrak{Z}$ ist ein (nicht metrisierbarer) vollständiger lokalkonvexer Hausdorff-Raum.

3.1.2 Das in 2.5.1 für $z \in \mathbb{C}$ definierte lineare Funktional A_z auf $\mathfrak{Z}$ ist stetig.

Für $z \in \mathbb{R}_+$ werde A_z mit dem in 2.5.3 definierten Maß identifiziert. Ich schreibe bisweilen $A_z =: e^{zA} = \mu_z$.

3.2 Faltung der e^{zA}. A und B seien fast positive straffe Funktionale auf $\mathbb{T}$. u und v seien komplexe Zahlen. Sei $f \in L^1_*(\hat{G})$. Dann zeigt eine kurze Rechnung:

3.2.1 $(e^{vB} * \hat{f}^o)(x) := \langle e^{vB}, (\hat{f}^o)_x \rangle = (e^{v\hat{B}} f)^\wedge(x)$.

Aus 3.2.1 folgt:

3.2.2 Konvergiert das Netz (f_α) in $L^1_*(\hat{G})$ gegen 0, so konvergiert das Netz $(e^{vB} * \hat{f}^o_\alpha)$ in $\mathfrak{Z}$ gegen 0.

Wegen 3.2.1 ist durch

3.2.3 $\langle e^{uA} * e^{vB}, \hat{f} \rangle := \langle e^{uA}, (e^{vB} * \hat{f}^o)^o \rangle = \langle (e^{uA})_x, \langle (e^{vB})_y, \hat{f}(x+y) \rangle \rangle$

ein lineares Funktional auf $\mathfrak{Z}$ definiert, das nach 3.2.2 stetig ist. Es werde Faltung von e^{uA} und e^{vB} genannt.

3.3 Definition: μ sei ein unbegrenzt teilbares positives beschränktes Radon-Maß auf $\mathfrak{G}$, A sei das infinitesimale Funktional einer Faltungshalbgruppe (μ_t) mit $\mu_1 = \mu$. Dann seien $(\mu_z = e^{zA} : z \in \mathbb{C})$ bzw. $(\mu_s = e^{sA} : s \in \mathbb{R})$ die zu μ bzw. A <u>assoziierten Faltungsgruppen</u> mit komplexer bzw. reeller Indexmenge genannt.

Wegen 2.5.2 ergibt der Satz von der dominierten Konvergenz:

3.3.1 Die Faltungsgruppe $(\mu_z : z \in \mathbb{C})$ ist schwach stetig.

3.4 <u>Die Potenzreihe der Faltungsgruppe:</u> Sei $f \in \mathfrak{Z}$. μ_f bezeichne die Abbildung $z \longmapsto \langle e^{zA}, f\rangle$. Dann gilt nach 2.5.1:

3.4.1 $$\mu_f(z) = \sum_{j=0}^{\infty} \langle A^{*j}, f\rangle \frac{z^j}{j!} \quad \text{für alle } z \in \mathbb{C}. ^{+)}$$

μ_f ist also eine ganze analytische Funktion; oder gleichwertig: Die Faltungsgruppe (e^{zA}) ist ganz-analytisch im schwachen Sinne. Insbesondere gilt:

3.4.2 $$\langle A^{*n}, f\rangle = \frac{d^n}{dz^n} \langle e^{zA}, f\rangle \Big|_{z=0} \quad \text{für alle } f \in \mathfrak{Z}.$$

Dieses Resultat läßt sich je nach Regularität der e^{zA} weiter verschärfen; beispielsweise:

3.4.3 $$\langle A^{*n}, f\rangle = \frac{d^{+n}}{dt^n} \langle \mu_t, f\rangle \Big|_{t=0} \quad \text{für alle } f \in \mathcal{C}^{2n} \ (t \in \mathbb{R}_+).$$

<u>Beweis:</u> Im Falle n = 1 ist 3.4.2 ein Spezialfall von 2.7.1. Für den Fall n = 2 sei $f \in \mathcal{C}^4$; dann gilt $A * f \in \mathcal{C}^2$. Mit Hilfe des Darstellungssatzes 1.3 überzeugt man sich leicht von der Beziehung

$$\langle A * A, f\rangle = \langle A_x, \langle A_y, f(x+y)\rangle\rangle$$

(die in 1.4.3 nur für Funktionen aus $\mathcal{C}_*^\infty$ benutzt wurde); jetzt zeigt 2.7.1:

$$\langle A * A, f\rangle = \lim_{s \downarrow 0} \int \frac{\mu_{s_m} - \varepsilon_o}{s_m} (dx) \lim_{t_n \downarrow 0} \int \frac{\mu_{t_n} - \varepsilon_o}{t_n} (dy) f(x+y).$$

+) Da $\mathfrak{Z}$ tonnerliert ist, konvergieren die Partialsummen gleichmäßig auf den kompakten Teilmengen von $\mathfrak{Z}$, jedoch besteht i.a. keine Konvergenz bezüglich der starken Topologie.

Sei $$g_t(x) := \int f(x+y)\, \frac{\mu_t - \varepsilon_0}{t}\,(dy).$$

Dann gilt nach 2.7.1: $\lim_n g_{t_n}(x) = \langle A_y, f(x+y)\rangle$. Andererseits sind, ebenfalls nach 2.7.1, die $A_{t_n} := \frac{1}{t_n}(\mu_{t_n} - \varepsilon_0)$ in T' schwach-$*$-konvergent gegen A, also normbeschränkt; setzt man ${}_xf(y) := f(x+y)$ und $d := \sup_n \|A_{t_n}\|_{T}$, so folgt mit 2.3.1:

$$|g_{t_n}(x)| \leqslant d\,\|{}_x f\|_T \leqslant cd\|f\|_2,$$

d. h. die g_{t_n} sind gleichgradig beschränkt. Also ist der Satz von der majorisierten Konvergenz anwendbar; man erhält:

$$\langle A*A,f\rangle = \lim_{s_m \downarrow 0}\ \lim_{t_n \downarrow 0} \left\langle \frac{1}{s_m t_n}(\mu_{s_m+t_n} - (\mu_{s_m} + \mu_{t_n}) + \varepsilon_0, f\right\rangle,$$

woraus 3.4.3 folgt. Analog führt man den Beweis für den Fall $n \geqslant 2$. □

3.5 Definition: Ein schwach analytischer Vektor der Faltungsgruppe (e^{zA}) ist ein $f \in \mathcal{C}^0$ mit der Eigenschaft, daß die Reihe 3.4.1 für alle $z \in \mathbb{C}$ konvergiert. (In diesem Fall ist die Gleichung 3.4.1 für $t \geqslant 0$ stets erfüllt.) Der Analytizitätsbereich von (e^{zA}) ist die Menge der schwach analytischen Vektoren von (e^{zA}). Er werde mit $\mathcal{A}(A)$ bezeichnet.

$\mathcal{A}(A)$ ist ein linearer Unterraum von $\mathcal{C}^0$, der $\mathfrak{Z}$ enthält; es folgt:

3.6 $\mathcal{A}(A)$ liegt dicht bezüglich $\mathcal{C}^m_\infty$, m = 0,1,2 ... .

3.7.1 Beispiel: A sei ein fast positives straffes Maß, d.h. stetig bezüglich der Supremumsnorm. (Wegen 1.1.4 liegt T_∞ dicht in $\mathcal{C}^0_\infty$, so daß sich die Restriktion von A auf T eindeutig zu einem beschränkten Radon-Maß auf $\mathcal{C}^0_\infty$ und damit eindeutig zu einem endlichen straffen Maß $\bar{A}$ auf $\mathcal{C}^0$ fortsetzen läßt; die Restriktion von $\bar{A}$ auf T ist eine straffe Fortsetzung von $A|_{T_\infty}$ und stimmt wegen der Eindeutigkeit dieser Fortsetzung (siehe 1.5) mit A überein, was die obige Bezeichnungsweise legitimiert.) Man erhält $\mathcal{A}(A) = \mathcal{C}^0$ (e^{zA} ist sogar normkonvergent).

3.7.2 Beispiel: $G = \mathbb{R}$, $A = -\varepsilon_0'$; dann ist A erzeugendes Funktional der Translationshalbgruppe $(\varepsilon_t : t \in \mathbb{R}_+)$. Man erhält:

$\mathcal{A}(A) = \{f \in \mathcal{C}^0 : f$ ist ganz-analytisch$\}$.

3.7.3 Beispiel: $G = \mathbb{R}$, $A = \frac{\varepsilon_0''}{2}$; dann ist A erzeugendes Funktional der Brownschen Faltungshalbgruppe $(\nu_{0,t} : t \in \mathbb{R}_+)$. Bezeichnet $M_n(\nu_{0,t})$ das n-te Moment der Gaußverteilung $\nu_{0,t}$, so erhält 3.4.1 für $z = t \in \mathbb{R}_+$ und $f \in \mathcal{A}(A)$ folgende Form:

3.7.4 $$\int f(x)\nu_{0,t}(dx) = \sum_{j=0}^{\infty} M_j(\nu_{0,t}) \frac{f^{(j)}(0)}{j!}.$$

Ein derartiges Momentenproblem ist typisch für die Bestimmung der schwach-analytischen Vektoren. Setzt man speziell $f(x) = e^{-2x^2}$ und $t = 1$, so steht in 3.7.4 linker Hand $\frac{1}{\sqrt{2\pi}} \int e^{-\frac{5}{2}x^2} dx = \frac{1}{\sqrt{5}}$, während rechter Hand die divergente Reihe $\sum_{j=0}^{\infty} (-1)^j \frac{(2j)!}{(j!)^2}$ steht. Also ist die Reihe 3.4.1 nicht für alle ganz-analytischen Funktionen f aus $\mathcal{S}$ konvergent, selbst nicht für $z \in \mathbb{R}_+$.

4 Verallgemeinerung auf lokalkompakte abelsche Gruppen.

4.1 G sei abelsche Lie-Gruppe. Dann ist G direktes Produkt zweier abelscher Gruppen G' und G'', wobei G' zusammenhängende Lie-Gruppe und G'' diskret ist ([2], ch.I,4,Prop.2). Für $f \in \mathcal{C}^0(G)$ und $x'' \in G''$ bezeichne $f^{x''}$ die Funktion $x' \longmapsto f(x',x'')$ auf G'. Ist $f^{x''}$ für alle $x'' \in G''$ m-mal differenzierbar, so setze ich (mit $x=(x',x'') \in G$):

4.1.1 $$\|f\|_m(x) := \|f^{x''}\|_m(x'), \quad \|f\|_m := \sup_{x' \in G', x'' \in G''} \|f^{x''}\|_m(x').$$

$\mathcal{C}^m = \mathcal{C}^m(G)$ bezeichne den Vektorraum aller derartigen f mit $\|f\|_m < \infty$, versehen mit der Norm 4.1.1.

Die Definitionen von $\mathcal{C}^m_\infty$, $\mathcal{C}^m_*$, $\mathcal{C}^\infty$, $\mathcal{C}^\infty_\infty$, $\mathcal{C}^\infty_*$ können jetzt übernommen werden, ebenso die Definitionen 1.2 der integrierbaren Distribution und 1.6 der straffen Distribution, sowie 1.7 ihrer Fouriertransformierten. Die Darstellung 1.3.1 einer integrierbaren Distribution A erhält die Form

$$\langle A,f\rangle = \sum_{|l|\leq m} \int \partial^l_{x'} f^{x''}(x')\,\mu_l(d(x',x'')).$$

Jetzt läßt sich nachprüfen, daß alle Resultate des ersten Paragraphen bestehen bleiben, insbesondere die Ausführungen über die Faltung integrierbarer Distributionen, die Eindeutigkeit der Fortsetzung integrierbarer zu straffer Distributionen sowie der Satz von Parseval und der Faltungssatz. Allerdings bedarf 1.7.2 bzw. 1.7.3 einer Präzisierung des Begriffs "Polynom", auf die hier verzichtet sei.

$e=(e',e'')$ bezeichne die Einheit der Gruppe $\mathfrak{G}$. $T=T(\mathfrak{G})$ bezeichne den Vektorraum aller $f\in\mathcal{C}^0(\mathfrak{G})$ mit $f^{e''}\in T(\mathfrak{G}')$. Die Definition 2.1.1 der Norm $\|\ \|_T$ überträgt sich (wobei $(f^{e''}\cdot\vartheta)'(o)$ bzw. $(f^{e''}\cdot\vartheta)''(o)$ statt $f'(o)$ bzw. $f''(o)$ steht und man eine Koordinatenabbildung ξ bzw. eine Majorantenfunktion Ψ auf $\mathfrak{G}'$ wählt und sie vermöge $\tilde{\xi}(x',x''):=\xi(x')$, $\tilde{\Psi}(x',x''):=\Psi(x')$ auf $\mathfrak{G}$ fortsetzt). Jetzt können alle Definitionen und Resultate des zweiten Paragraphen übernommen werden. Dasselbe trifft auf den dritten Paragraphen zu (abgesehen von den Beispielen 3.7.2 und 3.7.3).

4.2 <u>$\mathfrak{G}$ sei lokalkompakte abelsche Gruppe.</u> Dann ist $\mathfrak{G}$ <u>Lie-projektiv</u>, d.h. es existiert ein absteigend filtrierendes System $(K_\alpha)_{\alpha\in I}$ kompakter Untergruppen von $\mathfrak{G}$ mit $\bigcap_{\alpha\in I} K_\alpha = \{e\}$, so daß jede Quotientengruppe $\mathfrak{G}_\alpha := \mathfrak{G}/K_\alpha$ eine (abelsche) Lie-Gruppe ist. π_α bezeichne die Surjektion $\mathfrak{G}\to\mathfrak{G}_\alpha$, $\mathcal{C}^m_\alpha$ den Vektorraum der Funktionen $f\circ\pi_\alpha$ mit $f\in\mathcal{C}^m(\mathfrak{G}_\alpha)$, versehen mit der mit Hilfe von 4.1.1 definierten Norm

4.2.1 $\|f\circ\pi_\alpha\|_m := \|f\|_m.$

Sind K_α und K_β zwei Untergruppen des Systems (K_α) mit $K_\alpha \supseteq K_\beta$, so gilt $\mathcal{E}^m_\alpha \subseteq \mathcal{E}^m_\beta$, und die (durch die Norm 4.2.1) auf $\mathcal{E}^m_\alpha$ gegebene Topologie stimmt mit der von $\mathcal{E}^m_\beta$ induzierten Topologie überein. $\mathcal{E}^m = \mathcal{E}^m(G)$ bezeichne den induktiven Limes der $\mathcal{E}^m_\alpha$. Ist K eine kompakte Untergruppe von G derart, daß $G/_K$ eine Lie-Gruppe ist, so existiert ein $\alpha \in I$ mit $K \supseteq K_\alpha$ ([3], lemme 1), woraus folgt, daß die Definition des lokalkonvexen Raumes $\mathcal{E}^m$ nicht von der Wahl des filtrierenden Systems (K_α) abhängt. Analog seien die $\mathcal{C}^m_{\infty,\alpha}$ und $\mathcal{C}^m_\infty$ definiert. Man erhält ([7], VII, §1, Satz 3):

4.2.2 $\mathcal{C}^m$ bzw. $\mathcal{C}^m_\infty$ ist eine folgenvollständige Hausdorffsche lokalkonvexe seperat-stetige topologische Algebra (und vollständig, wenn das definierende System $(K_\alpha)_{\alpha \in I}$ abzählbar gewählt werden kann).

Bemerkung: Die strikte induktive Topologie dient hier nur dazu, die Stetigkeit linearer Funktionale aus der Stetigkeit ihrer Restriktionen auf die definierenden Teilräume zu sichern. Kann $(K_\alpha)_{\alpha \in I}$ nicht abzählbar gewählt werden, so läßt sich der lokalkonvexe induktive Limes vorteilhaft durch eine LB-Struktur ersetzen (man gibt die Bornologie des Raumes durch ein überdeckendes System „beschränkter" absolutkonvexer Teilmengen an; siehe [1]).

Ist $A \in (\mathcal{C}^m_\infty)'$, so sei A_α die Restriktion von A auf $\mathcal{C}^m_{\infty,\alpha}$. Ist $B \in (\mathcal{C}^m_\infty)'$, so sei $A * B$ durch $(A*B)_\alpha := A_\alpha * B_\alpha$ definiert. Aus den definierenden Gleichungen 1.4.1 und 1.4.3 folgt zusammen mit 1.1.4, daß $K_\alpha \supseteq K_\beta$ impliziert: $(A_\beta * B_\beta)\big|_{\mathcal{C}^m_{\infty,\alpha}} = A_\alpha * B_\alpha$, so daß die Definition von $A*B$ sinnvoll ist. Ebenso läßt sich die straffe Fortsetzung $\bar{A} \in (\mathcal{C}^m)'$ aus den straffen Fortsetzungen $\bar{A}_\alpha \in (\mathcal{C}^m_\alpha)'$ zusammensetzen.

Bekanntlich sind die Charaktere von G Elemente von $\mathcal{C}^m$ (siehe z.B. [3], Seite 50), so daß sich die Definition 1.7 der Fouriertransformierten von $A \in (\mathcal{C}^m)'$ übernehmen läßt. 1.7.1 bleibt richtig, 1.7.2 wird im Folgenden nicht benötigt.

<u>Satz:</u> $\mathfrak{Z} := (L^1_*(\hat{G}))^\wedge \subseteq \mathcal{C}^\infty_\infty$.

<u>Beweis:</u> Sei $f \in L^1_*(\hat{G})$ und K die vom Träger f erzeugte Untergruppe; K liegt in einer kompakt erzeugten offenen Untergruppe H.
Da H offen, also $\hat{G}|_H$ diskret ist, ist $H^\perp \cong (\hat{G}/_H)^\wedge$ ([6], prop. 3.2) kompakt; wegen $f = f1_H$ gilt $\hat{f} = \hat{f} * \omega_{H^\perp}$, wenn $\omega_{H^\perp}$ das normierte Haarsche Maß auf $H^\perp \subseteq G$ ist. Also ist $\hat{f}$ $H^\perp$-invariant. Da H kompakt erzeugt ist, gilt $H \cong \mathbb{R}^p \times \mathbb{Z}^q \times \mathbb{F}$ ([6], cor. 3.1) ($\mathbb{F}$ kompakt), also $\hat{H} = \mathbb{R}^p \times \mathbb{T}^q \times \hat{\mathbb{F}}$ ([6], cor. 3.2), und da $\hat{\mathbb{F}}$ diskret ist, ist $G|_{H^\perp}$ Lie-Gruppe. Es folgt $\hat{f} \in \mathcal{C}^\infty_\infty(G/_{H^\perp}) \subseteq \mathcal{C}^\infty_\infty(G)$. □

$T_\alpha := T(G_\alpha)$ trage stets die mit Hilfe von 2.1.1 definierte Norm (siehe letzten Absatz von 4.1). Ist $K_\alpha \supseteq K_\beta$, so induziert $T_\beta = T(G/K_\beta)$ auf dem linearen Teilraum T_α wegen 2.1.2 die Topologie von T_α, d.h. $(T_\alpha)_{\alpha \in I}$ ist ein striktes induktives Spektrum; T bezeichne den lokalkonvexen induktiven Limes dieses Spektrums. Damit sind die Darlegungen der Paragraphen 2 und 3 übertragbar; in 2.7.5 und 2.8 muß "unbegrenzt teilbar" durch "stetig einbettbar" ersetzt werden.

<u>Nachbemerkung:</u> Ein Teil der Resultate läßt sich auf gewisse nicht-kommutative Gruppen (Moore-Gruppen) und symmetrische Räume übertragen. Diese Resultate (und Anwendungen) werden an anderer Stelle veröffentlicht.

Literatur

1 G. Allan, H. Dales, J. Mc Clure — Pseudo-Banach Algebras. Studia Math. 40, 59-63 (1971)

2 J. Braconnier — Sur les groupes topologiques localement compacts. J. Math. Pures Appl. 27, 1-85 (1948)

3 F. Bruhat — Distributions sur un groupe localement compac et applications à l'étude des représentation des groupes p-adiques. Bull. Soc. math. France 89, 43-75 (1961)

4 P. Courrège — Générateur infinitésimal d'un semi-groupe de convolution sur $\mathbb{R}^n$ et formule de Lévy-Khinchine. Bull. Sc. math. 88, 3-30 (1964)

5 Th. Drisch — Ein neuer Zugang zur Theorie schwach stetiger Faltungshalbgruppen. Eingereicht bei der Z. Wahrscheinlichkeitstheorie verw. Geb.

6 A. Guichardet — Analyse harmonique commutative. Dunod , Paris 1968

7 A. Robertson, W. Robertson — Topologische Vektorräume. Bibliographisches Institut, Mannheim 1967

8 L. Schwartz — Théorie des distributions I. Hermann, Paris 1950.

Thomas Drisch
Universität Dortmund
Abteilung Mathematik
Postfach 500 500
D-4600 Dortmund 50

QUELQUES PROPRIETES DU NOYAU POTENTIEL ASSOCIE A UNE MARCHE ALEATOIRE

par Laure ELIE

L'objet de cet exposé est de rassembler quelques résultats de compacité du noyau potentiel. Soient G un groupe localement compact à base dénombrable (LCD) et μ une mesure de probabilité sur G ayant les propriétés suivantes:

- μ est adaptée, c'est à dire que le sous-groupe fermé engendré par le support de μ est G.

- la mesure $U = \sum_{n\geq 0} \mu^n$ est une mesure de Radon, μ^n désignant la $n^{\text{ème}}$ puissance de convolution de μ ; en d'autres termes, la marche aléatoire droite $(X_n)_{n\in N}$ de loi μ - chaîne de Markov d'espace d'états G et de probabilité de transition $P(g,.)$ égale à $\varepsilon_g * \mu$ - est transiente.

Le noyau potentiel associé à la marche $(X_n)_{n\in N}$ sera encore désigné par U et, pour tout élément g et tout borélien A de G, nous avons

$$U(g,A) = \sum_{n\geq 0} P_n(g,A) = \sum_{n\geq 0} \varepsilon_g * \mu^n(A) = \varepsilon_g * U(A).$$

Il est facile de démontrer à l'aide du principe du maximum que l'ensemble des mesures $\{ \varepsilon_g * U , g \in G \}$ est vaguement relativement compact et nous pouvons nous poser les deux questions suivantes :

1) Quel est l'ensemble I_μ des valeurs d'adhérence vague de $\{ \varepsilon_g * U , g \in G \}$ lorsque g tend vers δ, point à l'infini du groupe dans la compactification d'Alexandroff ? Cet ensemble I_μ , qui contient nécessairement la mesure nulle, est-il réduit à cette seule mesure ? La résolution de ce problème est souvent appelée étude du renouvellement.

2) L'ensemble $\{ \varepsilon_g * U * \varepsilon_{g'} , (g,g') \in G\times G \}$ est-il vaguement relativement compact ?

Il découle du principe du maximum que cette question est équivalente à la suivante: l'ensemble $\{ \varepsilon_g * U * \varepsilon_{g^{-1}} , g \in G \}$ est-il vaguement relativement compact ?

Les réponses à ces questions dépendent fondamentalement de la structure du groupe G et nous chercherons à regrouper, à titre comparatif, les divers résultats relatifs à ces deux problèmes et à donner l'idée des démonstrations. Nous serons amenés à utiliser les définitions suivantes, la mesure de probabilité μ étant désormais supposée adaptée.

Définition 1 : Une marche aléatoire droite transiente de loi μ sur un groupe G (LCD) est dite de type I si l'ensemble I_μ est réduit à la mesure nulle. Elle est dite de type II dans le cas contraire.

Un groupe G (LCD) est dit de type I si toute marche aléatoire droite transiente de loi étalée sur G (i.e. possédant une puissance de convolution non étrangère à une mesure de Haar) est de type I. Il est dit de type II dans le cas contraire.

Définition 2 : Une marche aléatoire de loi μ sur un groupe G (LCD) est dite avoir la propriété (P) si l'ensemble $\{ \varepsilon_g * U * \varepsilon_{g'} , (g,g') \in G \times G \}$ est vaguement relativement compact.

Un groupe G (LCD) est dit de type (P) si toute marche aléatoire transiente de loi étalée sur G a la propriété (P).

Nous allons dans cet exposé expliciter la structure des groupes, extension compacte de leur composante connexe, qui sont de type I ou de type (P). Ces résultats sont donnés en détail dans (10) pour les groupes de type I et dans un article (4) écrit avec P. Bougerol pour les groupes de type (P).

Il est possible d'établir un certain lien entre les questions 1) et 2) par le fait (4) que toute marche transiente de loi symétrique sur un groupe LCD et qui possède la propriété (P) est de type I.

Le problème du renouvellement a d'abord été résolu sur les groupes abéliens (cf.(13)), puis sur les groupes nilpotents (11) et sur les groupes de Lie connexes de type (R) ((6), (11)) ; les résultats obtenus étaient les suivants:

Un de ces groupes G est de type II si et seulement si il existe un sous-groupe K compact distingué dans G tel que le groupe quotient G/K soit isomorphe à $\mathbb{R}$ ou $\mathbb{Z}$. Alors, si une marche de loi μ est de type II, I_μ contient deux éléments: la mesure nulle et une mesure de Haar.

En fait ces résultats se généralisent aux groupes unimodulaires. Par contre sur les groupes non unimodulaires, les résultats sont très différents. L'exemple le plus simple de groupe de type II non unimodulaire est le groupe affine de la droite réelle (9) et sur un tel groupe, si une marche de loi μ est de type II, alors I_μ contient une infinité d'éléments. Des résultats récents ont été obtenus sans l'hypothèse que le groupe est extension compacte de sa composante connexe, mais de ce fait les théorèmes de structure sont moins précis (14).

Les groupes abéliens sont évidemment de type (P) et on peut voir (6) qu'il en est de même des groupes unimodulaires. Considérons par contre le groupe affine connexe de la droite réelle. Ce groupe peut s'écrire comme le produit semi-direct $\mathbb{R} \times \mathbb{R}^{+*}$ où le produit de deux éléments (h,a) et (h',a') est défini par (h+ah',aa').

Comme ce groupe est non unimodulaire, toute marche sur ce groupe est transiente (5); mais toute marche de loi symétrique n'a pas la propriété (P). En effet si K_1 et K_2 sont respectivement des compacts de $\mathbb{R}$ et $\mathbb{R}^{+*}$, nous avons, pour tout élément a de $\mathbb{R}^{+*}$

$$\varepsilon_{(0,a)} * U * \varepsilon_{(0,a)^{-1}} (K_1 \times K_2) = U(a^{-1}K_1 \times K_2).$$

On peut choisir K_1 de façon à ce que, lorsque a décroit vers 0, $U(a^{-1}K_1 \times K_2)$ croisse vers $U(\mathbb{R} \times K_2)$. Comme la mesure μ est symétrique, $U(\mathbb{R} \times K_2)$ est infini si K_2 est d'intérieur non vide (la marche projection sur $\mathbb{R}^{+*}$ est alors récurrente). Par suite l'ensemble $\{ \varepsilon_{(0,a)} * U * \varepsilon_{(0,a)^{-1}} , a \in \mathbb{R}^{+*} \}$ n'est pas vaguement relativement compact, et le groupe affine n'est pas de type (P).

Rappelons enfin que les groupes non moyennables sont de type (I) ((2),(7)) et de type (P) (3).

Il est souvent intéressant puisque le groupe $G/[\overline{G,G}]$ est abélien, $[\overline{G,G}]$ étant le sous-groupe fermé de G engendré par les commutateurs, de considérer les marches projections sur $G/[\overline{G,G}]$. On notera π la projection de G sur $G/[\overline{G,G}]$.
Soit $(X_n)_{n\in\mathbb{N}}$ une marche aléatoire droite de loi μ sur G, il découle de l'inégalité suivante, vraie pour tout borélien A de G et tout couple (g,g') de G×G,

$$\varepsilon_g * U * \varepsilon_{g'} (A) \leq \varepsilon_{\pi(g)} * \pi(U) * \varepsilon_{\pi(g')} (\pi(A))$$

que - si la marche $\pi(X_n)$ est transiente sur $G/[\overline{G,G}]$, alors la marche X_n est transiente et a la propriété (P).

- si la marche $\pi(X_n)$ est transiente et de type I, alors la marche X_n est transiente et de type I.

Nous savons que si le groupe G est compactement engendré, le groupe abélien $G/[\overline{G,G}]$ s'écrit $R^{d_1} \oplus Z^{d_2} \oplus K$ où d_1 et d_2 sont deux entiers et K un groupe compact abélien. L'entier d_1+d_2 s'appelle le rang du groupe abélien $G/[\overline{G,G}]$.

- Si le rang de $G/[\overline{G,G}]$ est supérieur ou égal à trois, toute marche sur $G/[\overline{G,G}]$ est transiente et de type I (13) et il découle des affirmations ci-dessus que le groupe G est de type I et de type (P).

- Si le rang de $G/[\overline{G,G}]$ est deux, alors toute marche transiente sur $G/[\overline{G,G}]$ est de type I, mais il existe aussi sur un tel groupe des marches récurrentes (13). Par suite si X_n est une marche sur G dont la projection $\pi(X_n)$ sur $G/[\overline{G,G}]$ est transiente, alors X_n est transiente de type I et a la propriété (P). Mais que se passe-t-il si $\pi(X_n)$ est récurrente ?

Il est possible de montrer (10) qu'un groupe G, extension compacte de sa composante neutre G_o, est de type I si et seulement si toute marche sur G de loi

étalée et de projection sur $G/\overline{[G,G]}$ transiente est de type I .
Nous ferons donc désormais l'hypothèse que G/G_o est un groupe compact et alors si le rang de $G/\overline{[G,G]}$ est deux, le groupe G est de type I .

- Si le rang de $G/\overline{[G,G]}$ est nul, alors le groupe G est unimodulaire. Nous savons qu'il est de type (P). Comme toute marche sur un tel groupe est telle que sa projection sur le groupe $G/\overline{[G,G]}$ alors compact est récurrente, il découle de la remarque ci-dessus qu'un tel groupe est de type I .

Nous sommes donc amenés pour étudier les groupes de type II à considérer le cas où le rang de $G/\overline{[G,G]}$ est un, et pour les groupes qui ne sont pas de type (P) le cas où le rang de $G/\overline{[G,G]}$ est un ou deux.

La proposition suivante qui met en évidence la dissymétrie des groupes non unimodulaires sera un outil fondamental dans la démonstration des théorèmes de structure que nous donnerons ensuite.

<u>Proposition</u> : Soient G un groupe LCD non unimodulaire et μ une mesure de probabilité sur G étalée. Alors

1) $\varepsilon_g * U$ converge vaguement vers la mesure nulle lorsque g tend vers le point à l'infini δ de manière à ce que le module $\Delta(g)$ reste borné supérieurement.

2) pour tout réel positif M, l'ensemble $\{ \varepsilon_g * U * \varepsilon_{g^{-1}} , g \in G, \Delta(g) \leq M \}$ est vaguement relativement compact.

<u>Démonstration</u> : Si m_D est une mesure de Haar à droite sur G, la fonction module Δ est définie pour $g \in G$ par $\varepsilon_g * m_D = \Delta(g)\, m_D$.

1) Si σ_g est l'opérateur de translation à gauche par g dans G, alors pour toute fonction de $C_K^+(G)$ (i.e. continue positive à support compact), nous avons $U\sigma_g f(y) = Uf(gy)$ sachant que pour tout y de G, $\sigma_g f(y) = f(gy)$.

Soit ν une valeur d'adhérence de $\{ \varepsilon_g * U , g \in G \}$ lorsque g tend vers δ de manière à ce que $\Delta(g)$ reste borné supérieurement. Comme μ est étalée, il existe (10) une suite $(g_n)_{n\in N}$ d'éléments de G telle que $\overline{\lim_n} \Delta(g_n)$ soit fini, telle que $\varepsilon_{g_n} * U$ converge vaguement vers ν et surtout telle que $U\sigma_{g_n} f$ converge simplement pour tout f de $C_K^+(G)$.

Par dualité, nous obtenons, pour tout h de $C_K^+(G)$,

$$<U\sigma_g f,h>_{m_D} = <\sigma_g f,\hat{U}h>_{m_D} = \Delta(g) <f,\hat{U}\sigma_{g^{-1}}h>_{m_D} ,$$

si $\hat{U}$ désigne le noyau potentiel associé à la mesure $\hat{\mu}$,image de μ par l'application qui à $g \in G$ associe g^{-1}. Par suite,

$$<U\sigma_g f,h>_{m_D} \leq \Delta(g)\, m_D(f)\, \|\hat{U}h\| .$$

Il est alors clair que si $\underline{\lim}_n \Delta(g_n)$ est nul, $U\sigma_{g_n} f$ converge vers 0 pour tout f de $C_K^+(G)$ et donc que la mesure ν est nulle.

Si $\underline{\lim}_n \Delta(g_n)$ est non nul, $\Delta(g_n)$ reste dans un compact. L'égalité de dualité ci-dessus permet encore de conclure lorsque la marche $\Delta(X_n)$ sur $\mathbb{R}^{+*}$ est récurrente. Si la marche $\Delta(X_n)$ sur $\mathbb{R}^{+*}$ est transiente, il suffit de remarquer que pour tout compact K de G, la suite $g_n^{-1}K$ reste dans un borélien A de G d'image $\Delta(A)$ compacte et donc de potentiel U(A) fini. Il résulte alors du théorème de lebesgue que $\varepsilon_{g_n} * U(K)$ tend vers 0 lorsque $n \to \infty$.

2) En utilisant les méthodes de réductions de l'article (1), on montre qu'il suffit de vérifier l'énoncé 2) de la proposition lorsque la probabilité μ est symétrique et admet une densité ϕ continue à support compact par rapport à m_D. Alors

$$\varepsilon_g * U * \varepsilon_{g^{-1}}(f) = f(e) + \varepsilon_g * \phi.m_D * U * \varepsilon_{g^{-1}}(f)$$

$$\leq f(e) + \Delta(g) \; \|U\phi\| \; m_D(f)$$

si $f \in C_K^+(G)$. D'où le résultat.

Théorème 1 : Soit G un groupe de Lie tel que G/G_o soit fini. Alors G est de type II si et seulement si il est produit semi-direct de deux sous-groupes fermés N et A tels que

- N est distingué nilpotent simplement connexe.
- A est extension d'un sous-groupe compact distingué par $\mathbb{R}$.
- si l'algèbre de Lie $\mathcal{N}$ de N est non réduite à {0} , il existe un élément x de A tel que les valeurs propres de Adx restreint à $\mathcal{N}$ soient toutes de module strictement supérieur à un.

Théorème 2 : Soit G un groupe de Lie tel que G/G_o soit fini. Alors les trois assertions suivantes sont équivalentes :

1) G n'est pas de type (P).
2) G_o n'est pas de type (P).
3) G_o est produit semi-direct de deux sous-groupes fermés non triviaux N et A tels que

- N est distingué nilpotent simplement connexe.
- A est extension d'un sous-groupe compact distingué par $\mathbb{R}$ ou $\mathbb{R}^2$.
- si $\mathcal{N}$ désigne l'algèbre de Lie de N, il existe un élémenr x de A tel que les valeurs propres de Adx restreint à $\mathcal{N}$ soient toutes de module strictement supérieur à un.

Dans ce cas les marches sur G_o n'ayant pas la propriété (P) sont celles dont la projection sur G_o/N est récurrente.

Remarque : Il découle du théorème 1 qu'un groupe de type II est unimodulaire si et seulement si le groupe N est trivial. Un tel groupe est alors extension d'un sous-groupe compact distingué par $\mathbb{R}$. Dans le théorème 2 le groupe N n'est jamais trivial puisqu'un groupe unimodulaire est de type (P).

Démonstration des théorèmes 1 et 2 :

Soit G un groupe de Lie connexe qui soit de type II (respectivement ne soit pas de type (P)). Il est alors moyennable. Notons R le radical de G et C le sous-groupe compact maximum du radical nilpotent $N = \overline{[G,R]}$; ce groupe est central dans G et quitte à remplacer G par G/C , il suffit de montrer les théorèmes lorsque N est simplement connexe, ce que nous supposerons désormais. Le groupe G admet alors (cf.(12)) la décomposition G = NPS où P est un sous-groupe fermé nilpotent et S un sous-groupe compact semi-simple tels que $[P,S] = \{0\}$ et $\overline{[G,G]} = NS$.

Si le groupe N est trivial, le groupe G est de type (R). Il ne peut donc être de type non (P) et il est de type II si et seulement si il est extension d'un sous-groupe compact distingué par $\mathbb{R}$. Nous allons donc supposer que N est non trivial.

Soit $(g_n)_{n\in\mathbb{N}}$ une suite d'éléments de G tendant vers δ . Si il n'existe pas d'élément x de P tel que les valeurs propres de Adx restreint à l'algèbre de Lie $\mathcal{N}$ de N soient toutes de module strictement supérieur à un, on peut toujours construire (10) un sous-groupe fermé B distingué dans G, strictement inclus dans N, associé à la suite g_n tel que

- soit G/B est non unimodulaire et si s désigne la surjection de G sur G/B et $\dot{\Delta}$ le module sur G/B , la suite $\dot{\Delta}(s(g_n))$ est bornée.

- soit G/B est unimodulaire et de plus de type (R).

Si le groupe G/B est non unimodulaire, la proposition précédente appliquée au groupe G/B permet de montrer que pour toute mesure de probabilité étalée sur G la suite $\varepsilon_{g_n} * U$ converge vaguement vers la mesure nulle (respectivement la suite $\varepsilon_{g_n} * U * \varepsilon_{g_n^{-1}}$ est vaguement relativement compacte).

Supposons donc que G/B est unimodulaire. Alors si G est de type II , il existe une marche $(X_n)_{n\in\mathbb{N}}$ sur G de type II dont la projection $\pi(X_n)$ sur $G/\overline{[G,G]}$ est transiente. Nous savons de plus que le rang de $G/\overline{[G,G]}$ est nécessairement un. Comme B est strictement inclus dans N, lui-même inclus dans $\overline{[G,G]}$, le groupe G/B de type (R) est de type I ((6),(11)). La marche projection $s(X_n)$ transiente sur G/B est de type I et X_n ne peut donc être de type II.

Si G n'est pas de type (P), nous obtenons le même type de contradiction lorsque $G/\overline{[G,G]}$ est de rang deux. Si $G/\overline{[G,G]}$ est de rang un, le problème est plus délicat car la projection $s(X_n)$ peut être récurrente (cf.(4)).

En utilisant cette construction pour toute suite g_n de G tendant vers δ ,on prouve que si G est de type II (respectivement de type non (P)), il existe un élément x de P tel que les valeurs propres de Adx restreint à $\mathfrak{S}$ soient toutes de module strictement supérieur à un. Alors (12) le groupe G est produit semi-direct du groupe distingué N et du groupe A = PS. Comme nécessairement le groupe $G/\overline{[G,G]}$ est de rang un (resp. un ou deux), le groupe A est extension d'un sous-groupe compact distingué par $\mathbb{R}$ (resp. $\mathbb{R}$ ou $\mathbb{R}^2$).

Pour démontrer que de tels groupes sont de type II (resp. non (P)), on généralise les méthodes utilisées sur le groupe affine de la droite réelle ((9),(4),(10)).

Remarque : Le type (I,II,(P),non(P)) des groupes LCD est stable si on quotiente par un sous-groupe compact distingué et on peut donc déterminer le type des groupes LCD, extension compacte de leur composante neutre : un tel groupe contient des sous-groupes compacts distingués K (arbitrairement petits) tels que G/K soit un groupe de Lie ayant un nombre fini de composantes connexes.

On peut en fait montrer que si G est un groupe LCD tel que G/G_0 soit compact,

$$G_0 \text{ est de type (P)} \Longleftrightarrow G \text{ est de type (P)}$$

$$G_0 \text{ est de type I} \Longrightarrow G \text{ est de type I}$$

Mais la réciproque de cette dernière assertion est fausse. Il suffit de considérer le groupe des symétries et translations de $\mathbb{R}$, groupe de type I, mais dont la composante neutre (isomorphe à $\mathbb{R}$) est de type II.

BIBLIOGRAPHIE

1 - P. BALDI, N. LOHOUE, J. PEYRIERE - Sur la classification des groupes récurrents. CRAS 285 (1977) p.1103.

2 - C. BERG et J.P.R. CHRISTENSEN - On the relation between amenability of locally compact groups and the norms of convolution operators. Math. Ann. 208 (1974), p.149-153.

3 - P. BOUGEROL - Fonctions de concentration sur certains groupes localement compacts. Z. für Wahrscheinlichkeitstheorie verw. Gebiete 45, 135-157 (1978).

4 - P. BOUGEROL et L. ELIE - Sur une propriété de compacité du noyau potentiel associé à une probabilité sur un groupe. (A paraître).

5 - A. BRUNEL, P. CREPEL, Y. GUIVARC'H, M. KEANE - Marches aléatoires récurrentes sur les groupes localement compacts. CRAS 275 (1972).

6 - A. BRUNEL et D. REVUZ - Sur la théorie du renouvellement pour les groupes non abéliens. Israël J. Math. 20 n°1 (1975) p.46 .

7 - Y. DERRIENNIC et Y. GUIVARC'H - Théorème de renouvellement pour les groupes non moyennables. CRAS 277 (1973) p.613 .

8 - L. ELIE - Etude du renouvellement pour certains groupes résolubles. CRAS 280 (1975) p.1149 .

Etude du renouvellement sur les groupes moyennables. CRAS 284 (1977) p.555 .

9 - L. ELIE - Etude du renouvellement sur le groupe affine de la droite réelle. Ann. Univ. Clermont 85 (1977) p.47-62 .

10 - L. ELIE - Etude du renouvellement sur les groupes G (LCD) tels que G/G_o soit compact. (A paraître).

11 - Y. GUIVARC'H, M. KEANE, B. ROYNETTE - Marches aléatoires sur les groupes de Lie. Lecture Notes n°624 (1977).

12 - A. RAUGI - Fonctions harmoniques sur les groupes moyennables. CRAS 280 (1975) p.1309.

13 - D. REVUZ - Markov Chains. North-Holland Mathematical Library Vol.11 (1975).

14 - C. SUNYACH - Sur la théorie du renouvellement pour les groupes non unimodulaires CRAS 284 (1977) p.547.

Récurrence et renouvellement des chaînes de Markov. Thèse Université Paris VII.

Laure ELIE
UER DE MATHEMATIQUES
Université Paris VII
2 Place Jussieu
75005 - PARIS

FONCTIONS HARMONIQUES POSITIVES SUR LE GROUPE AFFINE

Laure ELIE

La détermination des fonctions harmoniques positives et leur "représentation intégrale" est un problème qui a été examiné sur certains groupes à croissance polynomiale ((3), (4), (10)) et sur les groupes semi-simples((5), (9)). Il semble qu'il y ait principalement deux méthodes pour aborder ce sujet ; la première consiste à déterminer, à l'aide d'une "propriété de droite fixe", les génératrices extrémales du cône convexe des fonctions harmoniques positives ; la deuxième vise à expliciter la frontière de Martin. C'est cette frontière que nous allons rechercher sur le groupe affine, exemple le plus simple de groupe résoluble à croissance exponentielle. Le compactifié de Martin sera ici un espace compact G^* obtenu en complétant l'espace des états G de façon à pouvoir prolonger par continuité le "noyau de Martin" et la frontière de Martin désignera l'ensemble des éléments du compactifié G^* n'appartenant pas à G. Il va donc s'agir d'étudier le comportement asymptotique du "noyau de Martin" et, par suite de déterminer les "directions" du groupe suivant lesquelles ce noyau converge. La connaissance du renouvellement sur ce groupe (7) nous apportera d'utiles renseignements sur la nature de ces limites.

On sait que par la construction de cette frontière de Martin que sous des hypothèses convenables (assurant une dualité entre les fonctions harmoniques et les mesures invariantes) toute fonction harmonique positive admet une représentation intégrale sur cette frontière. Il est souvent délicat de déterminer précisément la frontière, mais comme toute solution harmonique extrémale est alors représentée par un point de cette frontière, il est plus simple de rechercher les éléments extrémaux de la frontière et d'en déduire les génératrices extrémales du cône des fonctions harmoniques positives.

I - Hypothèses et notations :

Le groupe affine connexe de la droite réelle - groupe des transformations réelles $x \longrightarrow ax + b$ - sera désigné par G et μ sera une mesure de probabilité sur G ayant les propriétés suivantes :

Ha) Le semi-groupe topologique engendré par le support de μ est G.

Hb) Il existe une fonction ϕ positive continue à support compact telle que, si m_D désigne une mesure de Haar à droite sur G, la mesure μ s'écrive $\phi.m_D$.

Si E est un espace localement compact, nous noterons $\mathcal{M}^+(E)$ le cône des mesures de Radon positives non nulles sur E. Supposons que E soit un G-espace à gauche, alors si ν et m sont respectivement des éléments de $\mathcal{M}^+(G)$ et $\mathcal{M}^+(E)$, nous appellerons $\nu * m$ la mesure sur E définie, si A est un borélien de E, par

$$\nu * m(A) = \iint_{G \times E} 1_A(g.x)d\nu(g)dm(x).$$

A la mesure de probabilité μ sur G, nous associerons le cône des <u>mesures μ-invariantes</u> défini par

$$\mathcal{H}_\mu = \{\nu \in \mathcal{M}^+(G),\ \nu * \mu = \nu\}\ .$$

Il découle des hypothèses faites sur μ que ce cône est à base compacte (4).

Nous désignerons par $\hat{\mu}$ l'image de μ par l'application qui à $g \in G$ associe g^{-1} et une fonction borélienne positive h est dite <u>$\hat{\mu}$-harmonique</u> si, pour tout $g \in G$,

$$h(g) = \int_G h(gg')\hat{\mu}(dg').$$

Il résulte de Hb) que tout élément ν de $\mathcal{H}_\mu$ admet une densité par rapport à m_D continue positive $\hat{\mu}$-harmonique et réciproquement toute fonction $\hat{\mu}$-harmonique positive est densité d'une mesure μ-invariante. Nous pourrons donc identifier $\mathcal{H}_\mu$ et le cône des fonctions $\hat{\mu}$-harmoniques positives.

Comme le groupe G est non unimodulaire, la mesure $U = \sum_{n \geq 0} \mu^{*n}$ est une mesure de Radon et nous noterons encore U le noyau potentiel défini pour tout élément g de G et toute fonction borélienne positive f sur G par

$$U(g,f) = Uf(g) = \sum_{n \geq 0} \varepsilon_g * \mu^n(f) = \varepsilon_g * U(f).$$

Si $C_K(C_K^+)$ désigne l'ensemble des fonctions continues à support compact (positives) sur G, alors pour tout $r \in C_K^+$, la fonction Ur est continue bornée. Il découle de plus de Ha) que si r est non nulle, Ur est strictement positive.

Nous appellerons si r est un élément non nul de C_K^+, noyau de Martin associé à la fonction de référence r le noyau K_r défini par

$$K_r(g,.) = U(g,.)/Ur(g) = \varepsilon_g * U/Ur(g).$$

Nous savons (11) que si $f \in C_K^+$, la fonction $K_r f = K_r(.,f)$ est continue bornée et l'ensemble de mesures $A_r = \{K_r(g,.), g \in G\}$ est par suite vaguement relativement compact.

La compactification de Martin G_r^* associée à la fonction de référence r sera par définition la fermeture de A_r dans $\mathcal{M}^+(G)$ pour la topologie de la convergence vague. Des hypothèses faites sur μ, il résulte que les différents espaces A_r obtenus lorsque r décrit C_K^+ sont homéomorphes à l'espace topologique G et que les différents espaces G_r sont homéomorphes entre eux. La Frontière de Martin M sera l'un quelconque des espaces $G_r^* \setminus A_r$ et sera donc constitué des valeurs d'adhérence vague de A_r lorsque g tend vers le point à l'infini du groupe dans la compactification d'Alexandrov. Comme μ est à support compact, ces valeurs d'adhérence sont μ-invariantes.

II - Principaux résultats :

Nous représenterons le groupe G par le produit semi-direct $\mathbb{R}^{+*} \times \mathbb{R}$ muni du produit $(a,b)(a',b') = (aa',ab'+b)$. Le sous-groupe des translations de G est alors isomorphe à $\mathbb{R}$ et le sous-groupe des homothéties à $\mathbb{R}^{+*}$. Nous désignerons par a et b les projections respectives de G sur $\mathbb{R}^{+*}$ et $\mathbb{R}$ et tout élément g de G sera donc noté $(a(g), b(g))$. La mesure $a(\mu)$, image de μ par a, est une probabilité sur $\mathbb{R}^{+*}$ ayant les propriétés Ha) et Hb) de μ et nous savons (3) que les éléments extrémaux du cône des mesures $a(\mu)$-invariantes sur $\mathbb{R}^{+*}$ sont des mesures de la forme $h.m_1$ où m_1 est une mesure de Haar sur $\mathbb{R}^{+*}$ et h une exponentielle $\widehat{a(\mu)}$-harmonique, c'est-à-dire une fonction du type $h(x) = x^\alpha$, $\alpha \in \mathbb{R}$, telle que

$$\int_{\mathbb{R}^{+*}} h(x) d\widehat{a(\mu)}(x) = 1.$$

On vérifie aisément en étudiant la fonction convexe F définie sur $\mathbb{R}$ par $F(\alpha) = \int_{\mathbb{R}^{+*}} x^\alpha d\widehat{a(\mu)}(x)$ qu'il y a au plus deux exponentielles $\widehat{a(\mu)}$-harmoniques sur $\mathbb{R}^{+*}$

En outre il découle de Ha) et Hb) que $F(\alpha)$ tend vers l'infini quand $|\alpha|$ tend vers l'infini ; on en déduit qu'il existe exactement deux exponentielles harmoniques si et seulement si la fonction F n'atteint pas son minimum en 0, c'est-à-dire si et seulement si $\int_{\mathbb{R}^{+*}} \text{Log } x \, da(\mu)(x)$ est non nul. Dans ce cas, si α désigne le réel non nul vérifiant $F(\alpha) = 1$, nous noterons h_α l'exponentielle harmonique non constante prolongé à G.

Il est alors clair que si m_2 est une mesure de Haar sur $\mathbb{R}$, les mesures produit définies sur G par $m_1 \otimes m_2$ (mesure de Haar à droite sur G) et $h_\alpha(m_1 \otimes m_2)$ sont μ-invariantes. Il est possible de construire une autre classe de mesures μ-invariantes de la façon suivante : Le sous-groupe $\mathbb{R}$ de G peut être muni d'une structure de G-espace par l'application qui à (g,x) associe $g.x = a(g)x + b(g)$ et si on connait une mesure m sur $\mathbb{R}$ telle que $\hat{\mu} * m = m$ (produit au sens de G-espace), alors la mesure $\widehat{m_1 \otimes m}$ est μ-invariante ainsi que toutes ses translatées à gauche.

Nous démontrerons le théorème suivant :

<u>Théorème</u> 1 - Soit μ une mesure de probabilité sur G vérifiant Ha) et Hb). On notera h_α l'exponentielle $\hat{\mu}$-harmonique non constante lorsqu'elle existe.

1) Si $\int \text{Log } a(g) \, d\mu(g) > 0$, il existe une unique mesure de probabilité m sur $\mathbb{R}$ telle que $\hat{\mu} * m = m$ et les éléments extrémaux du cône $\mathcal{H}_\mu$ sont, si m_1 et m_2 désignent les mesures de Lebesgue respectives sur $\mathbb{R}^{+*}$ et $\mathbb{R}$, les mesures proportionnelles

- à la mesure $h_\alpha \, m_1 \otimes m_2$
- et aux mesures $\varepsilon_{(1,z)} * \widehat{m_1 \otimes m}$, z décrivant $\mathbb{R}$.

2) Si $\int \text{Log } a(g) \, d\mu(g) = 0$, il existe, à une constante multiplicative près, une unique mesure m (nécessairement non bornée) telle que $\hat{\mu} * m = m$ et les éléments extrémaux du cône $\mathcal{H}_\mu$ sont les mesures proportionnelles

- à la mesure de Haar à droite $m_1 \otimes m_2$
- et aux mesures $\varepsilon_{(1,z)} * \widehat{m_1 \otimes m}$, z décrivant $\mathbb{R}$.

3) Si $\int \text{Log } a(g) \, d\mu(g) < 0$, il existe une unique mesure de probabilité m sur $\mathbb{R}$ telle que $(h_\alpha \hat{\mu}) * m = m$ et les éléments extrémaux du cône $\mathcal{H}_\mu$ sont les mesures proportionnelles

- à la mesure de Haar à droite $m_1 \otimes m_2$
- et aux mesures $\varepsilon_{(1,z)} * h_\alpha \, \widehat{m_1 \otimes m}$, z décrivant $\mathbb{R}$.

<u>Remarque</u> 1 Il est aisé de voir que l'application qui à la mesure ν associe la mesure $h_\alpha^{-1}\nu$ est une bijection de $\mathcal{H}_\mu$ sur $\mathcal{H}_{h_\alpha^{-1}\mu}$. Les cas 1) et 3) sont alors duaux l'un de l'autre car si μ vérifie la condition 1), $h_\alpha^{-1}\mu$ vérifie la condition 3) et vice-versa. En effet il résulte de la convexité de la fonction F définie ci-dessus que

les tangentes au graphe de F aux points 0 et α ont des pentes opposées et donc que $\int \log a(g)\, d\mu(g)$ et $\int \log a(g)\, h_\alpha^{-1}(g)\, d\mu(g)$ sont de signe contraire.

De la bijection existant entre les mesures μ-invariantes et les fonctions $\hat{\mu}$-harmoniques positives, nous déduisons du théorème 1 et du théorème de représentation intégrale de Choquet le résultat suivant :

<u>Théorème</u> 2 - Soit μ une mesure de probabilité sur G vérifiant Ha) et Hb). On notera h_β l'exponentielle μ-harmonique non triviale lorqu'elle existe.

1) Si $\int \operatorname{Log} a(g)\, d\mu(g) < 0$, il existe une unique mesure de probabilité m sur $\mathbb{R}$ telle que $\mu * m = m$ et les fonctions μ-harmoniques positives extrémales sont les fonctions proportionnelles

. à l'exponentielle h_β

. aux fonctions k_z, $z \in \mathbb{R}$, définies pour tout $g \in G$ par

$$k_z(g) = \frac{dg.m}{dm_2}(z) \qquad \text{où } g.m = \varepsilon_g * m.$$

Par suite toute fonction μ-harmonique positive h s'écrit, si $g \in G$, $h(g) = ch_\beta(g) + \int_{\mathbb{R}} \frac{dg.m}{dm_2}\, d\rho$ où c est un élément de $\mathbb{R}^+$ et ρ une mesure de Radon sur $\mathbb{R}$ uniquement déterminés par h.

2) Si $\int \operatorname{Log} a(g)\, d\mu(g) = 0$, il existe, à une constante multiplicative près, une unique mesure m sur $\mathbb{R}$ telle que $\mu * m = m$ et les fonctions μ-harmoniques positives extrémales sont les fonctions proportionnelles

. à la fonction constante 1

. aux fonctions k_z, $z \in \mathbb{R}$, définies comme ci-dessus.

Toute fonction μ-harmonique positive h s'écrit donc $h(g) = c + \int_{\mathbb{R}} \frac{dg.m_2}{dm_2}\, d\rho$ où $c \in \mathbb{R}^+$ et $\rho \in \mathcal{M}(\mathbb{R})$.

3) Si $\int \operatorname{Log} a(g)\, d\mu(g) > 0$, il existe une unique mesure de probabilité m sur $\mathbb{R}$ telle que $h_\beta\mu * m = m$ et les fonctions μ-harmoniques positives extrémales sont les fonctions proportionnelles

. à la fonction constante 1

. aux fonctions $h_\beta\, k_z$, $z \in \mathbb{R}$; où les fonctions k_z sont définies comme ci-dessus.

Toute fonction μ-harmonique positive h s'écrit donc $h(g) = c + h_\beta(g)\int_{\mathbb{R}} \frac{dg.m}{dm_2}\, d\rho$ où $c \in R^+$ et $\rho \in \mathcal{M}(\mathbb{R})$.

Remarque 2 - 1) Si la mesure m admet pour densité ϕ, alors $k_z(g) = \frac{dg.m}{dm_2}(z) = a(g)^{-1}\phi(g^{-1}.z) = \Delta(g)\phi(g^{-1}z)$, Δ étant le module sur G.

2) On retrouve les résultats obtenus sur les fonctions harmoniques bornées, à savoir que dans les cas 2) et 3) elles sont constantes puisque 1 est extrémale.

Dans le cas 1) on sait qu'une fonction harmonique positive h est bornée si et seulement si la constante c est nulle et si la mesure ρ admet une densité bornée f. Alors h s'écrit, pour $g \in G$, $h(g) = g.m(f)$ (cf.(8)).

III - Propriétés générales du noyau de Martin et périodes des mesures limites :

Nous allons commencer par étudier les propriétés de compacité du noyau de Martin.

Proposition 1 - Soit μ une mesure de probabilité sur G vérifiant Ha) et Hb) et r et f deux éléments non nuls de $C_K^+(G)$. On note R_g et H_g les fonctions de G dans R définies pour tout $(g,x) \in G \times G$ par

$$R_g(x) = Uf(xg) \;/\; Ur(g)$$

$$H_g(x) = Uf(gx) \;/\; Ur(g).$$

Alors les ensembles de fonctions $\{R_g,\ g \in G\}$ et $\{H_g,\ g \in G\}$ sont équicontinus en tout point x de G.

Démonstration - Remarquons tout d'abord que comme la fonction Uf/Ur est bornée, il suffit en fait de prouver la proposition lorsque f = r.

Si τ_x désigne la translation à gauche dans G par l'élément x de G, alors $\tau_x f$ sera la fonction valant f(xg) au point g, et par suite

$$R_g(x) = \varepsilon_{xg} * U(f) \;/\; Uf(g) = U(\tau_x f) \;/\; Uf(g).$$

On considère l'expression

$$R_g(yx_0) - R_g(x_0) = U(\tau_{yx_0} f - \tau_{x_0} f)(g) \;/\; Uf(g).$$

pour x_o fixé dans G et y appartenant à un voisinage compact V de e. Vérifions alors qu'il existe une constante C_f telle que pour tout $g \in G$ et $y \in V$

$$U(\tau_{yx_o} f - \tau_{x_o} f)(g) \leq C_f \, \| \tau_{yx_o} f - \tau_{x_o} f \| \, Uf(g)$$

si $\| \ \|$ désigne la norme sup.

En effet, comme f est à support compact, la fonction $\tau_{yx_o} f - \tau_{x_o} f$ a son support pour $y \in V$ dans un compact K.

D'après Ha), Uf est strictement positive et est par suite supérieure à une constante $a > 0$ sur K. D'où

$$U(\tau_{yx_o} f - \tau_{x_o} f)(g) \leq \| \tau_{yx_o} f - \tau_{x_o} f \| \, \| U1_K \| \, Uf(g)/a$$

sur K et donc partout d'après le principe du maximum.

Comme f est uniformément continue à gauche, il est alors clair que, pour tout $\varepsilon > 0$, on peut trouver un voisinage compact W de e tel que pour tout $g \in G$ et $y \in W$

$$|R_g(yx_o) - R_g(x_o)| < \varepsilon$$

et l'ensemble de fonctions $\{R_g, \, g \in G\}$ est équicontinu au point x_o.

L'hypothèse Hb) qui pour l'instant n'a pas servi va nous permettre de prouver l'équicontinuité de $\{H_g, \, g \in G\}$. Il va s'agir d'étudier l'expression

$$H_g(x_o y) - H_g(x_o) = [Uf(gx_o y) - Uf(gx_o)]/Uf(g)$$

lorsque x_o est fixé, g décrit G et y appartient à un voisinage compact V' de e. La démonstration va reposer sur la dualité existant entre U et $\hat{U}$. On sait, puisque $\mu = \phi . m_D$, que pour $(g,x) \in G \times G$,

$$\begin{aligned} Uf(gx) &= \varepsilon_{gx} * \phi m_D * U(f) + f(gx) \\ &= \iint f(gxzt)\, \phi(z)\, m_D(dz)\, U(dt) + f(gx) \\ &= \Delta(x) \iint f(gz)\, \phi(x^{-1} z t^{-1})\, m_D(dz) U(dt) + f(gx) \\ &= \Delta(x) \int f(gz)\, \hat{U}\phi(x^{-1} z)\, m_D(dz) + f(gx) \\ &= \Delta(x) \; < \tau_g f, \, \hat{U} \tau_{x^{-1}} \phi > + f(gx) \end{aligned}$$

si Δ désigne la fonction module définie par $\varepsilon_x * m_D = \Delta(x) \, m_D$.

Par suite on obtient

$$|H_g(x_o y) - H_g(x_o)| \leq |f(gx_o y) - f(gx_o)| \;/\; Uf(g)$$
$$+ \Delta(x_o) < \tau_g f, \hat{U}(\Delta(y)\tau_{y^{-1}x_o^{-1}}\phi - \tau_{x_o^{-1}}\phi) > / < \tau_g f, \hat{U}\phi >$$

Comme f est à support compact, la fonction $f(.x_o y) - f(.x_o)$ a son support dans un compact K' si $y \in V'$ et il existe alors, puisque Uf est strictement positive, une constante C telle que

$$|f(g\, x_o y) - f(gx_o)| \leq C \| f(.x_o y) - f(.x_o)\| \, Uf(g).$$

Par hypothèse ϕ est un élément de C_K^+, et il découle donc de la première partie de la démonstration qu'il existe une constante C_ϕ telle que, pour tout $z \in G$ et $y \in V'$, on ait

$$\hat{U}(\tau_{y^{-1}x_o^{-1}}\phi - \tau_{x_o^{-1}}\phi)(z) \leq C_\phi \|\tau_{y^{-1}x_o^{-1}}\phi - \tau_{x_o^{-1}}\phi\| \; \hat{U}\phi(z).$$

Comme la fonction $\hat{U}\tau_{x_o^{-1}}\phi \,/\hat{U}\phi$ est bornée par une constante M, on en déduit que

$$\hat{U}(\Delta(y)\, \tau_{y^{-1}x_o^{-1}}\phi - \tau_{x_o^{-1}}\phi)(z) \leq C_\phi \Delta(y) \|\tau_{y^{-1}x_o^{-1}}\phi - \tau_{x_o^{-1}}\phi\| \; \hat{U}\phi(z)$$
$$+ |\Delta(y)-1| \; M \; \hat{U}\phi(z)$$

et donc que

$$H_g(x_o y) - H_g(x_o) \leq C \| f(.x_o y) - f(.x_o)\| + C_\phi \Delta(y) \|\tau_{y^{-1}x_o^{-1}}\phi - \tau_{x_o^{-1}}\phi\|$$
$$+ M|\Delta(y)-1|.$$

Il résulte alors de l'uniforme continuité de f et de ϕ et de la continuité de Δ que l'ensemble $\{H_g, g \in G\}$ est équicontinu au point x_o.

A l'aide du théorème d'Ascoli, on obtient facilement le corollaire suivant :

<u>Corollaire</u> 1 - Sous les hypothèses de la proposition, les ensembles de fonctions $\{R_g, g \in G\}$ et $\{H_g, g \in G\}$ sont relativement compacts pour la topologie de la convergence uniforme sur tout compact.

On en déduit alors le

Corollaire 2 - Soit μ une mesure de probabilité sur G vérifiant Ha) et Hb) et r un élément non nul de $C_K^+(G)$. De toute suite tendant vers Δ, on peut extraire une sous-suite g_n telle que $\varepsilon_{yg_n} * U / Ur(g_n)$ et $\varepsilon_{g_n y} * U / Ur(g_n)$ convergent vaguement pour tout $y \in G$. De plus, si $f \in C_K^+$, les suites de fonctions $Uf(.g_n) / Ur(g_n)$ et $Uf(g_n.) / Ur(g_n)$ convergent uniformément sur tout compact.

Démonstration - Comme G est un groupe localement compact à base dénombrable, il existe une suite $(f_n)_{n \in \mathbb{N}}$ dense dans C_K^+ pour la topologie de la convergence compacte, c.a.d. que si $f \in C_K^+$, il existe une sous-suite $\{f_k\}$ de $\{f_n\}$ convergeant uniformément vers f et telle que les ensembles $\{f_k > 0\}$ soient inclus dans un compact fixé. On peut alors montrer par des arguments analogues à ceux de la proposition 1 que les ensembles de fonctions $\{Uf_n(.y) / Ur(.)\}_{n \in \mathbb{N}}$ et $\{Uf_n(y.) / Ur(.)\}_{n \in \mathbb{N}}$ sont denses respectivement dans les ensembles $\{Uf(.y) / Ur(.)\}_{f \in C_K^+}$ et $\{Uf(y.) / Ur(.)\}_{f \in C_K^+}$ pour la topologie de la convergence uniforme et ceci uniformément pour y appartenant à un compact de G. Le procédé diagonal permet ensuite de conclure.

Ces résultats techniques acquis, nous allons pouvoir étudier les mesures limites et leurs périodes.

Définition 1 - Soit $\nu \in \mathcal{M}^+(G)$. Un élément y de G est appelé période de la mesure ν si $\varepsilon_y * \nu = \nu$.

Théorème 3 - Soient μ une mesure de probabilité sur G vérifiant Ha) et Hb) et r un élément non nul de $C_K^+(G)$. On considère un élément ν de la frontière de Martin M et $(g_n)_{n \in \mathbb{N}}$ une suite d'éléments de G telle que $K_r(g_n,.)$ converge vaguement vers ν. Supposons qu'il existe un élément y de G et une suite h_n dans G tels que

- la suite h_n converge vers y,
- la suite $g_n^{-1} h_n g_n$ admette e pour valeur d'adhérence,

alors la mesure ν admet y comme période.

Démonstration - Une fois remarqué, quitte à remplacer la suite g_n par une suite extraite, qu'il existe deux suites k_n et t_n tendant vers e telles que $yg_n = k_n g_n t_n$, nous obtenons

$$\varepsilon_y * \nu = \lim_n (\varepsilon_{yg_n} * U / Ur(g_n)) = \lim_n (\varepsilon_{k_n g_n t_n} * U / Ur(g_n)).$$

Nous déduisons alors du corollaire 2, en réextrayant au besoin une sous-suite que

$$\lim_n (\varepsilon_{k_n g_n t_n} * U / Ur(g_n)) = \lim_n \varepsilon_{g_n} * U / Ur(g_n) = \nu \text{ et donc que}$$

$$\varepsilon_y * \nu = \nu.$$

On peut remarquer que ce résultat est en fait vrai sur un groupe G quelconque localement compact à base dénombrable, mais nous ne l'utiliserons dans cet article que sur le groupe affine.

IV - Frontière de Martin du groupe affine.

Proposition 2 - Soient μ une mesure de probabilité sur G vérifiant Ha) et Hb) et r un élément non nul de $C_K^+(G)$. On considère un élément ν de la frontière de Martin M et $(g_n)_{n \in \mathbb{N}}$ une suite d'éléments de G telle que $K_r(g_n,.)$ converge vaguement vers ν. Alors si $\overline{\lim_n}\, a(g_n) = \infty$ ou si $\overline{\lim}|b(g_n)| = \infty$, la mesure ν admet pour période tout élément du sous-groupe des translations de G.

Démonstration - On peut extraire de la suite g_n une sous-suite $(g'_n)_{n \in \mathbb{N}}$ telle que $a(g'_n) \longrightarrow \infty$ (respectivement $|(b(g'_n)| \longrightarrow \infty$) si $\overline{\lim_n}\, a(g_n) = \infty$ (resp. $\overline{\lim_n}\, |b(g_n)| = \infty$).

. Supposons que $a(g'_n) \longrightarrow \infty$ lorsque $n \longrightarrow \infty$. Pour tout $y \in \mathbb{R}$, la suite $g_n'^{-1}(1,y)g'_n = (1,a(g'_n)^{-1}y)$ tend vers e et il découle alors du théorème 3 que (1,y) est un période de ν.

. Supposons que $|b(g'_n)| \longrightarrow \infty$. Pour tout $y \in \mathbb{R}$, nous pouvons écrire si n est suffisamment grand

$$(1,y)g'_n = [a(g'_n),\ b(g'_n) + y] = [a(g'_n),\ b(g'_n)\ (1 + \frac{y}{b(g'_n)})]$$

$$= [1 + \frac{y}{b(g'_n)},0]\ [a(g'_n),\ b(g'_n)]\ [(1 + \frac{y}{b(g'_n)})^{-1},0]$$

$$= k_n g'_n k_n^{-1}$$

où la suite k_n converge vers e lorsque $n \longrightarrow \infty$. Il est alors clair que la suite $h_n = k_n^{-1} y$ obéit aux conditions du théorème 3 et donc que (1,y) est une période de ν.

<u>Lemme</u> 1 - Soit μ une mesure de probabilité sur G. Si ν est un élément de $\mathcal{H}_\mu$ admettant pour période tout élément du sous-groupe des translations de G, alors ν s'écrit $\rho \otimes m_2$ où ρ est une mesure $a(\mu)$-invariante sur $\mathbb{R}^{+*}$ et m_2 la mesure de Lebesgue sur $\mathbb{R}$.

<u>Démonstration</u> - Puisque ν admet pour période tout élément du sous-groupe $\mathbb{R}$, nous avons

$$\forall f_1 \in C_K(\mathbb{R}^{+*}), \quad \forall f_2 \in C_K(\mathbb{R}), \quad \forall y \in \mathbb{R},$$

$$\nu(f_1 \times f_2) = \varepsilon_{(1,y)} * \nu(f_1 \times f_2) = \nu(f_1 \times \tau_y f_2).$$

Par suite si la fonction f_1 est fixée, nous obtenons une mesure sur $\mathbb{R}$ invariante par translation et qui est donc proportionnelle à la mesure de Lebesgue m_2. De ce fait il existe, pour tout $f_1 \in C_K(\mathbb{R}^{+*})$, un réel positif $\rho(f_1)$ tel que

$$\nu(f_1 \times f_2) = \rho(f_1)\, m_2(f_2).$$

L'application de $C_K(\mathbb{R}^{+*})$ dans $\mathbb{R}$ qui à f_1 associe $\rho(f_1)$ est en fait une mesure de Radon et, comme de plus $\nu * \mu = \nu$, on vérifie aisément que $\rho * a(\mu) = \rho$.

Nous pouvons alors obtenir à l'aide de résultats connus sur le renouvellement (7) le théorème suivant qui va en partie décrire la frontière de Martin.

<u>Théorème</u> 4 - Soient μ une mesure de probabilité sur G vérifiant Ha) et Hb) et r un élément non nul de $C_K^+(G)$. Alors si $\int \log a(g)\, d\mu(g) \geq 0$, la frontière de Martin M est incluse dans l'ensemble B formé des mesures

. $\rho \otimes m_2 / \rho \otimes m_2(r)$ où ρ est une mesure $a(\mu)$ invariante sur $\mathbb{R}^{+*}$ et m_2 la mesure de Lebesgue sur $\mathbb{R}$.

. $\varepsilon(1,z) * \widehat{m_1 \otimes m} / \varepsilon(1,z) * \widehat{m_1 \otimes m}(r)$, $z \in \mathbb{R}$, où m_1 est la mesure de Lebesgue sur $\mathbb{R}^{+*}$ et m est, à une constante multiplicative près, l'unique mesure sur $\mathbb{R}$ telle que $\hat{\mu} * m = m$.

<u>Remarque</u> 3 - Par dualité (cf. Remarque1) on peut obtenir un théorème analogue lorsque $\int \log a(g)\, d\mu(g)$ est négatif.

<u>Démonstration du théorème</u> - Considérons un élément ν de M et g_n une suite d'éléments de G telle que $K_r(g_n,.) = \varepsilon_{g_n} * U / Ur(g_n)$ converge vaguement vers ν.

Nous venons de voir que si $\overline{\lim}\ a(g_n) = \infty$ ou si $\overline{\lim}|b(g_n)| = \infty$, alors $\nu = \rho \otimes m_2$ avec ρ mesure $a(\mu)$-invariante. Il reste donc à étudier le cas où ni $\overline{\lim}\ a(g_n) = \infty$, ni $\overline{\lim}|b(g_n)| = \infty$. On peut dans ce cas extraire une sous-suite g'_n telle que $a(g'_n)$ tende vers 0 dans $\mathbb{R}^{+*}$ et $b(g'_n)$ converge vers un élément b de $\mathbb{R}$.

Or on sait (7) que si $\int \log a(g)\, d\mu(g)$ est ≥ 0, il existe, à une constante multiplicative près, une unique mesure m sur $\mathbb{R}$ telle $\hat{\mu} * m = m$. Cette mesure m est de masse finie (respectivement infinie) si $\int \log a(g)\, d\mu(g)$ est > 0 (resp. $\int \log a(g)\, d\mu(g) = 0$). De plus les valeurs d'adhérence du noyau potentiel $\varepsilon_g * U$ sont non nulles lorsque $a(g) \longrightarrow 0$ et $b(g)$ reste dans un compact. Elles sont par contre nulles dans tout autre direction. Et plus précisément si $a(g'_n)$ tend vers 0 et $b(g'_n)$ vers b, alors

$$\varepsilon_{g'_n} * U \longrightarrow \varepsilon_{(1,b)} * \widehat{m_1 \otimes m}.$$

On en déduit aisément que

$$K_r(g'_n,.) \longrightarrow \varepsilon_{(1,b)} * \widehat{m_1 \otimes m} \,/\, \varepsilon_{(1,b)} * \widehat{m_1 \otimes m}(r).$$

<u>Remarque</u> 4 - Tout élément λ de $\mathcal{H}_\mu$ tel que $\lambda(r) = 1$ admet (cf. (11)) une représentation intégrale sur la frontière M ; il existe une probabilité γ_λ sur M associée à λ telle que

$$\lambda = \int_M \nu \, d\gamma_\lambda(\nu).$$

En conséquence, si λ appartient à une génératrice extrémale de $\mathcal{H}_\mu$, nécessairement λ est un point de la frontière M. On en déduit que le cône $\mathcal{S}$ des génératrices extrémales de $\mathcal{H}_\mu$ est inclus dans le cône engendré par M et donc à fortiori par B. La recherche des éléments de B qui engendrent des génératrices extrémales de $\mathcal{H}_\mu$ va être l'objet du paragraphe V.

V - <u>Génératrices extrémales</u>

Nous pouvons remarquer tout de suite que si la mesure $\rho \otimes m_2$ engendre une génératrice extrémale de $\mathcal{H}_\mu$ alors la mesure ρ engendre une génératrice extrémale du cône des mesures $a(\mu)$-invariantes sur $\mathbb{R}^{+*}$. De ce fait ρ est soit la mesure m_1, soit la mesure $h_\alpha m_1$ (cf. paragraphe II) où h_α est, lorqu'elle existe, l'exponentielle $a(\hat{\mu})$-harmonique sur R^{+*} non constante. On note encore h_α l'esponentielle $h_\alpha \circ a$ prolongée à G. Par suite le cône $\mathcal{S}$ des génératrices extrémales est inclus si $\int \log a(g)\, d\mu(g)$ est ≥ 0 dans le cône B' engendré par les mesures :

. $m_1 \otimes m_2$

. $h_\alpha . m_1 \otimes m_2$

. $\varepsilon_{(1,z)} * \widehat{m_1 \otimes m}$, $z \in \mathbb{R}$.

a) La mesure $\widehat{m_1 \otimes m}$ est nécessairement un élément extrémal de $\mathcal{H}_\mu$. Si elle ne l'était pas, il en serait de même des mesures $\varepsilon_{(1,z)} * \widehat{m_1 \otimes m}$, pour tout $z \in \mathbb{R}$. Par suite les éléments extrémaux de $\mathcal{H}_\mu$ ne pourraient être que les mesures proportionnelles à $m_1 \otimes m_2$ et $h_\alpha\, m_1 \otimes m_2$; et la mesure $\widehat{m_1 \otimes m}$ s'écrirait en fait $\rho \otimes m_2$ avec ρ mesure non nulle $a(\mu)$-invariante sur $\mathbb{R}^{+*}$. Ceci est impossible. En effet si $\int \log a(g)\, d\mu(g)$ est > 0, la mesure m est de masse finie et pour tout compact K_1 de $\mathbb{R}^{+*}$, $\widehat{m_1 \otimes m}\,(K_1 \times \mathbb{R}) = m_1(K_1)$ est fini, ce qui n'est pas le cas de $\rho \otimes m_2(K_1 \times \mathbb{R})$. Si $\int \log a(g)\, d\mu(g)$ est nul, la mesure m_1 sur $\mathbb{R}^{+*}$ est la seule mesure $a(\mu)$-invariante et $\widehat{m_1 \otimes m}$ devrait s'écrire $m_1 \otimes m_2$. Or la mesure $\widehat{m_1 \otimes m}$ est invariante par translation à gauche par tout élément (x,o), $x \in \mathbb{R}^{+*}$, ce qui est faux pour la mesure de Haar à droite $m_1 \otimes m_2$.

b) Il s'agit maintenant de chercher si les mesures $m_1 \otimes m_2$ et $h_\alpha(m_1 \otimes m_2)$ sont extrémales.

Lemme 2 - On se place sous les hypothèses du théorème 4. Alors pour toute mesure β sur $\mathbb{R}$ et toute suite x_n dans $\mathbb{R}^{+*}$ tendant vers $+\infty$, nous avons si $f_1 \in C_K^+(\mathbb{R}^{+*})$ et $f_2 \in C_K^+(\mathbb{R})$,

$$\underline{\lim_n}(\varepsilon_1 \otimes \beta) * \widehat{m_1 \otimes m} * \varepsilon_{(x_n,0)}(f_1 \times f_2) \geq m_1(f_1)\beta(f_2)\, m(\mathbb{R}).$$

Démonstration du lemme

$$(\varepsilon_1 \otimes \beta) * \widehat{m_1 \otimes m} * \varepsilon_{(x_n,0)}(f_1 \times f_2)$$

$$= \iiint (f_1 \times f_2)\,[(1,b)(x^{-1},-yx^{-1})(x_n,0)]\, d\beta(b)\, dm_1(x)\, dm(y)$$

$$= \iiint f_1(x^{-1}x_n)\, f_2(b-yx^{-1})\, d\beta(b)\, dm_1(x)\, dm(y)$$

$$= \iiint f_1(x)\, f_2(b-yx_n^{-1}x)\, d\beta(b)\, dm_1(x)\, dm(y)$$

puisque m_1 est invariante par symétrie et translation.

Si $x_n \longrightarrow +\infty$ quand $n \longrightarrow +\infty$, alors $\lim_n f_2(b-yx_n^{-1}x) = f_2(b)$ et il résulte du lemme de Fatou que

$$\iiint f_1(x)\, f_2(b)\, d\beta(b)\, dm_1(x)\, dm(y) \le \underline{\lim}_n\, (\varepsilon_1 \otimes \beta) * \widehat{m_1 \otimes m} * \varepsilon_{(x_n,0)}(f_1 \times f_2).$$

D'où le lemme.

. Supposons que $\int \log a(g)\, d\mu(g)$ est nul. Si la mesure $m_1 \otimes m_2$ n'est pas extrémale, alors les mesures extrémales sont exactement les mesures proportionnelles aux mesures $\varepsilon_{(1,z)} * \widehat{m_1 \otimes m}$ avec $z \in \mathbb{R}$. Par suite il existe une mesure β non nulle sur $\mathbb{R}$ telle que $m_1 \otimes m_2 = (\varepsilon_1 \otimes \beta) * \widehat{m_1 \otimes m}$. Comme $m_1 \otimes m_2$ est une mesure de Haar à droite, on déduit du lemme 2 que

$$m_1(f_1)\beta(f_2)m(\mathbb{R}) \le \underline{\lim}_n\, m_1 \otimes m_2 * \varepsilon_{(x_n,0)}(f_1 \times f_2) = m_1(f_1)\, m_2(f_2)$$

ce qui est impossible puisque m est de masse infinie. Par conséquent les éléments extrémaux de $\mathcal{H}_\mu$ sont dans ce cas les mesures proportionnelles à

$$\begin{cases} \cdot\ m_1 \otimes m_2 \\ \cdot\ \varepsilon_{(1,z)} * \widehat{m_1 \otimes m},\ z \in \mathbb{R}. \end{cases}$$

. Supposons que $\int \log a(g)\, d\mu(g)$ est positif. Il est facile de voir, puisque m est une mesure bornée, que

$$m_1 \otimes m_2 = (\varepsilon_1 \otimes m_2) * (\widehat{m_1 \otimes m})$$

et donc que $m_1 \otimes m_2$ n'est pas un élément extrémal. Si $h_\alpha m_1 \otimes m_2$ n'était pas une mesure extrémale, alors comme ci-dessus il existerait une mesure β non nulle sur $\mathbb{R}$ telle que

$$h_\alpha m_1 \otimes m_2 = (\varepsilon_1 \otimes \beta) * \widehat{m_1 \otimes m}$$

et il découlerait du lemme 2 que

$$m_1(f_1)\, \beta(f_2)\, m(\mathbb{R}) \le \underline{\lim}_n\, (h_\alpha m_1 \otimes m_2) * \varepsilon_{(x_n,0)}\, (f_1 \times f_2)$$

or $(h_\alpha m_1 \otimes m_2) * \varepsilon_{(x_n,0)} = h_\alpha[(x_n,0)]^{-1}\, (h_\alpha m_1 \otimes m_2)$ et comme $\int \log a(g)\, d\hat{\mu}(g)$ est négatif, $h_\alpha(g) = a(g)^\alpha$ avec α réel positif (cf. étude de F au paragraphe II).Par suite $m_1(f_1)\beta(f_2)\, m(\mathbb{R})$ serait ≤ 0, ce qui est impossible. On en conclut que les éléments extrémaux de $\mathcal{H}_\mu$ sont dans ce cas proportionnels aux mesures

. $h_\alpha m_1 \otimes m_2$

. $\varepsilon_{(1,z)} * \widehat{m_1 \otimes m}$ où $z \in \mathbb{R}$.

Le théorème 1 est alors démontré.

BIBLIOGRAPHIE

1 - R. AZENCOTT - Espaces de Poisson des groupes localement compacts. Lecture Notes n° 148 (1970).

2 - R. AZENCOTT et P. CARTIER - Martin Boundaries of Random Waks on locally compact Groups. Proceeding of the 6th Berkeley Symposium on Mathematical Statistics and Probability III (p. 87-129).

3 - G. CHOQUET et J. DENY - Sur l'équation de convolution $\mu = \mu * \sigma$. CRAS t. 250 (1960) p. 799.

4 - J.P. CONZE et Y. GUIVARC'H - Propriété de droite fixe et fonctions propres des opérateurs de convolution - Séminaire de Rennes - 1976.

5 - Y. DERRIENNIC - Marche aléatoire sur le groupe libre et frontière de Martin Z.Wahrscheinlichkeitstheorie verw. Gebiete 32 p. 261-276 (1975)

6 - L. ELIE - Etude du renouvellement sur les groupes moyennables CRAS 284 (1977) p. 555.

7 - L. ELIE - Etude du renouvellement sur le groupe affine de la droite réelle. Ann. Univ. Clermont 85 (1977) p. 47-62.

8 - ELIE - RAUGI - Fonctions harmoniques sur certains groupes résolubles - CRAS 280 (1975) p. 377.

9 - H. FURSTENBERG - Translation-invariant cones of functions on semi-simple groups. Bull. Amer. Math. Soc. 71 (1965), 271-326.

10 - G.A. MARGULIS - Positive harmonic functions on nilpotent groups. Doklady 1966 Tom. 166 n°5.

11 - D. REVUZ - Markov Chains - North-Holland - Mathematical Library Vol. 11 (1975).

UER DE MATHEMATIQUES

Université Paris VII

2 Place Jussieu

75005 - PARIS

INFINITELY DIVISIBLE POSITIVE FUNCTIONS ON SO(3) Ⓢ $\mathbb{R}^3$

B.-J. Falkowski

Introduction

The paper is divided into four parts. In § 1 we review the relevant definitions and give some examples. In § 2 we exhibit the connection between infinitely divisible σ-positive functions and first order cocycles. In § 3 the first order cocycles for SO(3) Ⓢ $\mathbb{R}^3$ are described. Finally, in § 4 we give a description of all infinitely divisible positive functions on SO(3) Ⓢ $\mathbb{R}^3$ associated with first order cocycles arising from irreducible representations. It is understood that all representations arising in this paper are continuous and unitary.

§ 1 *First and Second Order Cocycles*

Let G be a topological group. Let $g \to U_g$ be a representation of G in a Hilbert space H.

(1.1) Def.: A first order cocycle associated with U is a continuous map $\delta : G \to H$ satisfying

$$U_{g_1}\delta(g_2) = \delta(g_1 g_2) - \delta(g_1) \quad \forall (g_1, g_2) \in G \times G$$

Example: Let $v \in H$ be fixed. Then $\delta(g) = U_g v - v$ defines a trivial cocycle (coboundary) associated with U.

(1.2) Def.: A second order cocycle is a continuous map $\sigma : G \times G \to S^1$ satisfying

(i) $\sigma(g_1,g_2)\sigma(g_1g_2,g_3) = \sigma(g_1,g_2g_3)\sigma(g_2,g_3)$

(ii) $\sigma(g,e) = \sigma(e,g) = 1$

$\forall\, g,g_1,g_2,g_3 \in G$, where e is the identity of G.

Example: $\sigma(g_1,g_2) = a(g_1)a(g_2)a(g_1g_2)^{-1}$, where $a:G \to S^1$ is continuous, defines a trivial second order cocycle (coboundary).

It will be noted, that the examples given are of a trivial nature. The reason for this is that nontrivial examples are rather hard to come by in general.

§ 2 *Infinitely Divisible σ-positive Functions and Cocycles*

(2.1) Def.: A continuous function $f:G \to \mathbb{C}$ is called σ-positive if

(i) $\sum_{i=1}^{n}\sum_{j=1}^{n} \alpha_i\bar{\alpha}_j\, \sigma(g_j^{-1},g_i)f(g_j^{-1}g_i) \geq 0$

$\forall\, (\alpha_1,\ldots,\alpha_n) \in \mathbb{C}^n, \ \forall (g_1,\ldots,g_n) \in G^n$

(ii) $f(e) = 1$ (i.e. f is normalized)

Here σ is a second order cocycle.

(2.2) Def.: A pair (f,σ), with f σ-positive, is called infinitely divisible if $\forall n \in \mathbb{N}\, \exists\, (f_n,\sigma_n)$ with

(i) f_n is σ_n-positive

(ii) $f_n^n \equiv f$

(iii) $\sigma_n^n \equiv \sigma$

The connection between cocycles and infinitely divisible σ-positive functions is then established by the following

(2.3) Theorem (cf. [5], [1]): Let G be a connected, locally connected, locally compact, second countable group. Let (f,σ) be infinitely divisible as in (2.2). Then there exists a first order cocycle δ such that

(i) $f(g) = \exp[-\frac{1}{2} < \delta(g), \delta(g) > + i\, a(g)]$

(ii) $\sigma(g_1,g_2) = \exp i\,[\,\mathrm{Im} < \delta(g_2), \delta(g_1^{-1}) > + a(g_1)+a(g_2)-a(g_1g_2)]$

where $a: G \to \mathbb{R}$ is a continuous function satisfying $a(g^{-1})=-a(g)\ \forall g \in G$

(N.B. $<\cdot,\cdot>$ denotes the inner product in the Hilbert space H and Im the imaginary part of a complex number.)

§ 3 *First Order Cocycles for SO(3) Ⓢ $\mathbb{R}^3$*

(3.1) Def.: SO(3) Ⓢ $\mathbb{R}^3$ is SO(3) x $\mathbb{R}^3$ as a set with group operation given by

$$(h_1,\underline{x_1}) \cdot (h_2,\underline{x_2}) = (h_1h_2, \underline{x_1} + h_1(\underline{x_2}))$$

The SO(3)-action on $\mathbb{R}^3$ is here just the natural one.

This is a regular semi-direct product (cf. [3]). Hence all irreducible representations may be computed as induced representations and an analysis of the associated first order cocycles is possible (cf. [3], [2]). Some fairly tedious calculations yield:

(3.2) Theorem: The only nontrivial cocycles for SO(3) Ⓢ $\mathbb{R}^3$ are of the form

$$\delta(h,\underline{x}) = c\,\underline{x} \qquad c \in \mathbb{R}$$

These are associated with the adjoint representation of SO(3) in its Lie-algebra (unitary with respect to the Cartan-Killing form) which is identified with $\mathbb{R}^3$ in the obvious manner.

Remark: We have dealt only with cocycles associated with irreducible representations. The general case may be considered by using the decomposition theory given in [5].

§ 4 *Infinitely Divisible Positive (I.D.P.) Functions on $SO(3) \circledS \mathbb{R}^3$*

An I.D.P. function is described as in (2.2) but here $\sigma_n \equiv \sigma \equiv 1$. For these we now obtain

(4.1) Lemma: Let $a: SO(3) \circledS \mathbb{R}^3 \to \mathbb{R}$ be continuous with $a(g^{-1}) = -a(g)$ $\forall g \in SO(3) \circledS \mathbb{R}^3$. Then there is a bijection η between pairs (f,a) with f I.D.P. and certain σ-positive (Ψ,σ) given by

$$\eta : (f,a) \to (\Psi,\sigma) \quad \text{where}$$

(i) $\Psi(g) = \exp\,[ia(g)]\; f(g)$

(ii) $\sigma(g_1,g_2) = \exp\, i\,[a(g_1)+a(g_2)-a(g_1g_2)]$

Proof: η is injective:
Suppose $\exp\,[ia_1(g)]\; f_1(g) = \exp\,[ia_2(g)]\; f_2(g)$ and $\sigma_1 \equiv \sigma_2$. Then we have

$$f_2(g) = \exp\,[\,i(a_1(g)-a_2(g))]\; f_1(g) \quad \text{and}$$

$$a_1(g) - a_2(g) = b(g) \quad \text{where}$$

$b: SO(3) \circledS \mathbb{R}^3 \to \mathbb{R}$ is an additive continuous homomorphism. But $SO(3)$ is semi-simple. Hence it is its own commutator subgroup and thus $b \equiv 0$ on $SO(3)$. Then it is easily seen that $b \equiv 0$ on $SO(3) \circledS \mathbb{R}^3$.

η is sujective:
Let (Ψ,σ) be given with $\sigma(g_1,g_2) = \exp\, i\,[\,a(g_1)+a(g_2)-a(g_1g_2)]$
Define $f(g) = \exp\,[\,-ia(g)]\;\Psi(g)$, then we have
$\eta(f,a) = (\Psi,\sigma)$
Moreover one easily checks that
f I.D.P. $\Leftrightarrow$ $\eta(f,a)$ infinitely divisible σ-positive with σ as above. Q.E.D.

It is obvious that $SO(3) \circledS \mathbb{R}^3$ satisfies the conditions of (2.3). Hence using (2.3), (3.2), and (4.1) we now obtain:

(4.2) Theorem: All I.D.P. functions on $SO(3) \circledS \mathbb{R}^3$ arising from cocycles associated with irreducible representations are given by:

a) $f(h,\underline{x}) = \exp < U_{(h,\underline{x})} v - v, v >$ $\quad v \in H$ fixed

b) $f(h,\underline{x}) = \exp - b^2 < \underline{x},\underline{x} >$ $\quad b \in \mathbb{R}$

Those in case a) arise from trivial cocycles. Case b) results from the nontrivial cocycles associated with the adjoint representation.

References

[1] *B.-J. Falkowski:* Factorizable and Infinitely Divisible PUA representations of Locally Compact Groups. J. Math. Phys., Vol. 15, No. 7 (1974)

[2] *B.-J. Falkowski:* Cohomology for Certain Leibnitz Extensions. (Unpublished Manuscript)

[3] *G.W. Mackey:* Induced Representations and Quantum Mechanics. Benjamin (1968)

[4] *K.R. Parthasarathy, K. Schmidt:* Factorizable Representations of Current Groups and the Araki-Woods Imbedding Theorem. Acta Mathematica, Vol. 128 (1972)

[5] *K.R. Parthasarathy, K. Schmidt:* Positive Definite Kernels, Continuous Tensor Products, and Central Limit Theorems of Probability Theory. Springer (Lecture Notes in Mathematics), Vol. 272 (1972)

B.-J. Falkowski
Hochschule der Bundeswehr München
(Fachbereich Informatik)
8014 Neubiberg

MULTIPLES OF RENEWAL FUNCTIONS:

Remark on a result of D.J. Daley

Gunnar Forst

D.G. Kendall mentions, cf. [6] p. 27, a result of D.J. Daley [5] as an important contribution to the arithmetical study of renewal sequences and renewal densities, and it is the purpose of this note to obtain Daley's result as an easy consequence of simple properties of potential kernels on the half-line. Also a question posed by Daley is answered.

A positive measure λ on $\mathbb{R}_+$ of the form

$$\lambda = \sum_{n=1}^{\infty} \mu^{*n}$$

where μ is a probability measure on $\mathbb{R}_+$ with $\mu \neq \varepsilon_0$ (ε_0 is the point mass at 0) is called a renewal measure (clearly the series converges vaguely) with generating measure μ. Daley proves the following

Theorem 1. Suppose λ is a renewal measure. Then $c\lambda$ is a renewal measure for every $c \in \,]0,1]$.

Theorem 2. A renewal measure λ has the property that $c\lambda$ is a renewal measure for all $c \in \,]0,\infty[$ if and only if the Laplace trans form $L\lambda$ of λ can be written

$$\frac{1}{L\lambda(s)} = bs + \int_0^{\infty} (1-e^{-xs})\,d\nu(x) \quad \text{for} \quad s > 0,$$

where $(b,\nu) \neq (0,0)$ is a couple of a number $b \geq 0$ and a positive measure ν on $]0,\infty[$ such that

$$\int_0^{\infty} \frac{x}{1+x}\,d\nu(x) < \infty . \tag{1}$$

Remark. Daley's result in Theorem 2 has been slightly rephrased using the simple fact that the generating measure μ has zero mass at 0 if and only if "$b > 0$" or "$b = 0$ and $\int_0^{\infty} d\nu(x) = +\infty$", with the notation as above.

A renewal measure in "almost" an <u>elementary kernel</u> in the sense of potential theory, and the integral representation in Theorem 2 is analogous to the integral representation of <u>subordinator exponents</u> or <u>Bernstein functions</u>. More explicitly: A <u>potential kernel</u> on $\mathbb{R}_+$ is a positive measure κ on $\mathbb{R}$ of the form

$$\kappa = \int_0^\infty \eta_t dt ,$$

where $(\eta_t)_{t>0}$ is a vaguely continuous convolution semigroup on $\mathbb{R}$ with $\operatorname{supp}\eta_t \subseteq \mathbb{R}_+$ (i.e. $(\eta_t)_{t>0}$ is a <u>subordinator</u>) and $\eta_t \neq \varepsilon_0$ for all $t > 0$ (the integral then converges vaguely). A <u>Bernstein function</u> is a C^∞-function $f\colon]0,\infty[\to]0,\infty[$ such that Df is completely monotone, or equivalently that

$$f(s) = a + bs + \int_0^\infty (1-e^{-xs})\,d\nu(x) \quad \text{for} \quad s > 0 \tag{2}$$

for a uniquely determined triple $(a,b,\nu) \neq (0,0,0)$ of numbers $a,b \geq 0$ and a positive measure ν on $]0,\infty[$ such that (1) holds. The Laplace transformation determines a bijection between the set P of potential kernels on $\mathbb{R}_+$ and the set B of Bernstein functions: $\kappa \in P$ corresponds to $f \in B$ if $L\kappa = \frac{1}{f}$.

An <u>elementary kernel</u> on $\mathbb{R}_+$ is a measure of the form

$$\kappa = \sum_{n=0}^{\infty} \mu^{*n}$$

($\mu^{*0} = \varepsilon_0$), where μ is a subprobability on $\mathbb{R}_+$ with $\mu \neq \varepsilon_0$. Such a κ belongs to P and the corresponding Bernstein function is $f(s) = 1 - L\mu(s)$.

Details of the above facts and the following Lemma can e.g. be found in [2] (the arguments in [2] simplify in the present context of the half-line).

<u>Lemma 1</u>. (i) P is a cone ($\kappa \in P$, $c > 0 \Rightarrow c\kappa \in P$)
(ii) $\kappa \in P$, $c \geq 0 \Rightarrow \kappa + c\varepsilon_0 \in P$
(iii) $\kappa \in P$ is an elementary kernel if and only if $\kappa(\{0\}) \geq 1$.
(iv) P is vaguely closed in the set of non-zero measures on $\mathbb{R}_+$.

The following trivial observation is the key to Daley's result.

<u>Lemma 2</u>. A positive measure λ on $\mathbb{R}_+$ with infinite mass is a renewal measure if and only if $\lambda + \varepsilon_0 \in P$.

Proof of Theorem 1. By Lemma 2 it suffices to see that $c\lambda + \varepsilon_0 \in P$ for $c \in]0,1[$. But

$$c\lambda + \varepsilon_0 = c\left(\lambda + \frac{1}{c}\varepsilon_0\right) = c\left(\lambda + \varepsilon_0 + \left(\frac{1}{c} - 1\right)\varepsilon_0\right)$$

which belongs to P by Lemma 1 (i) and (ii). □

Proof of Theorem 2. In view of the integral representation (2) of Bernstein functions, it is enough to see the following biimplication:

$$\left[\forall c \in]0,\infty[: c\lambda + \varepsilon_0 \in P\right] \Leftrightarrow \lambda \in P ,$$

which is evident by Lemma 1 (i), (ii) and (iv) (letting $c \to \infty$). □.

Remark. Let λ be a renewal measure with the property from Theorem 2, i.e. $\lambda \in P$. Daley obtains, by a probabilistic argument, that the generating measure μ is infinitely divisible. For this it is perhaps simpler to note that μ is necessarily the resolvent measure of index 1 for $\lambda \in P$, cf. [2], hence a probability in P, and such a measure is infinitely divisible, cf. [3].

Remark. Let $\kappa \in P$ with $L\kappa = \frac{1}{f}$ for $f \in B$ and suppose that

$$f(s) = a + bs + \int_0^\infty (1-e^{-xs})\,d\nu(x) \qquad \text{for} \qquad s > 0$$

cf. (2). Then it follows by Kingman [7] that κ has a continuous density $k: [0,\infty[\to]0,\infty[$ with respect to Lebesgue measure if and only if $b > 0$, and that in the affirmative case k is proportional with a (standard) p-function. It should be remarked that the regenerative property of p-functions is equivalent with a very important property of potential kernels, namely that they satisfy the so-called balayage principle, cf. [2].

Remark. Let λ be a renewal measure with the property from Theorem 2, i.e. $\lambda \in P$, with representing couple (b,ν). If $b > 0$, then λ has a density which is continuous and bounded on $\mathbb{R}_+$, cf. Kingman [7], and the distribution function $\Lambda(x)$ for λ is then continuously differentiable on $]0,\infty[$. The question of differentiability of $\Lambda(x)$ in the case $b = 0$ is left open by Daley. However, when $b = 0$ the function $\Lambda(x)$ can be continuously differentiable on $]0,\infty[$ as e.g. for λ the potential kernel of the one-sided stable semigroup of order $\alpha \in]0,1[$, cf. [2], and $\Lambda(x)$ can also be non-differentiable. In fact, let λ be a potential kernel on $\mathbb{R}_+$ with infinite mass and such that λ is continuous singular with respect to Lebesgue measure, cf. Berg [1]. The distribution function $\Lambda(x)$ for this λ is not differentiable on $]0,\infty[$, because if it were it would be constant (e.g. by [4] p. 597) which is a contradiction.

References:

1. Berg, C.: Hunt convolution kernels which are continuous singular with respect to Haar measure. This volume.

2. Berg, C. & Forst, G.: Potential Theory on Locally Compact Abelian Groups. Berlin: Springer 1975.

3. Berg, C. & Forst, G.: Infinitely divisible probability measures and potential kernels. This volume.

4. Carathéodory, C.: Vorlesungen über reelle Funktionen. Leipzig, Berlin: Teubner 1918.

5. Daley, D.J.: On a class of renewal functions. Proc. Camb. Phil. Soc. 61, 519 - 526 (1965).

6. Kendall, D.G.: An introduction to stochastic analysis. In: Stochastic Analysis. Ed. D.G. Kendall & E.F. Harding. London: Wiley 1973.

7. Kingman, J.F.C.: Regenerative Phenomena. London: Wiley 1972.

Gunnar Forst
Matematisk Institut
Universitetsparken 5
DK-2100 København Ø
Denmark

EIN KONVERGENZSATZ FÜR FALTUNGSPOTENZEN

Peter Gerl

1. Der Satz

Es sei G eine unendliche diskrete Gruppe mit Einheitselement e. P sei ein Wahrscheinlichkeitsmaß auf G (d.h. $P : G \to \mathbb{R}$ mit $P(x) \geq 0$ für alle $x \in G$ und $\sum_{x\in G} P(x) = 1$), welches <u>irreduzibel</u> ist (d.h. zu jedem $x \in G$ gibt es eine natürliche Zahl n, so daß $P^n(x) > 0$ ist, wobei P^n die n-te Faltungspotenz von P bedeutet). In [1], Th. 4. 14 wurde gezeigt (sogar viel allgemeiner, wenn G eine nichtkompakte, vollständig einfache, lokal kompakte Halbgruppe ist, die dem zweiten Abzählbarkeitsaxiom genügt), daß dann P^n für $n \to \infty$ vag gegen Null konvergiert; in unserem Fall heißt das gerade, daß $\lim_{n\to\infty} P^n(x) = 0$ ist für alle $x \in G$. Dieses Ergebnis soll hier im Fall diskreter Gruppen verallgemeinert werden.

Es bedeute

$$\sigma = \sigma(P) = R^{-1} = \limsup_{n\to\infty} (P^n(e))^{\frac{1}{n}}.$$

Dann ist also $R \geq 1$. Es gilt nun der

<u>Satz</u>: *Ist* P *ein irreduzibles Wahrscheinlichkeitsmaß auf der unendlichen diskreten Gruppe* G *, so ist*

$$\lim_{n\to\infty} \frac{P^n(x)}{\sigma^n} = 0 \quad \text{für alle } x \in G$$

(d.h. $R^n P^n$ *konvergiert für* $n \to \infty$ *vag gegen Null).*

Für $R = 1$ ergibt sich gerade das oben zitierte Ergebnis in [1], für $R > 1$ ist dieser Satz eine Verschärfung.

2. Beweis des Satzes

Wir führen den Beweis in einzelnen Schritten durch.

a) Nach [4] ist $\sum_n P^n(x) R^n$ für alle $x \in G$ entweder konvergent oder divergent; das Wahrscheinlichkeitsmaß P heißt dann R-transient oder R-rekurrent. Im R-rekurrenten Fall ist für alle $x \in G$ entweder

$$\lim_{n\to\infty} P^n(x) R^n = 0 \quad \text{(P heißt dann R-null)}$$

oder

$$0 < \lim_{n\to\infty} P^n(x)\, R^n < \infty \quad \text{(P heißt dann R-positiv)}$$

(wenn P periodisch ist, d.h. wenn $g.g.T.\{n \mid P^n(e) > 0\} = d > 1$ ist, so muß im letzten Grenzwert n in einer geeigneten, von x abhängigen Restklasse von d gegen unendlich streben; wegen der Irreduzibilität von P gibt es ja eine natürliche Zahl n_0, so daß $P^{n_0}(x) > 0$. Dann ist aber auch $P^{n_0+nd}(x) > 0$ für alle natürlichen Zahlen n und n_0 legt die richtige Restklasse für x fest). Es ist klar, daß der Satz richtig ist, wenn P R-transient oder R-null ist.

b) Wir setzen also von jetzt an P als R-positiv voraus und zeigen, daß das unmöglich ist.

Nach [2] gibt es dann eine eindeutig bestimmte Funktion

$$f : G \to \mathbb{R},$$

welche $P * f = \sigma f$ und $f(e) = 1$ erfüllt ($*$ bedeutet Faltung); es ist dann sogar $f(x) > 0$ für alle $x \in G$. Wir setzen

$$Q(x,y) = \frac{P(xy^{-1})}{\sigma} \frac{f(y)}{f(x)} \quad \text{für alle } x,y \in G.$$

Dann gilt:

$$Q(x,y) \geq 0 \quad \text{für alle } x,y \in G,$$

$$\sum_{y\in G} Q(x,y) = 1 \quad \text{für alle } x \in G.$$

Q ist also eine Übergangswahrscheinlichkeit auf G. Da

(1) $$Q^n(x,y) = \frac{P^n(xy^{-1})}{\sigma^n} \frac{f(y)}{f(x)} \quad \text{für alle } x,y \in G,$$

so folgt sofort

$$\limsup_{n\to\infty} (Q^n(x,y))^{\frac{1}{n}} = 1 \quad \text{für alle } x,y \in G$$

(dabei ist Q^n die n-te Faltungspotenz von Q, d.h.

$$Q^1(x,y) = Q(x,y), \quad Q^n(x,y) = \sum_{z\in G} Q(x,z)\, Q^{n-1}(z,y) \;).$$

Weil P R-positiv ist, ist also wegen (1) die Übergangswahrscheinlichkeit Q 1-positiv. Nach [4] haben wir daher:

(2) $$\lim_{n\to\infty} Q^n(x,y) \text{ exist.} > 0 \quad \text{für alle } x,y \in G$$

(wobei im Fall, daß $g.g.T.\{n \mid Q^n(e,e) > 0\} = d > 1$ ist, n wieder in einer geeigneten, von x und y abhängigen Restklasse von d gegen unendlich strebt).

c) Wir definieren nun für $x,y \in G$:

$$F_0(x,y) = 0, \quad F_1(x,y) = Q(x,y)$$

$$F_n(x,y) = \sum_{y\neq z\in G} Q(x,z)F_{n-1}(z,y) = \sum_{y\neq z_i\in G} Q(x,z_1)Q(z_1,z_2)\ldots Q(z_{n-1},y).$$

Dann gelten folgende Beziehungen für $n \geq 1$ und $x,y \in G$:

(3)
$$\sum_{k=1}^{n} F_k(x,y) \leq 1$$

(4)
$$Q^n(x,y) = \sum_{k=1}^{n} F_k(x,y)\, Q^{n-k}(y,y)$$

(Beweis wie in [3], P 1. 2 (b) und (c))

(5)
$$Q^n(x,x) = Q^n(e,e) \quad \text{(wegen (1))}$$

(6)
$$F_n(x,x) = F_n(e,e) .$$

<u>Beweis</u> von (6) durch Induktion nach n: Es ist ja

$$F_n(x,x) = \frac{1}{\sigma^n} \sum_{x\neq z_i\in G} P(x\, z_1^{-1})\, P(z_1\, z_2^{-1})\ldots P(z_{n-1}\, x^{-1}) =$$

$$= \frac{1}{\sigma^n} \Big(\sum_{x\neq z_2,\ldots,z_{n-1}} P^2(x\, z_2^{-1})\, P(z_2^{-1} z_3)\ldots P(z_{n-1}x^{-1}) -$$

$$- P(e) \sum_{x\neq z_2,\ldots,z_{n-1}} P(x\, z_2^{-1})\ldots P(z_{n-1}\, x^{-1})\Big)$$

$$= \frac{1}{\sigma^n} \Big(\sum_{x\neq z_2,\ldots,z_{n-1}} P^2(x\, z_2^{-1})P(z_2^{-1} z_3)\ldots P(z_{n-1}x^{-1}) - P(e)\sigma^{n-1} F_{n-1}(x,x)\Big)$$

$$= \frac{1}{\sigma^n} \Big(\sum_{x\neq z_3,\ldots,z_{n-1}} P^3(x\, z_3^{-1})P(z_3 z_4^{-1})\ldots P(z_{n-1}x^{-1}) -$$

$$- P^2(e) \sum_{x\neq z_3,\ldots,z_{n-1}} P(x\, z_3^{-1})\ldots P(z_{n-1}x^{-1}) - \text{(von } x \text{ unabhängig)}\Big)$$

......

$$= \frac{1}{\sigma^n} \left(P^n(e) - P^{n-1}(e)\, P(e) - \ldots \right) = \text{von } x \text{ unabhängig} = F_n(e,e).$$

d) Wir definieren für $x,y \in G$ und $n \geq 0$:

$$G_n(x,y) = \sum_{k=0}^{n} Q^k(x,y) \qquad (Q^0(x,y) = \delta(x,y))$$

$$F(x,y) = \sum_{k=1}^{\infty} F_k(x,y).$$

Für $x,y \in G$ und $n \geq 0$ gelten dann folgende Beziehungen:

(7)
$$G_n(x,y) \leq G_n(e,e)$$

(Beweis wie in [3], P 1. 3)

(8) $$F(x,x) = F(e,e) = 1$$

(das erste Gleichheitszeichen folgt aus (6); $F(e,e) = 1$ wird wie in [3], P 1. 4 gezeigt, wobei man berücksichtigt, daß wegen der vorausgesetzten 1-Positivität von Q ja $\sum_{k=0}^{\infty} Q^k(e,e) = \infty$ ist).

(9) $$\lim_{n\to\infty} \frac{G_n(x,y)}{G_n(e,e)} = F(x,y)$$

(Beweis wie in [3], P 1. 5)

(10) $$\sum_{t\in G} Q(x,t)\, F(t,y) = F(x,y)$$

(Beweis wie in [3], P 2. 4 (b))

(11) $$F(x,y) = 1$$

<u>Beweis</u> von (11) (ähnlich zu [3], P 2. 5): Aus (10) folgt

$$\sum_{t\in G} Q^2(s,t)\, F(t,y) = \sum_{x,t\in G} Q(s,x)\, Q(x,t)\, F(t,y) = \sum_{x\in G} Q(s,x)\, F(x,y) = F(s,y)$$

und Iteration liefert

$$\sum_{t\in G} Q^m(x,t)\, F(t,y) = F(x,y) \quad \text{für alle } m \geq 0 .$$

Für $x = y$ ergibt sich daraus nach (8):

$$\sum_{t\in G} Q^m(x,t)\, F(t,x) = F(x,x) = F(e,e) = 1 \quad \text{für alle } m \geq 0.$$

Für feste $x, t_0 \in G$ gibt es wegen der Irreduzibilität von P und (1) ein $m_0 \geq 0$, so daß $Q^{m_0}(x,t_0) > 0$. Daher wird

$$1 = Q^{m_0}(x,t_0)\, F(t_0,y) + \sum_{t_0 \neq t\in G} Q^{m_0}(x,t)\, F(t,x)$$

$$\overset{(3)}{\leq} Q^{m_0}(x,t_0)\, F(t_0,y) + \sum_{t_0\neq t\in G} Q^{m_0}(x,t) = 1 + Q^{m_0}(x,t_0)(F(t_0,x) - 1).$$

Da $Q^{m_0}(x,t_0) > 0$, so muß $F(t_0,x) - 1 \geq 0$ sein, also folgt (11) aus (3).

Aus (9) und (11) ergibt sich die im weiteren wichtige Beziehung

(12) $$\lim_{n\to\infty} \frac{G_n(x,y)}{G_n(e,e)} = 1 \quad \text{für alle } x,y \in G .$$

e) Wir beweisen nun den Satz, wenn Q (d.h. P) <u>nicht periodisch</u> ist, d.h. wenn gilt:

Für alle $x,y \in G$ gibt es ein $n_0 = n_0(x,y)$, so daß

$$Q^n(x,y) > 0 \quad \text{für alle } n \geq n_0.$$

Aus (2) folgt, daß für alle $x,y \in G$

(13) $$\lim_{n\to\infty} \frac{Q^n(x,y)}{Q^n(e,e)} \text{ exist.} = a(x,y).$$

Da nun $Q^n(e,e) > 0$ für alle genügend großen n (Q ist irreduzibel und nicht periodisch) und $\sum_n Q^n(x,y) = \infty$ (nach (2)), so folgt aus (12) und (13), daß $a(x,y) = 1$, also daß

$$\lim_{n\to\infty} Q^n(x,y) = \alpha > 0 \quad \text{für alle } x,y \in G.$$

Bezeichnet H_k eine Teilmenge von G mit k Elementen, so gilt für natürliche Zahlen m,n und $x,y \in G$:

$$Q^{m+n}(x,y) \geq \sum_{t\in H_k} Q^m(x,t)\, Q^n(t,y)$$

und Grenzübergang $m,n \to \infty$ liefert

$$\alpha \geq k\,\alpha^2,$$

also wegen $\alpha > 0$

$$1 \geq k\,\alpha \quad \text{für alle natürlichen Zahlen } k.$$

Das ist aber ein Widerspruch zu $\alpha > 0$, daher kann P nicht R-positiv sein.

f) Es sei Q (d.h. P) periodisch mit Periode $d > 1$. Dann ist also

(14) $$Q^{nd}(e,e) > 0, \text{ wenn } n \text{ genügend groß und}$$
$$Q^{nd+j}(e,e) = 0 \quad \text{für } 1 \leq j \leq d-1 \text{ und alle } n.$$

Zu $x,y \in G$ wählen wir ein j mit $0 \leq j \leq d-1$, so daß

$$Q^{nd+j}(x,y) > 0 \quad \text{für alle genügend großen } n.$$

Dann folgt aus (2), daß

(15) $$\lim_{n\to\infty} \frac{Q^{nd+j}(x,y)}{Q^{nd}(e,e)} \text{ exist.} = a(x,y).$$

Aus (14) und $\sum_n Q^{nd}(e,e) = \infty$ folgt wegen (12) (n geht längs der Teilfolge $(m+1)d-1$ gegen ∞) und (15) wie vorher, daß $a(x,y) = 1$, also daß

(16) $$\lim_{n\to\infty} Q^{nd+j}(x,y) = \alpha > 0 \quad \text{für alle } x,y \in G$$

(dabei ist j durch $x,y \in G$ und $0 \leq j \leq d-1$ festgelegt).

Zu jeder natürlichen Zahl k gibt es sicher $k \cdot d$ Elemente $t_r \in G$ und ein i, so daß

$$Q^{nd+i}(e,t_r) > 0 \quad \text{für alle } n \geq n_1;$$

unter diesen $k \cdot d$ Elementen t_r gibt es nach dem Schubfachprinzip sicher k (etwa $t_1,\dots,t_k$) und ein $j\ (\equiv -i \pmod d)$, so daß

$$Q^{nd+j}(t_r,e) > 0 \quad \text{für alle} \quad n \geq n_1 \quad \text{und} \quad r = 1,\dots,k.$$

Daher erhalten wir

$$Q^{2nd+i+j}(e,e) \geq \sum_{r=1}^{k} Q^{nd+i}(e,t_r)\, Q^{nd+j}(t_r,e)\,.$$

Geht man in dieser Ungleichung mit $n \to \infty$, so ergibt sich wegen (16) für jede natürliche Zahl k:

$$\alpha \geq k\,\alpha^2\,,$$

und das stellt einen Widerspruch zu $\alpha > 0$ dar. Damit ist der Satz bewiesen.

LITERATUR

[1] A. Mukherjea - N. A. Tserpes, Measures on topological Semigroups. L. N. in Mathem. 547, Springer-Verlag 1976.

[2] W. E. Pruitt, Eigenvalues of nonnegative matrices. Ann. Math. Stat. 35 (1966), 1797 - 1800.

[3] F. Spitzer, Principles of random walk. Springer-Verlag 1976.

[4] D. Vere-Jones, Ergodic properties of nonnegative matrices I. Pac. J. Math. 22 (1967), 361 - 386.

Mathematisches Institut
der Universität Salzburg
Salzburg / Austria

EIN GLEICHVERTEILUNGSSATZ AUF F_2

Peter Gerl

1. Das Ergebnis

Es sei $F_2 = \langle a,b | \ \rangle$ die freie Gruppe mit den zwei Erzeugenden a,b und e das leere Wort (= Einheitselement in F_2). P bezeichnet im folgenden stets ein Wahrscheinlichkeitsmaß auf F_2 mit Träger $T = \{a, b, a^{-1}, b^{-1}\}$, d.h. $P(x) > 0$ für $x \in T$, $P(x) = 0$ für $x \notin T$ und $P(a) + P(b) + P(a^{-1}) + P(b^{-1}) = 1$. P^n bedeutet die n-te Faltungspotenz von P. Wir setzen o.B.d.A. stets voraus, daß $P(a)\,P(a^{-1}) \leq P(b)\,P(b^{-1})$.

In [2], [3] wurde das asymptotische Verhalten von $P^{2n}(e)$ für $n \to \infty$ untersucht. Hier soll allgemeiner das asymptotische Verhalten von $P^n(w)$ (wo $w \in F_2$ ist) studiert werden.

Insbesondere beweisen wir den

Satz: *Ist* P *ein Wahrscheinlichkeitsmaß auf* F_2 *mit Träger* T, *so gilt für Elemente* $w \in F_2$ *für* $n \to \infty$,

a) wenn die reduzierte Länge von w *gerade ist:*

$$P^{2n+1}(w) = 0$$

$$P^{2n}(w) \sim h(w) \frac{1}{n\sqrt{n}\, x_0^n} + O\left(\frac{1}{n^2 x_0^n}\right),$$

b) wenn die reduzierte Länge von w *ungerade ist:*

$$P^{2n}(w) = 0$$

$$P^{2n+1}(w) \sim h(w) \frac{1}{n\sqrt{n}\, x_0^n} + O\left(\frac{1}{n^2 x_0^n}\right),$$

wobei $x_0 = \lim_{n\to\infty} (P^{2n}(e))^{-\frac{1}{n}}$ *und* h(w) *durch* (6) *bzw.* (7) *gegeben ist.*

Aus diesem Satz ergibt sich sofort durch Bildung der Quotienten $\frac{P^{2n}(w)}{P^{2n}(e)}$, wenn die reduzierte Länge von w gerade ist (bzw. $\frac{P^{2n+1}(w)}{P^{2n}(e)}$, wenn die reduzierte Länge von w ungerade ist), und Grenzübergang für $n \to \infty$ ein Gleichverteilungssatz auf F_2.

Für den Spezialfall, daß P die Gleichverteilung auf T ist, wurde dieses Ergebnis bereits in [1] erhalten.

2. Der Beweis

Zunächst einige Bezeichnungen (wie in [3], nur schreiben wir hier

G bzw. G^* für das dort verwendete g; dieser Arbeit und [2] entstammen auch alle im weiteren benötigte Ergebnisse):

Liegt $w \in F_2$, so ist

$P^n(w)$ = Wahrscheinlichkeit, in der durch P definierten Irrfahrt auf F_2 (d.h. durch Rechtsmultiplikation mit Elementen aus T mit den durch P gegebenen Wahrscheinlichkeiten) in n Schritten von e nach w zu kommen.

Weiter sei

$Q^n(w)$ = Wahrscheinlichkeit, in der durch P definierten Irrfahrt auf F_2 in n Schritten das erste Mal von e nach w zu kommen.

a_{2n} (bzw. b_{2n}) = Wahrscheinlichkeit, in der durch P definierten Irrfahrt auf F_2 in 2n Schritten das erste Mal von e nach e zu kommen, wobei der erste Schritt Multiplikation mit a oder a^{-1} (bzw. b oder b^{-1}) ist.

Wir führen erzeugende Funktionen ein ($z^2 = x$ in [3]):

$$G_w^*(z) = \sum_{n=0}^{\infty} P^n(w)\, z^n , \quad G_e^*(z) = G^*(z) = G(x)$$

(1) $$F_w^*(z) = \sum_{n=1}^{\infty} Q^n(w)\, z^n ,$$

$$A(x) = \sum_{n=1}^{\infty} a_{2n}\, x^n , \quad B(x) = \sum_{n=1}^{\infty} b_{2n}\, x^n .$$

Es gilt dann der folgende

Hilfssatz: Die reduzierte Darstellung von $w \in F_2$ sei: $w = c_1 c_2 \dots c_k$ mit $c_i \in T$. Es bestehen folgende Beziehungen ($z^2 = x$):

a) $A(x)(1 - A(x) - 2B(x)) = P(a)\, P(a^{-1})\, x$

$B(x)(1 - B(x) - 2A(x)) = P(b)\, P(b^{-1})\, x$

b) $G(x) = (1 - 2(A(x) + B(x)))^{-1}$

c) $G_w^*(z) = G^*(z)\, F_w^*(z)$

d) $F_w^*(z) = F_{c_1}^*(z)\, F_{c_2}^*(z) \dots F_{c_k}^*(z)$

e) $$F_a^*(z) = \frac{A(z^2)}{P(a^{-1})\, z} , \quad F_{a^{-1}}^*(z) = \frac{A(z^2)}{P(a)\, z} ,$$

$$F_b^*(z) = \frac{B(z^2)}{P(b^{-1})\, z} , \quad F_{b^{-1}}^*(z) = \frac{B(z^2)}{P(b)\, z} .$$

Beweis: a) und b) finden sich in [2] bzw. [3].
c) und d) folgen sofort aus dem Graphen von F_2 nach der Methode von Howard [4] bzw. durch Studium von Rekursionsformeln, z.B. $P^n(w) = \sum_{k=1}^{n} Q^k(w)\, P^{n-k}(e)$ für c).

e) ergibt sich ebenso, wenn noch a) verwendet wird.

Wir verwenden nun die in [3] gefundenen Eigenschaften der Funktionen $A(x)$, $B(x)$, $G(x)$, um $P^n(w)$ asymptotisch auszuwerten. Dort wurde gezeigt, daß die Reihenentwicklungen (1) von $A(x)$, $B(x)$, $G(x)$ alle den gleichen Konvergenzradius $x_0 > 0$ besitzen (daher ist $x_0^{-1} = \limsup_{n\to\infty} (P^{2n}(e))^{\frac{1}{n}} = \lim_{n\to\infty} (P^{2n}(e))^{\frac{1}{n}}$, denn x_0 ist Konvergenzradius der Reihe $G(x) = \sum_{n=0}^{\infty} P^{2n}(e)\, x^n$) und daß am Rande des Konvergenzkreises (dort konvergieren noch die Reihenentwicklungen) nur die einzige Singularität x_0 liegt (x_0 ist ein algebraischer Verzweigungspunkt der Ordnung 1). Die PUISEUX-Reihen um x_0 lauten:

$$
\begin{aligned}
A(x) &= a_0 - a_1 \sqrt{x_0 - x} + \dots , \\
B(x) &= b_0 - b_1 \sqrt{x_0 - x} + \dots , \qquad (2) \\
G(x) &= g_0 - g_1 \sqrt{x_0 - x} + \dots ;
\end{aligned}
$$

dabei sind die Koeffizienten gegeben durch:

$a_0 = A(x_0)$, $b_0 = B(x_0)$, $g_0 = G(x_0) = (1 - 2(a_0 + b_0))^{-1}$,

a_1 = Formel (29) in [3], b_1 = Formel (25) in [3] und $g_1 = 2(a_1 + b_1) g_0^2$.

Es sei jetzt $w \in F_2$ ein Wort gerader reduzierter Länge, welches in der reduzierten Darstellung r-mal a, s-mal a^{-1}, t-mal b und u-mal b^{-1} enthält ($r + s + t + u$ = gerade). Weiter sei

$$
c(w) = (P(a))^s \, (P(a^{-1}))^r \, (P(b))^u \, (P(b^{-1}))^t . \qquad (3)
$$

Nach dem Hilfssatz gilt dann (da $P^{2n+1}(w) = 0$)

$$
\begin{aligned}
G_w(x) &= G_w^*(z) = G(x) (F_a^*(z))^r \, (F_{a^{-1}}^*(z))^s \, (F_b^*(z))^t \, (F_{b^{-1}}^*(z))^u = \qquad (4) \\
&= G(x) (A(x))^{r+s} \, (B(x))^{t+u} \, (c(w))^{-1} \, x^{-\frac{1}{2}(r+s+t+u)} .
\end{aligned}
$$

Daraus folgt sofort, daß auch $G_w(x)$ als Potenzreihe um den Nullpunkt entwickelt werden kann, welche den Konvergenzradius $x_0 > 0$ besitzt und daß am Rande des Konvergenzkreises nur die Singularität x_0 liegt (die ein algebraischer Verzweigungspunkt erster Ordnung ist). Für die PUISEUX-Reihe von $G_w(x)$ um x_0

$$
G_w(x) = g_0(w) - g_1(w) \sqrt{x_0 - x} + \dots
$$

ergibt sich daher aus (4) durch Koeffizientenvergleich, wenn (2) und $x^{\frac{1}{2}(r+s+t+u)} = x_0^{\frac{1}{2}(r+s+t+u)} - (*)(x_0 - x)$ berücksichtigt wird:

$$g_0(w) = G_w(x_0) = g_0\, a_0^{r+s}\, b_0^{t+u}\, (c(w))^{-1}\, x_0^{-\frac{1}{2}(r+s+t+u)}$$

(5)

$$g_1(w) = G_w(x_0) \left(\frac{(r+s)\, a_1\, b_0 + (t+u)\, a_0\, b_1}{a_0\, b_0} + \frac{g_1}{g_0} \right) .$$

Wie in [3] liefert nun die Methode von DARBOUX:

$$p^{2n}(w) \sim \frac{g_1(w)}{2\sqrt{\pi}} \; \frac{\sqrt{x_0}}{n\sqrt{n}\; x_0^n} + O\left(\frac{1}{n^2\; x_0^n} \right) \quad \text{für } n \to \infty$$

und daraus ergibt sich Teil a) des Satzes mit

$$h(w) = g_1(w)\, \frac{\sqrt{x_0}}{2\sqrt{\pi}} . \tag{6}$$

Für $w' \in F_2$ ein Wort ungerader Länge ergibt eine analoge Vorgangsweise (die reduzierte Darstellung von w' enthalte r-mal a, s-mal a^{-1}, t-mal b und u-mal b^{-1}, wobei $r+s+t+u =$ ungerade; $c(w')$ wie in (3)):

$$G_{w'}^{*}(z) = G(x)(A(x))^{r+s}\,(B(x))^{t+u}\,(c(w))^{-1}\, x^{-\frac{1}{2}(r+s+t+u)}$$

und daher

$$\sum_{n=0}^{\infty} p^{2n+1}(w')\, x^n = \frac{1}{\sqrt{x}}\, G_{w'}^{*}(\sqrt{x}) = g_0'(w') - g_1'(w')\,\sqrt{x_0 - x} + \ldots$$

Daraus folgt wieder

$$g_0'(w') = \frac{1}{\sqrt{x_0}}\, G_{w'}^{*}(\sqrt{x_0}) = g_0\, a_0^{r+s}\, b_0^{t+u}\,(c(w))^{-1}\, x_0^{-\frac{1}{2}(r+s+t+u+1)}$$

$$g_1'(w') = g_0'(w') \left(\frac{(r+s)\, a_1\, b_0 + (t+u)\, a_0\, b_1}{a_0\, b_0} + \frac{g_1}{g_0} \right)$$

und schließlich nach der Methode von DARBOUX:

$$p^{2n+1}(w') \sim \frac{g_1'(w')}{2\sqrt{\pi}} \; \frac{\sqrt{x_0}}{n\sqrt{n}\; x_0^n} + O\left(\frac{1}{n^2\; x_0^n} \right) \quad \text{für } n \to \infty$$

und daraus ergibt sich Teil b) des Satzes mit

$$h(w') = g_1'(w')\, \frac{\sqrt{x_0}}{2\sqrt{\pi}} . \tag{7}$$

LITERATUR

[1] P. Gerl, Über die Anzahl der Darstellungen von Worten. Mh. Math. 75 (1971), 205 - 214.

[2] P. Gerl, Irrfahrten auf F . Mh. Math. 84 (1977), 29 - 35.

[3] P. Gerl, Eine asymptotische Auswertung von Faltungspotenzen. Sitzungsber. der Österr. Akad. Wiss. (im Druck).

[4] R. A. Howard, Dynamic Probabilistic Systems. Vol. I. New York: John Wiley & Sons, Inc. 1971.

Mathematisches Institut
der Universität Salzburg
Salzburg / Austria

A RANDOM WALK ON THE GENERAL LINEAR GROUP RELATED TO A PROBLEM OF ATOMIC PHYSICS

Lutz Hantsch and Wilhelm von Waldenfels

Introduction

The object of the paper is the investigation of a problem in theoretical spectroscopy which can be modelled as the study of a random walk on the general linear group. We describe at first the random walk and then formulate the physical problem and its relation to this random walk.

Let g be a hermitian traceless $d\times d$-matrix. Let $t_0 = 0 < t_1 < t_2 < \dots$ be the jump points of a Poisson process with mean density c. Let $U_1, U_2, \dots$ be a sequence of independent unitary random $d\times d$-matrices, independent of the Poisson process. Each of the random matrices U_i is distributed with respect to the Haar measure on the unitary group.

If $t_n < t < t_{n+1}$ define

$$X(t) = e^{(g/2)(t-t_n)}\, U_n e^{(g/2)(t_n-t_{n-1})}\, U_{n-1} \dots e^{(g/2)(t_2-t_1)}\, U_1 e^{(g/2)t_1}.$$

Then $X(t)$ is a process with independent multiplicative increments on the general linear group $GL(d)$. We are interested in the behaviour of $X(t)X(t)^*$. If $g = 0$, then $X(t)$ is unitary and $X(t)X(t)^* = 1$. If the density c of the Poisson process is 0, then $X(t) = e^{(g/2)t}$ and $X(t)X(t)^* = e^{gt}$ and

$$\| X(t)X(t)^* - 1 \| < e^{\|g\|t} - 1$$

where the norm equals the maximal eigenvalue. As g is traceless, in the limit $c \longrightarrow \infty$ the quantity $X(t)X(t)^*$ approaches 1 and, what is essential, its behaviour can be estimated. One gets roughly for $t \gg \frac{1}{c}$

$$E\,\tau(X(t)X(t)^*) = \left(1 - \frac{\tau g^2}{c^2}\right) e^{(\tau g^2/c^2)ct}$$

where

$$\tau A = \frac{1}{d} \text{ trace } A$$

is the normalized trace (exact expression: prop. 1). It is, however, possible to state the stronger estimate

$$P\left\{ \max_{0 \leq s \leq t} \frac{1}{2}\tau(X(s)X(s)^* + X(s)^{*-1}X(s)^{-1} - 2) \geqslant \varepsilon \right\}$$

$$\leq \frac{2}{\varepsilon} \left(\left(1 - \frac{\tau g^2}{c^2}\right) e^{(\tau g^2/c^2)ct} - 1 \right).$$

(exact expression cf. prop. 4). So the deviation of $X(t)X(t)^*$ from 1 has a much slower increase than $e^{\|g\|t} - 1$.

The quantity

$$\frac{1}{2}\tau(X(t)X(t)^* + X(t)^{*-1}X(t)^{-1} - 2)$$

is somewhat unfamiliar. Introduce the eigenvalues $\xi_1(t), \ldots, \xi_d(t)$ of $X(t)X(t)^*$. Then the quantity is

$$\frac{1}{2d} \sum_{i=1}^{d} \frac{(\xi_i - 1)^2}{\xi_i}$$

and behaves like $\frac{1}{2d} \sum (\xi_i - 1)^2$ for $\xi_i \approx 1$.

We want now to discuss the physical application. Assume an atom with a highly degenerate energy level, e.g. the hydrogen atom with the level of principal quantum number 3. It consists of 18 linearly independent states, namely two 3s-states, six 3p-states and ten 3d-states. These states have different lifetimes, namely, 16×10^{-8} sec, 0.54×10^{-8} sec and 1.56×10^{-8} sec, resp. Assuming at time 0 each state occupied with the same probability one would guess that after a certain time only the state with the largest lifetime, namely 3s, survives. This is not true under usual experimental conditions. At all times all states are occupied with approximately the same probability and we have a mean lifetime corresponding to the arithmetic mean of the decay rates. This effect is due to collisions with neighboring atoms recharging the empty states from the full ones.

We want to make a mathematical model for that effect. We consider a d-dimensional state space $\mathfrak{H}$, where d is the number of states of the energy level, e.g. $d = 18$ in our case. The atom is at time t described by a positive definite $d \times d$-matrix $\rho(t)$. Equal probability for each state at $t = 0$ corresponds to

$$\rho(0) = \frac{1}{d} \mathbb{1}$$

If there are no collisions then

$$\rho(t) = e^{-\frac{\gamma}{2}t} \rho(0) e^{-\frac{\gamma}{2}t}$$

where γ is the matrix of the decay rates. γ is a positive definite matrix. Let us assume that a collision is represented by a random unitary matrix which we assume for simplicity to have a distribution given by the Haar measure.

If there are collisions at times $t_1 < t_2 < \ldots < t_n$, then

$$\rho(t) = e^{-(\gamma/2)(t-t_n)} U_n e^{-\frac{\gamma}{2}(t_n-t_{n-1})} \cdots$$
$$U_1 e^{-\frac{\gamma}{2}t_1} \rho(0) e^{-\frac{\gamma}{2}t_1} U_1^* \ldots e^{-\frac{\gamma}{2}(t_n-t_{n-1})}$$
$$U_n^* e^{-(\gamma/2)(t-t_n)}$$

Assuming still that the $t_1, t_2, \ldots, t_n, \ldots$ are jump points of a Poisson process of mean density c we would like to prove that

$$(*) \quad \rho(t) \approx e^{-\tau(\gamma)t} \cdot \mathbb{1}/d,$$

where

$$\tau(\gamma) = \frac{1}{d} \sum \gamma_i$$

is the average of the decay rates.

We introduce the traceless matrix $g = -\gamma + \tau(\gamma)\mathbb{1}$. Then with the previous notation

$$\rho(t) = \frac{1}{d} e^{-\tau(\gamma)t} X(t)X(t)^* .$$

An essential assumption is that there are many impacts during the lifetime of the energy level, so $\|\gamma\|/c \ll 1$, which implies $\|g\|/c \ll 1$. We consider times of the order of magnitude $\frac{1}{\|g\|}$, the lifetime of the level. The probability that $X(t)X(t)^*$ differs from $\mathbb{1}$ by ε for $0 \leq t \leq \frac{1}{\|g\|}$ is then

$$\leq \frac{2}{\varepsilon}\left(\left(1 - \frac{\tau g^2}{c^2}\right) e^{\frac{\tau g^2}{c^2}\frac{c}{\|g\|}} - 1\right)$$

which is still of the order of magnitude $\|g\|/c$ and very small. So a least during times not exceeding greatly $1/_{\|g\|}$ the relation (*) is fulfilled. The times $t \gg \frac{1}{\|g\|}$, however, are not interesting because then the line has died off.

This paper is a simplified version of the thesis of L. Hantsch |2| There one finds more sophisticated estimates. The key for them is lemma 2 by which one can prove as in proposition 3 submartingale properties of $p(X(t)X(t)^*)$.

We have to thank H. Rost, Heidelberg, for many discussions and suggestions.

Proof of results

At first we state a well-known lemma denoting by dU the Haar measure on $U(d)$ and as before by τ the normalized trace.

Lemma 1: Let A be a $d \times d$-matrix, then

$$\int_{U(d)} dU \quad U^* A U = \tau(A).$$

Proposition 1: One has

$$E\,\tau\, X(t)X(t)^* = \frac{1}{c} F(t)$$

with $F(t) = k + k * k + k * k * k + \ldots$ and

$$k(t) = \begin{cases} \tau(c\, e^{-ct+gt}) & \text{for } t \geq 0 \\ 0 & \text{for } t < 0. \end{cases}$$

Proof: One has

$$c\; E\tau X(t)X(t)^* = c e^{-ct} \quad (e^{(g/2)t}\; e^{(g/2)t}) +$$

$$+ \sum_{n=1}^{\infty} e^{-ct} c^n \int_{0 \leqslant t_1 \leqslant \ldots \leqslant t_n \leqslant t} \cdots \int dt_1 \ldots dt_n \int \ldots \int dU_1 \ldots dU_n$$

$$\tau\Big(e^{(g/2)(t-t_n)} U_n e^{(g/2)(t_n-t_{n-1})} U_{n-1} \ldots$$

$$e^{(g/2)(t_2-t_1)} U_1 e^{(g/2)t_1} e^{(g/2)t_1} U_1^* \ldots$$

$$U_{n-1}^* e^{(g/2)(t_n-t_{n-1})} U_n^* e^{(g/2)(t-t_n)} \Big) .$$

Integrating successively over $U_1, U_2, \ldots$ and applying lemma 1 one obtains

$$ce^{-ct} \tau(e^{gt}) + \sum_{n=1}^{\infty} e^{-ct} c^n \int_{0 \leqslant t_1 \leqslant \ldots \leqslant t_n} \cdots \int dt_1 \ldots dt_n \, \tau(e^{g(t-t_n)})$$

$$\tau\, e^{g(t_n-t_{n-1})} \ldots \tau(e^{gt_1})$$

and this is just the proposition.

In the following we assume always that $\|g\|/c$ is sufficiently small, which is surely the case if $\|g\|/c < \frac{1}{3}$. But as we are interested in small $\|g\|/c$ an exact evaluation of the validity of the formulas is not interesting. In the same way the determination of the constants in the O-terms is not very useful.

<u>Proposition 2:</u> For $p \geqslant 0$ the equation

$$\tau \frac{c}{p_0 + c - g} = 1$$

has only one solution p_0

$$\frac{p_0}{c} = \frac{\tau g^2}{c^2} + O\left(\frac{\|g\|^3}{c^3}\right).$$

One has

$$E \tau X(t) X(t)^* = (1-\alpha) e^{p_0 t} + O(e^{-\tilde{c} t})$$

with

$$1-\alpha = \left(\tau \frac{c^2}{(p_0+c-g)^2}\right)^{-1} = 1- \frac{\tau g^2}{c^2} + 0\left(\frac{\|g\|^3}{c^3}\right)$$

$$\tilde{c} = c\left(1- \frac{\|g\|}{c} - \left(\frac{2\,\tau g^2}{c^2}\right)^{1/3}\right) .$$

Proof: We apply the Laplace transform

$$\hat{k}(p) = \int_0^\infty e^{-pt}k(t)dt$$

$$= \tau \frac{c}{c+p-g}$$

$$\hat{F}(p) = \int_0^\infty e^{-pt}F(t)dt = \frac{\hat{k}}{1-\hat{k}} = \hat{k} + \frac{\hat{k}^2}{1-\hat{k}}$$

As $k^2(\xi + i\eta)$ is integrable with respect to η for sufficiently large ξ one has for sufficiently large ξ

$$F(t) = k(t) + \frac{1}{2\pi i}\int_{\xi-i\infty}^{\xi+i\infty} dp \frac{\hat{k}^2}{1-\hat{k}} e^{+pt} .$$

We look for the poles of $\hat{k}^2/(1-\hat{k})$. Let us assume that g is diagonal

$$g = \begin{pmatrix} g_1 & & 0 \\ & \ddots & \\ 0 & & g_d \end{pmatrix}, \quad \text{then}$$

$$\hat{k}(p) = \frac{1}{d} \sum_i \frac{c}{p+c-g_i} .$$

All the poles of $\hat{k}$ are real and $\leqslant -c + \|g\|$. In the interval $-c + \|g\| \leqslant p < \infty$, the function $\hat{k}(p)$ is monotonically decreasing; one has

$$\frac{c}{p+c} \leqslant \hat{k}(p) \leqslant \frac{c}{p+c+\|g\|} .$$

The first inequality comes from the convexity of the function $x \longmapsto x^{-1}$, $x > 0$. For real $p \geqslant -c + \|g\|$ there exists only one value p_0 such that $k(p_0) = 1$, and one has $0 \leqslant p_0 \leqslant \|g\|$.

We want to calculate p_0 more exactly. One has

$$(c+p)(1-\hat{k}(p)) = c+p- \tau \frac{c}{1-\frac{g}{c+p}}$$

$$= c + p - c - \tau \frac{cg}{c+p} - \varphi(p)$$

$$= c - \varphi(p)$$

as $\tau g = 0$, where

$$\varphi(p) = \frac{1}{c+p} \tau \frac{cg^2}{c+p-g} .$$

One has

$$\varphi'(p) = -\frac{1}{(c+p)^2} \tau \frac{cg^2}{c+p-g} - \frac{1}{c+p} \tau \frac{cg^2}{(c+p-g)^2} .$$

In the interval $0 \leqslant p \leqslant \|g\|$ one has

$$|\varphi'(p)| < 2 \frac{\tau g^2}{(c-\|g\|)^2} = \beta .$$

We solve the equation $p = \varphi(p)$ in an iterative way

$$\tilde{p}_0 = \varphi(0)$$

$$p_1 = \varphi(\tilde{p}_0)$$

$$p_2 = \varphi(p_1)$$

$$\vdots \qquad \vdots$$

As $|p_{i+1} - p_i| = |\varphi(p_i) - \varphi(p_{i-1})| \leqslant \beta |p_i - p_{i-1}|$ one has

$$|p_0 - \tilde{p}_0| \leqslant \frac{\beta}{1-\beta} \tilde{p}_0$$

and

$$\tilde{p}_0 = \tau \frac{g^2}{c-g}$$

and finally

$$\frac{p_0}{c} = \frac{\tau g^2}{c^2} + 0 \left(\frac{\|g\|^3}{c^3} \right) .$$

From the formula for $\varphi'(p)$ above one obtains

$$|\varphi'(p)| \leqslant \frac{2c\,\tau g^2}{(c+\mathcal{Re}\,p-\|g\|)^3} - \varepsilon \; ; \; \varepsilon > 0$$

for $\mathcal{Re}\, p \geqslant \|g\| - c$. From this,

$$|\varphi'(p)| \leqslant 1 - \delta \; ; \; \delta > 0$$

for $\mathcal{Re}\, p \geqslant \|g\| - c + (2c\,\tau g^2)^{1/3}$. This implies that in this half-plane the equation $p - \varphi(p) = 0$ has only one simple solution p_0. For if $\varphi(p_1) = p_1$, then $|p_0 - p_1| = |\varphi(p_0) - \varphi(p_1)| =$

$= |\int_{p_0}^{p_1} \varphi'(p)dp| \leqslant |(1-\delta)p_1 - p_0|$ and hence $p_1 = p_0$. So the equation $p - \varphi(p)$ has only one solution and the relation

$$(p - \varphi(p))' = 1 - \varphi'(p)$$

implies that p_0 is a simple zero. Concluding these results one obtains that the function $\hat{k}(p)^2/_{1-\hat{k}}$ has in the halfplane

$$\mathcal{Re}\, p \geqslant \|g\| - c + (2c\,\tau g^2)^{1/3}$$

only one simple pole p_0 and that it is continuous and integrable on the boundary.

So

$$F(t) = k(t) + F_0(t) + \int_{\xi_0 - i\infty}^{\xi_0 + i\infty} dp \frac{\hat{k}^2}{1-\hat{k}} e^{+pt} = O(e^{\xi_0 t}) + F_0(t)$$

where $\xi_0 = -c + \|g\| + (2c\tau g^2)^{1/3} = \tilde{c}$ and $F_0(t)$ is the residue of $(\hat{k}^2/_{1-\hat{k}})e^{-ipt}$ at p_0. One has

$$F_0(t) = -\frac{1}{\hat{k}'(p_0)} e^{p_0 t}$$

with

$$-\frac{1}{\hat{k}'(p_0)} = \left(\tau \frac{c}{(c+p_0-g)^2}\right)^{-1} = c\left(1 - \frac{\tau g^2}{c^2} + O\left(\frac{\|g\|^3}{c^3}\right)\right)$$

This yields the proposition.

The following lemma is more general than is needed, but it has some interest in itself and has been applied to more sophisticated estimations of our process |2|.

__Lemma 2:__ Let p be a sublinear function on all positive definite complex $d \times d$-matrices such that

(i) $0 \leqslant A \leqslant B \Rightarrow 0 \leqslant p(A) \leqslant p(B)$

(ii) $p(U^*AU) = p(A)$ for all unitary U, where $\geqslant 0$ means positive definite. Then for any matrix $B \geqslant 0$ of determinant 1 and for any $A \geqslant 0$

$$\int p(BUAU^*B)dU \geqslant p(B)$$

where dU denotes the Haar measure.

__Proof:__ We want to apply the lemma for $p = \tau$. Then the proof becomes very easy. Without loss of generality we may assume that B is orthogonal $B = \mathrm{diag}(\beta_1, \ldots, \beta_d)$ with $\Pi\beta_i = 1$. Then by lemma 1

$$\int \tau(BUAU^*B)dU = \tau(B\int UAU^*dUB) = \tau(A)\tau(B^2)$$

and $\tau(B^2) = \frac{1}{d}\sum \beta_i^2 \geqslant 1$ as the geometric mean is not bigger than the arithmetic mean.

In the general case one uses the sublinearity of p and the invariance of p and dU and obtains

$$\int p(BUAU^*B)dU \geqslant \int dU p\left(\frac{1}{d!}\sum_{\pi\in\mathcal{S}_d} \pi B \pi^{-1} UAU^* \pi B \pi^{-1}\right)$$

where $\mathcal{S}_d$ denotes the permutation group of d elements. For any $C \geqslant 0$ the inequality (*) holds

$$(*)\quad \frac{1}{d!}\sum \pi B \pi^{-1} C \pi B \pi^{-1} \geqslant C.$$

If (*) holds one concludes by (i) and (ii) that the last integral

$$\geqslant \int dU p(UAU^*) = p(A).$$

The inequality (*) has still to be proven. If $B = \mathrm{diag}(\beta_1, \ldots, \beta_d) = \sum \beta_i E_i$, where E_i is the matrix with 1 on the (i, i)-place and 0 elsewhere, then

$$\pi B \pi^{-1} = \sum_i \beta_{\pi(i)} E_i$$

and the left side of (*) becomes

$$\frac{1}{d!}\sum_{i,k,\pi} \beta_{\pi(i)} \beta_{\pi(k)} E_i C E_k.$$

We split the sum into the sums over all pairs (i, k) with $i = k$ and $i \neq k$:

$$\frac{1}{d}\sum \beta_i^2 \sum E_i C E_i + \frac{1}{d(d-1)}\sum_{i\neq k} \beta_i \beta_k \sum_{i\neq k} E_i C E_k.$$

Now

$$\sum_{i\neq k} E_i C E_k = \sum_{i,k} E_i C E_k - \sum_i E_i C E_i = C - \sum_i E_i C E_i.$$

So one has

$$\left(\frac{1}{d}\sum \beta_i^2 - \frac{1}{d(d-1)}\sum_{i\neq k} \beta_i \beta_k\right)\sum_i E_i C E_i + \frac{1}{d(d-1)}\sum_{i\neq k} \beta_i \beta_k C.$$

Now

$$\prod_{i\neq k} \beta_i \beta_k = \prod_{i,k} \beta_i \beta_k \left(\prod_i \beta_i^2\right)^{-1} = 1$$

and hence

$$\frac{1}{d(d-1)} \sum_{i\neq k} \beta_i \beta_k \geqslant 1.$$

Furthermore

$$\frac{1}{d}\sum \beta_i^2 - \frac{1}{d(d-1)} \sum_{i\neq k} \beta_i \beta_k = \frac{d}{d-1}\left(\frac{1}{d}\sum \beta_i^2 - \left(\frac{1}{d}\sum \beta_i\right)^2\right) \geqslant 0.$$

This proves (*) and hence the lemma.

Proposition 3: For any $t \geqslant 0$ the following inequality holds:

$$P\{\sup_{0\leqslant s\leqslant t} \tau\,(X(s)X(s)^* - 1) \geqslant \varepsilon\} \leqslant E(\tau X(t)X(t)^* - 1)/\varepsilon$$

and

$$\tau\,(X(t)X(t)^*) \geqslant 1 \quad \text{for all} \quad X.$$

Proof: Assume n fixed, and $t_o = 0 \leqslant t_1 \leqslant \dots \leqslant t_n \leqslant t = t_{n+1}$ numbers. Define for $t_k < s \leqslant t_{k+1}$

$$Y(s) = e^{(g/2)(s-t_k)} U_k e^{(g/2)(t_k - t_{k-1})} \dots U_1 e^{(g/2)t_1}$$

$$Z(s) = \tau\,(Y(s)Y(s)^*)$$

and $Y(0) = 1$, $Z(0) = 1$ and

$$Y_k = Y(t_{k+1}) = e^{(g/2)(t_{k+1}-t_k)} U_k \dots U_1 e^{(g/2)t_1}$$

$$Z_k = \tau\,(Y_k Y_k^*).$$

Then one observes:

(i) For any choice of $U_1, \dots, U_n$ the function $Z(s)$ is convex in the interval $[t_k, t_{k+1}]$, $k = 0, \dots, n$. So $Z(s)$ takes its maximum on one of the points $t_o, t_1, \dots, t_n, t$ and $\max_{0\leqslant s\leqslant t} Z(s) = \max_{k=0,\dots,n+1} Z_k$.

(ii) Assume as always that the U_k, $k = 1, \ldots, n$ are independent, identically distributed with respect to the Haar measure, then Z_k, $k = 0, 1, \ldots, n+1$ is a submartingale.

(iii) $Z(s) \geq 1$ for $0 \leq s \leq t$.

For any $A \geq 0$ the function $s \longmapsto \tau(e^{(g/2)s} A e^{(g/2)s})$ has a second derivative ≥ 0 and therefore it is convex. This yields (i). By lemma 2

$$E\{Z_k | U_1, \ldots, U_{k-1}\} \geq Z_{k-1}.$$

This shows (ii). As $Y(s)Y(s)^* \geq 0$ one has

$$Z(s) = \tau(Y(s)Y(s)^*) \geq \det Y(s)Y(s)^* = 1.$$

This gives (iii).

By (i), (ii) and (iii) we conclude

$$P\{\sup_{0 \leq s \leq t} (Z(s)-1) \geq \varepsilon\} < \frac{1}{\varepsilon}(EZ(t) - 1)$$

and finally assuming now $n, t_0, \ldots, t_n$ as random variables

$$P\{\sup_{0 \leq s \leq t} (\tau(X(s)X(s)^*) \geq \varepsilon\}$$

$$= EP\{\sup_{0 \leq s \leq t} (X(s)X(s)^*) \geq \varepsilon | n; t_1, \ldots, t_n\}$$

$$\leq \frac{1}{\varepsilon} E(E\{\tau(X(t)X(t)^*) - 1 | n; t_1, \ldots, t_n\}$$

$$= \frac{1}{\varepsilon} E(\tau(X(t)X(t)^*) - 1).$$

<u>Proposition 4:</u> There exist constants $\tilde{p}$ and $\tilde{\alpha}$

$$\frac{\tilde{p}}{c} = \frac{\tau g^2}{c^2} + 0\left(\frac{\|g\|^3}{c^3}\right)$$

$$\tilde{\alpha} = \frac{\tau g^2}{c^2} + 0\left(\frac{\|g\|^3}{c^3}\right)$$

such that

$$P\{\max_{0 \leq s \leq t} \frac{1}{2}(X(s)X(s)^* + X(s)^{-1*}X(s)^{-1} - 2) \geq \varepsilon\}$$

$$\leq \frac{2}{\varepsilon}((1-\tilde{\alpha})e^{\tilde{p}t} - 1 + 0(e^{-\tilde{c}t}))$$

with

$$\tilde{c} = c\left(1 - \frac{\|g\|}{c} - \left(\frac{2\tau g^2}{c^2}\right)^{1/3}\right)$$

<u>Proof:</u> If A is a matrix then denote $\hat{A} = A^{-1*}$. Then

$$\hat{X}(t) = e^{-(g/2)(t-t_n)} U_n e^{-(g/2)(t_n - t_{n-1})} \dots U_1 e^{-(g/2)t_1}.$$

So $\hat{X}(t)$ is a process of the same type as $X(t)$, its only difference is the sign of g. All previous results can be applied to $\hat{X}(t)$ as well as to $X(t)$.

By proposition 3

$$P\{\max_{0\leqslant s\leqslant t} \tau\,(\hat{X}(s)\hat{X}(s)^* - 1) \geqslant \varepsilon\} \leqslant \frac{1}{\varepsilon}\,(E\,\tau X(t)X(t)^* - 1)$$

and

$$P\{\max_{0\leqslant s\leqslant t} \frac{1}{2}\tau(X(s)X(s)^* + \hat{X}(s)\hat{X}(s)^* - 2) \geqslant \varepsilon\}$$

$$\leqslant \frac{2}{\varepsilon}\, E\, \frac{1}{2}\,(X(t)X(t)^* + \hat{X}(t)\hat{X}(t)^* - 2).$$

By proposition 1 one gets the result.

<u>Literature</u>

|1| E.U. Condon, G.H. Shortley: The theory of atomic spectra. Cambridge University Press 1953.

|2| L. Hantsch: Stochastische Mittelung von Abklingzeiten bei atomaren Niveauübergängen. Doktorarbeit, Heidelberg 1976.

|3| E. Hewitt, K.A. Ross: Abstract Harmonic Analysis II. (esp. § 29). Springer-Verlag 1970.

Institut für Angewandte Mathematik
der Universität Heidelberg
Im Neuenheimer Feld 294
69 Heidelberg, BRD

Subordination von Faltungs- und Operatorhalbgruppen

W. Hazod

Sei (Ω,Σ,P) ein Wahrscheinlichkeitsraum, $(\xi_n:\Omega \to \mathbb{R})$ eine Folge unabhängiger zufälliger Variabler mit identischer Verteilung $\xi_n(P) = \mu$, es sei $\eta_n := \sum_{i=0}^{n} \xi_i$ (mit Verteilung $\eta_n(P) = \mu^n$) und schließlich sei $Z:\Omega \to \mathbb{Z}_+$ eine von den (ξ_n) unabhängige Zufallsvariable. Dann sei $\Psi(\cdot) := \sum_{i=0}^{Z(\cdot)} \xi_i(\cdot) = \eta_{Z(\cdot)}(\cdot)$ die "untergeordnete" Zufallsvariable, deren Verteilung $\Psi(P)$ durch $\sum_{n=0}^{\infty} \mu^n \cdot P(\{Z=n\}) = \sum_{n\in\mathbb{Z}_+} \mu^n F(\{n\})$ gegeben ist, (dabei ist $F := Z(P)$).

Analog: Sei (Ω,Σ,P) gegeben, $(\eta_t:\Omega \to \mathbb{R})_{t>0}$ sei ein zeitlich homogener Prozess mit unabhängigen Zuwächsen, es sei $\mu_t := (\eta_{s+t}-\eta_s)(P)$. Wieder sei $Z:\Omega \to \mathbb{R}_+$ eine von (η_s) unabhängige Zufallsvariable mit Verteilung F, dann heißt $\Psi := \eta_{Z(\cdot)}(\cdot)$ "untergeordnet" und besitzt die Verteilung $\Psi(P) = \int_{\mathbb{R}_+} \mu_t dF(t)$. $t\in\mathbb{Z}_+$ resp. $\in\mathbb{R}_+$ wird in beiden Fällen als "zufälliger - diskreter oder stetiger - Zeitparameter" aufgefaßt (s.z.B. W. Feller [8] oder B.W. Huff [17 - 19]).

Nun sei zusätzlich $Z = Z_1$ in einen stationären Prozeß mit unabhängigen Zuwächsen (Z_t) (mit stetigem $(t\in\mathbb{R}_+)$ oder diskretem $(t\in\mathbb{Z}_+)$ Zeitparameter) eingebettet mit Verteilungen $Z_t(P) =: F_t (= F_1^t$ für $t\in\mathbb{Z}_+)$, sei weiter $\Psi_t(\cdot) = \eta_{Z_t(\cdot)}(\cdot)$ $(t\in\mathbb{Z}_+, \mathbb{R}_+)$ wie vorhin definiert mit Verteilung $\int_{\mathbb{R}_+} \mu_s dF_t(s) = \Psi_t(P) =: \lambda_t$. (Ist F_t auf $\mathbb{Z}_+$ konzentriert, so kann das Integral durch eine Summe ersetzt werden.) Dann heißt $(\Psi_t : t\in\mathbb{Z}_+$ oder $\in\mathbb{R}_+)$ untergeordneter Prozeß, genauer: der Prozeß, der (η_t) untergeordnet wird vermöge des zufälligen Zeitprozesses (Z_t). Die Verteilungen $(\mu_t$ resp. $\mu_1^t)$, $(F_t$ resp. $F_1^t)$, $(\lambda_t$ resp. $\lambda_1^t)$, $t\in\mathbb{R}_+$ oder $\in\mathbb{Z}_+$, bilden Faltungshalbgruppen, i.e. $\mathbb{Z}_+[\mathbb{R}_+] \ni t \to \mu_t$ resp. $\to F_t$ resp. $\to \lambda_t$ ist ein (stetiger) Homomorphismus der additiven Halbgruppen $\mathbb{Z}_+$ oder $\mathbb{R}_+$ in die Halbgruppe der Wahrscheinlichkeitsmaße $M^1(\mathbb{R})$.(F_t) heißt dann „Subordinationshalbgruppe."

Bei diesen Überlegungen geht nicht ein, daß (η_t) unabhängige Zuwächse hat, ganz analog läßt sich die Subordination von Übergangswahrscheinlichkeiten zeitlich-homogener Markoffprozesse erklären (s.z.B. W. Feller [8]).

Ersetzt man μ_t und λ_t durch die Faltungsoperatoren

$f \longrightarrow R_{\mu_t} f(\cdot) := \int f(y+\cdot)d\mu_t(y)$ $(=E(f\circ\eta_{t+s}|\eta_s = \cdot))$ resp.

$R_{\lambda_t} f(\cdot) := \int f(\cdot+y)d\lambda_t(y)$, $f\in C_0(\mathbb{R})$ und verzichtet man auf die Deutung durch Prozesse, so kommt man in natürlicher Weise zur Begriffsbildung der Subordination von Operatorhalbgruppen: Sei $(T_t)_{t\geq 0}$ eine stetige Halbgruppe von Kontraktionen auf einem Banachraum $\mathbf{B}$, sei $(F_t)\in M^1(\mathbb{R}_+)$ eine stetige Halbgruppe von Maßen, dann ist $(S_t := \int_{\mathbb{R}_+} T_s dF_t(s))_{t\geq 0}$ - zu verstehen als schwaches oder starkes Integral - eine stetige Halbgruppe von Kontraktionen (s. z.B. E.Hille, R. Phillips [14], R. Phillips [28]).

Eine wichtige Klasse solcher Operatorhalbgruppen wird im Zusammenhang mit Faltungshalbgruppen studiert: Sei $\mathcal{G}$ eine lokalkompakte Gruppe, $(\mu_t)_{t\geq 0}$ eine stetige Halbgruppe von Wahrscheinlichkeitsmaßen, $(R_{\mu_t} : f\in C_0(\mathcal{G}) \longrightarrow \int_{\mathcal{G}} f(\cdot y)d\mu_t(y))_{t\geq 0}$ die zugehörige Halbgruppe der Faltungsoperatoren auf $C_0(\mathcal{G}) =: \mathbf{B}$, dann ist $(\int_{\mathbb{R}_+} R_{\mu_s} dF_t(s))_{t\geq 0}$ von der Gestalt (R_{λ_t}) mit $\lambda_t = \int_{\mathbb{R}_+} \mu_s dF_t(s)$, (Für $\mathcal{G} = \mathbb{R}^n$ und Abelsche Gruppen s. S. Bochner [2], für kompakte Gruppen s. H. Carnal [4], allgemein s. J. Woll [37], resp. [11, 12].) In diesen Fällen lassen sich die neu gewonnenen Verteilungen durch ihre erzeugenden Distributionen resp. durch ihre Fouriertransformierten beschreiben. Für $\mathcal{G} = \mathbb{R}$ s. insbesondere auch B. W. Huff [17, 18, 19].

Ist $\mathcal{G}$ separabel, so könnte man - wie eingangs für $\mathcal{G} = \mathbb{R}$ geschildert - (μ_t) resp. (λ_t) als Verteilungen der Zuwächse eines zeitlich-homogenen $\mathcal{G}$-wertigen Prozesses mit unabhängigen Zuwächsen beschreiben, wobei der zu (λ_t) gehörige Prozeß durch Subordination mittels eines (unabhängigen) zufälligen Zeitparameters gewonnen wird. Und dies wiederum für stetigen $(Z \in \mathbb{R}_+)$ als auch für diskreten $(Z \in \mathbb{Z}_+)$ Zeitparameter.

Ersetzt man $\mathcal{G}$ durch eine vollständig-reguläre topologische Halbgruppe mit Einheit e und betrachtet die Faltungsoperatoren auf $C(\mathcal{G})$ oder betrachtet man anstelle der Operatoren R_μ Markoffsche Kerne auf einem Funktionenverband, dann sieht man sich veranlaßt, Subordinationen von Halbgruppen $(T_t)_{t \geq 0}$ zu betrachten, die in einem schwächeren Sinne stetig sind und nicht mehr notwendig auf einem Banachraum definiert sind. Daher scheint es geboten, allgemeiner Subordinationen von Homomorphismen $\Psi: \mathbb{R}_+ \longrightarrow \mathcal{Y}$ in affine Halbgruppen zu studieren.

Will man überdies Subordinationen von diskreten und stetigen Halbgruppen von einer gemeinsamen Warte aus betrachten, so ist es nützlich, allgemeinere Abbildungen $\Psi: X \longrightarrow \mathcal{Y}$ von einem topologischen Raum X [in den Anwendungen: $X = \mathbb{Z}_+$ oder $\mathbb{R}_+$] und Mischungsintegrale der Gestalt $\int_X \Psi(s)dF(s)$ zu betrachten, $F \in M^1(X)$.

Der erste Teil der Arbeit (§ 1) entwickelt einen Kalkül für Mischungen oben genannter Gestalt sowie Grenzwertsätze für derartige Mischungen. Überdies werden Beispiele von affinen Halbgruppen, die in diesen Rahmen passen, angegeben.

In § 2 betrachten wir spezielle affine Halbgruppen, nämlich $M^1(\mathbb{R}_+)$ und $M^1(\overline{\mathbb{R}}_+)$ ($\overline{\mathbb{R}}_+ := [0,\infty]$ mit ∞ als Nullelement). Mittels des Subordinationsbegriffs wird in den unendlich teilbaren Maßen dieser Halbgruppen eine neue Komposition (- verschieden von der Faltung -) eingeführt, die diese zu einer kompakten topologischen Halbgruppe macht.

Diese Struktur verwenden wir in § 3, um - zumindest für Faltungsoperatoren auf lokalkompakten Gruppen - nachzuweisen, daß durch den Subordinationsbegriff eine (partielle) Ordnungsrelation definiert wird: Man setzt $(S_t) \preccurlyeq_s (T_t)$, wenn es eine Subordinationshalbgruppe $(F_t) \subseteq M^1(\mathbb{R}_+)$ gibt mit $(\int T_s dF_t(s) = S_t)_{t \geq 0}$.

Wir nennen eine Subordination trivial, wenn es einen Automorphismus $\tau_c : t \longrightarrow c \cdot t$ von $\mathbb{R}_+$ gibt, so daß $(\int_{\mathbb{R}_+} T_t dF_s(t) = \int_{\mathbb{R}_+} T_t d\tau_c(\varepsilon_s)(t) = T_{ct}), t \geq 0$.

Wir zeigen, daß (zumindest) im Spezialfall der Subordination von Faltungshalbgruppen die Äquivalenzklassen $\{(T_t) \approx_s (S_t) \Longleftrightarrow (T_t) \preccurlyeq_s (S_t)$ und $(S_t) \preccurlyeq_s (T_t)\}$ durch triviale Subordinationen beschrieben sind.

In § 4 behandeln wir einige Klassen von Faltungshalbgruppen und deren Subordinationen, die von eigenem Interesse sind: Wir zeigen, daß die Halbgruppen lokalen Typs (Gaußhalbgruppen) und die elementaren Poissonhalbgruppen, maximal bezüglich $\preccurlyeq_s$ sind, d.h. nur in trivialer Weise als Subordinationen dargestellt werden können. (Damit werden bekannte Sätze von S. Bochner [2], H. Carnal [4], J. Woll [37] u.[11] leicht verallgemeinert, s. auch [12].) Daneben studieren wir Halbgruppen "vom Typ P+L", die dadurch beschrieben sind, daß sich die Generatoren aus einem primitiven und einem Poissonschen Anteil zusammensetzen. (Diese sind u.a. deshalb von Interesse, da die Vermutung naheliegt, daß dies die Klasse der Halbgruppen ist, bei denen jedes Maß μ_t invertierbar (in M(G)) ist.)

In § 5 behandeln wir schließlich eine weitere spezielle Klasse von Maßen, nämlich die Bochner-stabilen Verteilungen, i.e. jene Halbgruppen, die sich aus symmetrischen Gaußhalbgruppen durch Subordination nach einseitigen stabilen Verteilungen gewinnen lassen. Diese Konstruktion ist ein spezieller Fall der Konstruktion gebrochener Potenzen von Halbgruppengeneratoren mittels Subordinationen, und damit von Lösungen (verallgemeinerter) Wellengleichungen.

Es ist natürlich nicht möglich, einen umfassenden Literaturüberblick über Subordinationen und ihre Anwendungen zu geben. Ich gehe daher nur stichwortartig auf einige wenige Arbeiten ein: Subordinationen im Zusammenhang mit gebrochenen Potenzen werden z.B. in den Arbeiten von Komatsu [23] u. U. Westphal [35,36] behandelt, ein umfassender Kalkül wurde von Phillips [28] und E. Nelson [24] entwickelt. Subordination und Randomisierung von Wahrscheinlichkeitsverteilungen findet man bei W. Feller [8], S. Bochner [2], im Zusammenhang mit gebrochenen Potenzen sei auf V. Nollau [25,26] verwiesen. Subordination von Faltungshalbgruppen auf $\mathbb{R}_1$ (in der Sprechweise der stochastischen Prozesse) werden von B. W. Huff [17 - 19] eingehend studiert. Ohne dies näher hervorzuheben, werden in § 1 und § 4 einige Resultate von Huff verallgemeinert. Mischungen von Maßen $(\mu_t) \subseteq M^1(\mathbb{R})$ etwas allgemeinerer Art werden von A. Tortrat studiert [34] (hier darf das Mischungsmaß signiert sein!). Eine andere Verallgemeinerung wurde in [12] angegeben, dort werden Mischungen von Generatoren betrachtet. (Die zugehörigen Halbgruppen sind dann nicht mehr als Mischungen, sondern als (verallgemeinerte) Produktintegrale darstellbar).

F. Hirsch [15] untersucht eine spezielle Klasse von Subordinationen, die als Mischungen von Resolventen (und somit als Stieltjestransformationen von Halbgruppen) darstellbar sind. Dieser Gedanke wird aber hier nicht weiter verfolgt.

Im Text wird die Kenntnis der stetigen Halbgruppen von Wahrscheinlichkeitsmaßen auf lokalkompakten Gruppen und deren Erzeuger vorausgesetzt, insbesondere die kanonische Zerlegung von E. Siebert. Der Leser sei hierzu besonders auf die von H. Heyer [13] angegebene Darstellung verwiesen.

§ 1 Vorbereitungen

1.1

$\mathbb{E}$ sei ein lokalkonvexer topologischer Vektorraum, mit $\mathbb{E}'$ resp. $\mathbb{E}''$ resp. $\mathbb{E}'^*$ werden der topologische Dualraum resp. Bidualraum resp. der algebraische Dualraum des topologischen Dualraumes bezeichnet. $\mathbb{E}'^*$ wird mit der schwachen Topologie $\sigma(\mathbb{E}'^*,\mathbb{E}')$ ausgestattet. Es gilt die kanonische Einbettung $\mathbb{E}\hookrightarrow\mathbb{E}''\hookrightarrow\mathbb{E}'^*$.

Sei $\mathbb{F}\subseteq\mathbb{E}'$ ein Unterraum, so daß $\sigma(\mathbb{E},\mathbb{F})$ eine Hausdorfftopologie ist.

1.2

X sei ein topologischer Hausdorffraum, M(X) resp. $M_+(X)$ resp. $M^1(X)$ sei die Menge der straffen beschränkten Maße resp. positiven beschränkten Maße resp. der Wahrscheinlichkeitsmaße auf X. Wenn nichts anderes vereinbart wird, sind Maße im folgendem stets als straffe Maße zu verstehen.

1.3 Definition.

Sei $\mathcal{M}(X,\mathbb{E};\mathbb{F})$ die Menge aller Abbildungen $f:X\longrightarrow\mathbb{E}$, so daß $x\longrightarrow\langle f(x),g\rangle$ stetig und beschränkt ist für alle $g\in\mathbb{F}$. Wenn $\mathbb{F}=\mathbb{E}'$ ist, so schreiben wir $\mathcal{M}(X,\mathbb{E})$. Sei $\mu\in M(X)$, dann wird durch

$$\mathbb{F}\ni g\longrightarrow\int_X\langle f(x),g\rangle d\mu(x)$$

ein Element in $\mathbb{F}^*$ bestimmt, das mit $\int_X f d\mu$ bezeichnet werde.

1.4 Definition.

Sei $\mathcal{M}_0(X,\mathbb{E};\mathbb{F}):=\{f\in\mathcal{M}(X,\mathbb{E};\mathbb{F}),$ so daß $\int_X f d\mu\in\mathbb{E}$ für alle $\mu\in M(X)\}$.

In den Beispielen, die uns im folgenden interessieren, läßt sich stets leicht direkt verifizieren, daß für alle geeignet beschränkten f gilt: $\int f d\mu\in\mathbb{E}$. Eine ausführliche Diskussion findet man in N. Bourbaki [3], livre VI Ch.6 § 1 (für lokalkompakte Räume X) sowie in S. Khurana [22].

1.5 Bemerkung.

Sei $\mathscr{V} \subseteq \mathbb{E}$ abgeschlossen und konvex, $\mathcal{M}_0(X,\mathscr{V};\mathbb{F}) := \{f \in \mathcal{M}_0(X,\mathbb{E};\mathbb{F})$ mit $f(X) \subseteq \mathscr{V}\}$, so ist $\int_X f d\mu \in \mathscr{V}$ für alle $\mu \in M^1(X)$ und $f \in \mathcal{M}_0(X,\mathscr{V};\mathbb{F})$.

1.6 Bemerkung.

Sei μ diskret, $\mu = \sum_{x \in X} \alpha_x \varepsilon_x \in M(X)$ resp. $M^1(X)$.

Sei $\mathscr{V}$ wie in 1.5. Dann ist für jedes $f \in \mathcal{M}_0(X,\mathbb{E};\mathbb{F})$ resp. $\mathcal{M}_0(X,\mathscr{V};\mathbb{F})$
$\int f d\mu = \sum \alpha_x f(x) \in \mathbb{E}$ resp. $\in \mathscr{V}$.

1.7 Voraussetzung.

Sei $\mathscr{V}$ konvex und abgeschlossen, $\mathbb{F} \subseteq \mathbb{E}'$, so daß für jeden vollständig regulären Raum X $\{f \in \mathcal{M}(X,\mathbb{E};\mathbb{F}) : f(X) \subseteq \mathscr{V}\} \subseteq \mathcal{M}_0(X,\mathscr{V},\mathbb{F})$. Diese Voraussetzung werden wir vielfach der Bequemlichkeit halber machen. Insbesondere folgt daraus (wenn wir $X = \mathbb{N}$ als diskreten Raum ansehen), daß für jede Folge ($\alpha_n \geq 0$, $\sum \alpha_n = 1$) und für jede Punktfolge $(x_n) \subseteq \mathscr{V}$ auch $\sum \alpha_n x_n \in \mathscr{V}$ ist.

1.8 Beispiele.

1.8.1

Sei $\mathbb{B}$ ein Banachraum, $\mathbb{E} := \mathscr{L}(\mathbb{B})$ der Raum der stetigen linearen Operatoren mit der starken resp. schwachen Operatorentopologie, $\mathscr{V} := \{T \in \mathscr{L}(\mathbb{B}) : \|T\| \leq 1\}$. Weiter sei
$\mathbb{F} := \{L_{x,y} \in \mathbb{E}',\ x \in \mathbb{B},\ y \in \mathbb{B}' : L_{x,y}(T) = \langle T, L_{x,y}\rangle_{\mathbb{E}} := \langle Tx, y\rangle_{\mathbb{B}}\}$.

1.8.2

Sei $\mathbb{B}$ überdies ein Banachverband, und
$\mathscr{V} := \{T \in \mathscr{L}(\mathbb{B}) : T \geq 0\ \|T\| \leq 1\}$.

1.8.3 Speziell:

$\mathbb{B} = l^1$, $\mathscr{V} :=$ {Markoff-Matrizen} resp. {sub-Markoff-Matrizen} versehen mit der Topologie der koordinatenweisen Konvergenz.

Sei $\mathbb{F}:=\langle\{L_{ij}, i,j\in\mathbb{N} : T = (T_{kl}) \longrightarrow T_{ij}\}\rangle$.

Sei $f: x\in X \longrightarrow (f_{ij}(x)) = f(x)\in\mathcal{T}$ eine matrizenwertige Funktion, dann ist $\int f d\mu = (\int_X f_{ij}(x) d\mu(x))$.

In allen genannten Fällen ist die Voraussetzung 1.7 erfüllt, wie man unmittelbar einsieht.

1.8.4

Seien X,Y topologische Räume, $\mathbb{E} := M(Y)$, $\mathcal{T} := M^1(Y)$ die (konvexe) Menge der straffen Wahrscheinlichkeitsmaße, versehen mit der schwachen Topologie, erzeugt von den Funktionalen $(\mu \longrightarrow \int h d\mu : h \in C(Y))$.

$f: X \ni x \longrightarrow \sigma_x \in M^1(Y)$ sei stetig, beschränkt, weiter gelten

(i) Y sei vollständig regulär,

(ii) für alle kompakten $K \subseteq X$ sei $f(K) = \{\sigma_x : x\in K\}$ gleichmäßig straff

[Dies ist offenbar erfüllt, wenn Y ein Prohorov Raum ist!].

Dann ist für $\mu\in M^1(X)$ $\int_X f d\mu \in M^1(Y)$ und $(\int f d\mu)(h) = \int_X [\int_Y h(y) d\sigma_x(y) d\mu(x)]$

1.8.5

Sei Y vollständig regulär und Hausdorffsch, $\mathbb{B} := C(Y)$, $\mathcal{T} := \{T\in\mathcal{L}(\mathbb{B}) : T\geq 0$, $\|T\|\leq 1$, $T\mathbb{1} = \mathbb{1}$ resp. $T\mathbb{1}\leq\mathbb{1}\}$. $\mathcal{L}(\mathbb{B})$ werde mit einer lokalkonvexen Hausdorfftopologie versehen, bezüglich der die Funktionale $L_{g,y}$: $T \longrightarrow Tg(y)$, $g\in C(Y)$, $y\in Y$, stetig sind. Weiter sei $\mathbb{F} := \{L_{g,y}: g\in C(Y), y\in Y\}$ oder $\mathbb{F} := \{L_{g,y}, g\in C_0(Y), y\in Y\}$.

Letzteres ist nur sinnvoll, wenn Y lokalkompakt ist. Sei nun $T: X\ni x \longrightarrow T(x)\in\mathcal{T}$, so daß $X\ni x \longrightarrow T(x)g(y)$ stetig und beschränkt ist, für $y\in Y$, $g\in C(Y)$, sei weiter $\mu\in M(X)$, $h(y) := \int_X T(x)g(y)d\mu(x)$.

Dann ist, da $(x,y) \longrightarrow T(x)g(y)$ nicht notwendig simultan stetig ist, h nicht notwendig $\in C(Y)$. Also müssen zusätzliche Bedingungen erfüllt sein, die die Existenz der Mischungsintegrale garantieren. Da μ straff ist, darf man oBdA voraussetzen, daß X kompakt ist.

(i) Wenn X das 2-te Abzählbarkeitsaxiom erfüllt, dann ist auch $\int T d\mu \in \mathcal{T}$, i.e. dann ist für jedes g auch $(y \longrightarrow \int_X T(x)g(y)d\mu(x)) \in C(Y)$

(ii) Sei $\mathcal{V}_1 \subseteq \mathcal{V}$ eine abgeschlossene konvexe Teilmenge, so beschaffen, daß aus der Stetigkeit von $x \longrightarrow T(x)g(y)\ \forall\ g \in C(Y), y \in Y$ folgt, daß $x \longrightarrow T(x)g(\cdot)$ stetig ist in der kompakt-offenen Topologie, so folgt wieder die Existenz der Mischungsintegrale und $\int_X T(x)d\mu(x) \in \mathcal{V}_1$ für $\mu \in M^1(X)$.

Die später fast ausschließlich betrachteten Faltungsoperatoren erfüllen die Voraussetzungen (ii).

1.8.6 Speziell:

Seien die Operatoren T aus $\mathcal{V}[\mathcal{V}_1]$ dargestellt durch Übergangskerne - die mit dem selben Symbol $\bar{T}$ bezeichnet werden, also sei $Tf(y) = \int_Y f(y')\bar{T}(y,dy')$.

Dann bedeutet $T_\alpha \longrightarrow T$ in der in 1.8.5 beschriebenen Konvergenz einfach die schwache resp. vage Konvergenz der Maße $\bar{T}_\alpha(x,\cdot) \to \bar{T}(x,\cdot)$, $y \in Y$. Die Voraussetzung (ii) in 1.8.5 besagt also, daß aus der schwachen Konvergenz $\bar{T}_\alpha(x,\cdot) \longrightarrow \bar{T}(x,\cdot) \in \mathcal{V}_1$ punktweise für jedes $y \in Y$ folgt, daß die Konvergenz gleichmäßig ist auf kompakten Mengen in Y.

1.9 Definition.

Seien $\mathbb{E}, \mathcal{V}, X, \mathbb{F}$ wie vorhin (1.1 - 1.6) gegeben. Dann führen wir weitere Räume von Abbildungen ein:

(i) $\mathcal{M}_1(X,\mathcal{V};\mathbb{F}) := \{f: X \longrightarrow \mathbb{E}, f(X) \subseteq \mathcal{V}$ und beschränkt, so daß eine Zerlegung $X = \bigcup_{j=1}^{\infty} E_j$ in paarweise disjunkte meßbare Teile existiert, wobei jedes Kompaktum in X nur endlich viele E_j schneidet und f die Gestalt $f = \sum 1_{E_j}(\cdot)y_j, y_j \in \mathbb{E}$ hat. Dabei sei die Folge $\{y_j\}$ beschränkt$\}$.

(ii) Wir setzen voraus, daß die Mischungsintegrale $\int f d\mu = \sum \mu(E_j)y_j$ existieren und in $\mathcal{V}$ liegen. (Dies ist offenbar erfüllt, falls $\mathcal{V}$ bezüglich $\mathcal{G}(\mathbb{E},\mathbb{F})$ folgenvollständig ist.)

Unter der Voraussetzung 1.7 ist die Forderung stets erfüllt. $\mathcal{M}_1(X,\mathcal{T};\mathbb{F})$ ist eine konvexe Teilmenge von $\mathcal{M}_1(X,\mathbb{E};\mathbb{F})$, somit $\subseteq\{\varphi:X \longrightarrow \mathbb{E} :\langle\varphi(\cdot),g\rangle$ meßbar und beschränkt für $g\in\mathbb{F}\}$. Der letzte Raum enthält auch $\mathcal{M}_0(X,\mathbb{E};\mathbb{F})$, daher ist folgende Menge sinnvoll definiert:

(iii) $\mathcal{M}_2(X,\mathcal{T};\mathbb{F}) := co(\mathcal{M}_0(X,\mathcal{T};\mathbb{F}), \mathcal{M}_1(X,\mathcal{T};\mathbb{F}))$ (co := konvexe Hülle),

$\mathcal{M}_2(X,\mathbb{E};\mathbb{F}) := \mathcal{M}_0(X,\mathbb{E};\mathbb{F}) \oplus \mathcal{M}_1(X,\mathbb{E};\mathbb{F})$.

(iv) Die Räume $\mathcal{M}_i(X,\mathbb{E},\mathbb{F})$ werden nun stets versehen mit der Topologie $\mathcal{T}_{co}$ der kompakt-gleichmäßigen Konvergenz, i.e. $f_\alpha \xrightarrow{\mathcal{T}_{co}} f$, falls für alle $g\in\mathbb{F}$, für alle kompakten $K\subseteq X$

$$\sup_{x\in K} |\langle f_\alpha(x) - f(x),g\rangle| \longrightarrow 0 \quad .$$

(v) Schließlich führen wir ad hoc die abkürzende Schreibweise $\Phi(f,\mu) := \int f d\mu$, $\mu\in M(X)$, $f\in\mathcal{M}_2(X,\mathbb{E};\mathbb{F})$ ein.

1.10 Lemma.

$\Phi:\mathcal{M}_2(X,\mathbb{E};\mathbb{F}) \otimes M(X) \longrightarrow \mathbb{E}$ ist bezüglich $\mathcal{T}_{co}$ in der ersten Komponente stetig,

Φ ist linear in jeder Komponente,

Φ ist in der zweiten Komponente stetig, wenn man $M(X)$ mit der Normtopologie versieht.

Überdies ist $\Phi\{\mathcal{M}_2(X,\mathcal{T};\mathbb{F}) \otimes M^1(X)\} \subseteq \mathcal{T}$.

1.11 Lemma.

Die Einschränkung $\Phi:\mathcal{M}_0(X,\mathbb{E};\mathbb{F}) \otimes M(X) \longrightarrow \mathbb{E}$ ist in jeder normbeschränkten Teilmenge von $M(X)$ stetig in der zweiten Komponente, wenn man $M(X)$ mit der schwachen Topologie versieht. [Offensichtlich].

Eine etwas speziellere Aussage gilt auch für $\mathcal{M}_1(X)$:

1.12 Lemma.

Seien $(\mu_\alpha), \mu \in M(X)$, $\sup_\alpha \|\mu_\alpha\| < \infty$, $\mu_\alpha \longrightarrow \mu$ schwach. Sei $f \in \mathcal{M}_2(X, \mathbb{E}; \mathbb{F})$, $f = g+h$, $g \in \mathcal{M}_0(X)$, $h \in \mathcal{M}_1(X)$, $h = \sum_1^\infty 1_{E_j} x_j$, so daß alle E_j μ-Stetigkeitsmengen sind (d. h. ihr Rand hat μ-Maß 0). Dann gilt:

$$\Phi(f, \mu_\alpha) \longrightarrow \Phi(f, \mu).$$

Darüber hinaus gilt allgemeiner

1.13 Proposition.

a) $\Phi: \mathcal{M}_0(X) \otimes M(X) \to \mathbb{E}$ ist in gewissem Sinne simultan stetig auf beschränkten Teilen, genauer:

Seien (f_β), (μ_α) Netze in $\mathcal{M}_0(X, \mathbb{E}, \mathbb{F})$ resp. $M(X)$,

sei $f \in \mathcal{M}_0(X)$, $\mu \in M(X)$, $\sup_\alpha \|\mu_\alpha\| < \infty$, (μ_α) gleichmäßig straff und sei

$\sup_\beta \sup_{x \in X} |\langle f_\beta(x), y\rangle| < \infty$ für alle $y \in \mathbb{F}$,

weiter konvergieren $f_\beta \longrightarrow f$ in τ_{co} und $\mu_\alpha \longrightarrow \mu$ schwach,

so folgt $\Phi(f_\beta, \mu_\alpha) \longrightarrow \Phi(f, \mu)$ in $\mathcal{G}(\mathbb{E}, \mathbb{F})$.

b) Ein entsprechendes Resultat gilt für $(f_\beta), f \in \mathcal{M}_2(X)$, wenn man zusätzlich fordert, daß alle bei der Zerlegung von f auftretenden Mengen E_j

.[- Es sei $f_\beta = g_\beta + \Sigma 1_{E_j^\beta} z_j^\beta$, $f = g + \Sigma 1_{E_j} z_j$ -] μ-Stetigkeitsmengen sind.

Beweis: Es ist für $y \in \mathbb{E}'$ $|\langle \Phi(f_\beta, \mu_\alpha) - \Phi(f, \mu), y\rangle| \le |\langle \Phi(f_\beta - f, \mu_\alpha), y\rangle| + |\langle \Phi(f, \mu_\alpha - \mu), y\rangle|$. Wählt man ein kompaktes $K \subseteq X$, so daß $\mu_\alpha(X \setminus K) < \varepsilon$, $\mu(X \setminus K) < \varepsilon$, so ist dieser Ausdruck abzuschätzen durch

$$\le 4 \cdot \varepsilon \cdot \sup_\alpha \|\mu_\alpha\| \cdot \sup_\beta \sup_{x \in X} |\langle f_\beta(x), y\rangle| + \left|\int_K \langle f_\beta(x) - f(x), y\rangle \, d\mu_\alpha(x)\right| +$$

$$+ \left|\int_K \langle f(x), y\rangle \, d(\mu_\alpha - \mu)(x)\right| \le 5\varepsilon \sup_\alpha \|\mu_\alpha\| \cdot \sup_{x,\beta} |\langle f_\beta(x), y\rangle| +$$

$$+ \left|\int_K \langle g(x), y\rangle \, d(\mu_\alpha - \mu)(x)\right| + \left|\int_K \langle \sum_j 1_{E_j}(x) z_j, y\rangle \, d(\mu_\alpha - \mu)(x)\right|$$

für genügend große β. Nach Voraussetzung schneidet K nur endliche viele E_j, die E_j sind μ-Stetigkeitsmengen, daraus folgt die Behauptung. □

1.14 Korollar.

Sei $\mathbb{E}$ ein Banachraum versehen mit der schwachen Topologie, oder $\mathbb{E} = \mathcal{L}(\mathbb{B})$, wobei $\mathbb{B}$ ein Banachraum ist, versehen mit der schwachen Operatorentopologie. Seien $f_\beta \in \mathcal{M}_2(X, \mathbb{E}, \mathbb{F}), \mu_\alpha \in M^1(X), \| f_\beta(\cdot) \| \leq 1$. Weiter sei $f_\beta \longrightarrow f$ in τ_{co} mit $f \in \mathcal{M}_0(X, \mathbb{E}; \mathbb{F})$ und überdies konvergiere $\mu_\alpha \longrightarrow \mu$ schwach; die (μ_α) seien gleichmäßig straff und $\|\cdot\|$-beschränkt. Dann folgt $\Phi(f_\beta, \mu_\alpha) \longrightarrow \Phi(f, \mu)$.

1.15

Nun sei zunächst $\mathbb{E}$ ein beliebiger linearer Raum, $\mathcal{T} \subseteq \mathbb{E}$ eine konvexe Teilmenge. $\mathcal{T}$ heißt affine Halbgruppe, wenn eine Komposition $\mathcal{T} \otimes \mathcal{T} \dashrightarrow \mathcal{T}$ definiert ist, so daß

(i) $\mathcal{T}$ ein Monoid ist,

(ii) $x\cdot(py+(1-p)z) = px\cdot y + (1-p)x\cdot z$

$(px + (1-p)y)z = px\cdot z + (1-p)y\cdot z$

für $x,y,z \in \mathcal{T}$, $0 \leq p \leq 1$.

Nun sei wiederum $\mathbb{E}$ ein lokalkonvexer Vektorraum, $\mathcal{T} \subseteq \mathbb{E}$ sei eine affine Halbgruppe. $\mathcal{T}$ sei - als Teilmenge von $\mathbb{E}$ - abgeschlossen und die Multiplikation $(x,y) \longrightarrow x\cdot y$ sei stetig in jeder Komponente. Dann heißt $\mathcal{T}$ affine halbtopologische Halbgruppe. Ist $(x,y) \longrightarrow xy$ simultan stetig, so sprechen wir von einer affinen topologischen Halbgruppe. [Zu affinen Halbgruppen s. z.B. H. Collins [6]]. Wir setzen stets voraus, daß $0 \in \mathcal{T}$, dies ist OBdA möglich. $\mathbb{E}_\mathcal{T}$ sei der von $\mathcal{T}$ erzeugte Vektorraum, dann wird $\mathbb{E}_\mathcal{T}$ in natürlicher Weise zu einer (topologischen, lokalkonvexen) Algebra.

1.16 Lemma.

Für $x \in \mathbb{E}_\gamma$, $z \in \mathbb{E}'_\gamma$ seien $\varphi_{x,z}: \mathbb{E}_\gamma \ni y \longrightarrow \langle x \cdot y, z \rangle$ sowie $\Psi_{x,z}: \mathbb{E}_\gamma \ni y \longrightarrow \langle y \cdot x, z \rangle$ definiert. Wenn γ eine affine halbtopologische Halbgruppe ist, dann sind

$$\varphi_{x,z} \text{ und } \Psi_{x,z} \in \mathbb{E}'_\gamma \quad \text{für } x \in \mathbb{E}_\gamma,\ z \in \mathbb{E}'_\gamma .$$

Dies erlaubt es, folgende Operatoren einzuführen:

$$\mathbb{E}_\gamma \ni x \longrightarrow T_x : \mathbb{E}'_\gamma \longrightarrow \mathbb{E}'_\gamma,\ T_x z := \varphi_{x,z}$$

$$\mathbb{E}_\gamma \ni x \longrightarrow S_x : \mathbb{E}'_\gamma \longrightarrow \mathbb{E}'_\gamma,\ S_x z := \Psi_{x,z}.$$

Es gelten: $x \longrightarrow T_x$ ist ein Antihomomorphismus, $x \longrightarrow S_x$ ist ein Homomorphismus, d.h. für $x,y \in \mathbb{E}$ ist

$$T_x T_y = T_{yx},\ S_x S_y = S_{xy}, \text{ überdies}$$

sind $x \longrightarrow T_x$ und $x \longrightarrow S_x$ $\mathcal{G}(\mathbb{E}_\gamma, \mathbb{E}'_\gamma)$ stetig.

1.17 Voraussetzung.

Es sei $\mathbb{F} \subseteq \mathbb{E}' \subseteq \mathbb{E}'_\gamma$. γ sei so beschaffen, daß für alle $x \in \mathbb{E}$, $z \in \mathbb{F}$ $T_x z$ und $S_x z$ zu Elementen von $\mathbb{F}$ fortgesetzt werden können. Dies gilt offenbar, wenn $\mathbb{E}_\gamma$ abgeschlossen in $\mathbb{E}$ und $\mathbb{F} = \mathbb{E}'$ ist. Bei unseren Beispielen (1.8) ist dies stets der Fall.

1.18 Beispiel: Affine Maßhalbgruppen.

X sei eine vollständig reguläre hausdorffsche topologische Halbgruppe, OBdA mit Einheit e. Man setze $\gamma := M^1(X)$ die Menge der straffen Wahrscheinlichkeitsmaße, $\mathbb{E}_\gamma = \mathbb{E}$ ist dann der Raum M(X) der straffen beschränkten Maße, versehen mit der Toplogie $\mathcal{G}(M(X), C(X))$ der schwachen Konvergenz. Man definiert die Faltung durch

$$\mu\nu(f) := \int_{X \otimes X} f(xy)\, d\mu \otimes \nu(x,y), \quad f \in C(X).$$

a) Wenn X eine topologische Halbgruppe ist, so existiert die Faltung stets und $\mathcal{T} = M^1(X)$ ist selbst eine affine topologische Halbgruppe.

b) Wenn X eine lokalkompakte halbtopologische Halbgruppe ist, dann ist $M^1(X)$ eine halbtopologische Halbgruppe (s. Glicksberg [10]).

1.19 Proposition.

Seien nun X eine vollständig reguläre topologische Halbgruppe, seien $\mathbb{E} \supseteq \mathcal{T}$ eine affine halbtopologische Halbgruppe (mit Einheit I) und es sei wieder $\mathbb{F} \subseteq \mathbb{E}'$, so daß 1.7 und 1.17 gelten.

$\Psi : X \longrightarrow \mathcal{T}$ sei ein stetiger Halbgruppenhomomorphismus, so daß $\Psi \in \mathcal{M}_0(X, \mathcal{T}; \mathbb{F})$, dann ist Ψ fortsetzbar zu einem affinen stetigen Halbgruppenhomomorphismus, nämlich : $M^1(X) \ni \mu \longrightarrow \Phi(\Psi,\mu) = \int_X \Psi d\mu$, der sich zu einem Algebrenhomomorphismus $M(X) \ni \mu \longrightarrow \int_X \Psi \, d\mu$ fortsetzen läßt.

Beweis: Sei $z \in \mathbb{E}'$, $\mu, \nu \in M^1(X)$, so ist $\int \langle \Psi(\cdot), z \rangle d\mu\nu =$

$$= \int_{X \otimes X} \langle \Psi(x \cdot y), z \rangle d\mu \otimes \nu(x,y)$$

$$= \int \langle \Psi(x)\Psi(y), z \rangle d\mu(x) d\nu(y) = \int [\int \langle \Psi(x), S_{\Psi(y)} z \rangle d\mu(x)] d\nu(y) =$$

$$= \int \langle \Phi(\Psi,\mu), S_{\Psi(y)} z \rangle d\nu(y) = \int \langle \Phi(\Psi,\mu)\Psi(y), z \rangle d\nu(y) = \ldots$$

$$= \langle \Phi(\Psi,\mu) \cdot \Phi(\Psi,\nu), z \rangle.$$

Die Erweiterung zum Algebrenhomomorphimus ist offensichtlich. □

1.20 Korollar.

Y sei nun eine weitere Halbgruppe [lokalkompakt und halbtopologisch oder vollständig regulär und topologisch]. Es sei $\varphi : Y \longrightarrow M^1(X)$ ein stetiger Halbgruppenhomomorphismus, der die "Prohorov-Eigenschaft" 1.8.4(ii) besitzt. Dann ist $\xi : M^1(Y) \longrightarrow M^1(X) \longrightarrow \mathcal{T}$:

$$M^1(Y) \ni \lambda \longrightarrow \Phi(\varphi,\lambda) = \int_Y \varphi d\lambda \, (\in M^1(X)) \longrightarrow \Phi(\Psi, \Phi(\varphi,\lambda)) = \int_X \Psi d(\int_Y \varphi d_\lambda)$$

ein stetiger affiner Halbgruppenhomomorphismus.

1.21 Proposition.

Sei $X = (o,\infty)$ mit „+" als Komposition, $\mathcal{T}$ sei eine affine topologische Halbgruppe mit Einheit I. $\varphi:(o,\infty) \longrightarrow \mathcal{T}$ sei ein schwach stetiger Homomorphismus $\in \mathcal{M}_b(X,\mathcal{T};\mathbb{F})$. Man setzt $\varphi(0) := I$. Es sei 1.17 erfüllt. Weiter sei $\mathbb{R}_+ \ni t \longrightarrow F_t \in M^1(\mathbb{R}_+)$ ein stetiger Homomorphismus - wir nennen (F_t) eine stetige Halbgruppe von Wahrscheinlichkeitsmaßen - mit der Lévy-Hinčin-Darstellung von (F_t)

$$\frac{d^+}{dt} F_t(f)\Big|_{t=0} = cf'(o) + \int (f(t)-f(e)d\eta(t)), \text{ s. § 2 .}$$

Man setzt $\Psi: \mathbb{R}_+ \ni t \longrightarrow \Phi(\varphi,F_t) = \int_{\mathbb{R}_+} \varphi(x)dF_t(x)$.

Dann gelten:

(i) $\Psi: \mathbb{R}_+^{\times} := (o,\infty) \longrightarrow \mathcal{T}$ ist ein stetiger Homomorphismus.

(ii) Sei $S(\varphi) := \{z \in \mathbb{E}' : \lim_{t\to o} \langle \varphi(t),z\rangle = \langle I,z\rangle\}$,

$S_1(\varphi) := \{x \in \mathbb{E} : \lim_{t\to o} \varphi(t)\, x = x\}$. Analog seien $S(\Psi)$, $S_1(\Psi)$ erklärt. Dann ist $S(\Psi) \supseteq S(\varphi)$, $S_1(\Psi) \supseteq S_1(\varphi)$.

(iii) Sei $D(\varphi) := \{z \in \mathbb{E}' : \frac{d^+}{dt}\Big|_{t=0} \langle\varphi(t),z\rangle \doteq D_\varphi(z)$ existiert$\}$,

analog seien D_Ψ, $D(\Psi)$ erklärt, dann ist $D(\Psi) \supseteq D(\varphi)$ und es ist $\qquad D_\Psi(z) = cD_\varphi(z) + \int_{\mathbb{R}_+^{\times}} \langle\Psi(t)-I,z\rangle d\eta(t)$ für $z \in D_\varphi$.

Beweis: (Für Operatorhalbgruppen s.z.B. Hille-Phillips [14]).
Setzt man $\varphi(o) := I$, so ist $t \longrightarrow \varphi(t)$ ein stetiger Homomorphismus von $X := \{o\} \cup (o,\infty)$ mit 0 als isoliertem Punkt. Daher ist auch $\Psi(t) = \Phi(\varphi,F_t)$ erklärbar und $t \longrightarrow \Psi(t)$ ist ein Homomorphismus von $\{o\} \cup (o,\infty) \longrightarrow \mathcal{T}$. Es bleibt also das Verhalten in $t = o$:
Sei $z \in S(\varphi)$, so ist $g: s \longrightarrow \langle\varphi(s),z\rangle \in C(\mathbb{R}_+)$, also $\langle\Psi(t),z\rangle =$

$$= \int_{\mathbb{R}_+} \langle\varphi(s),z\rangle dF_t(s) \xrightarrow[t\to o]{} \int g dF_0 = g(o) = \langle I,z\rangle$$

Entsprechend: Sei $\varphi(t)x \xrightarrow[t\to o]{} x, x\in\mathbb{E}$, so ist für jede stetige Halb-

norm q: $t \longrightarrow q(\varphi(t)x-x) \in C(\mathbb{R}_+)$ und man zeigt wieder,

daß $q(\psi(t)x-x) \leq \int q(\varphi(s)x-x)dF_t(s) \xrightarrow[t\to o]{} 0.$

Die letzte Aussage folgt unmittelbar aus der Lēvy-Hinčin-Darstellung. □

1.22 Definition.

Wir führen die folgenden Sprechweisen ein: Ein stetiger Homomorphismus $\varphi:(o,\infty) \longrightarrow \gamma$ heißt „stetige Halbgruppe", ein stetiger Homomorphismus $t \longrightarrow F_t \in M^1(\mathbb{R}_+)$ heißt Subordinationshalbgruppe und $(\psi(t) := \Phi(\varphi, F_t))$ heißt „untergeordnete Halbgruppe".

Um Operatorhalbgruppen, die in $t = o$ nicht stetig sind, studieren zu können, wurden allgemeinere Stetigkeitsbedingungen eingeführt ([14]).

Die Fragen, wie solche Stetigkeitsbedingungen sich von φ auf die untergeordneten Halbgruppen ψ übetragen, ist naheliegend, wird aber einer anderen Arbeit vorbehalten. Wir beschränkten uns hier auf einen (offensichtlichen) Spezialfall:

Sei (F_t) eine stetige Halbgruppe in $M^1(\mathbb{R}_+)$, sei $(\Lambda_{(\tau)})_{\tau\geq o} \subseteq M^1(\mathbb{R}_+)$ eine Familie von Maßen mit $\lim_{\tau\to o} \Lambda_{(\tau)} = \varepsilon_0$ und sei $\Sigma_{(\tau)} := \int F_t d\Lambda_{(\tau)}(t)$, so ist wiederum $\Sigma_{(\tau)} \longrightarrow \varepsilon_0$.

Für jeden (beschränkten, stetigen) Homomorphismus $\varphi:(o,\infty) \longrightarrow \gamma$ sei

$S(\varphi,\Lambda) := \{x\in\mathbb{E} : \lim_{\tau\to o} \Phi(\varphi,\Lambda_{(\tau)})x = x\}$. Analog sei $S(\varphi,\Sigma)$ definiert.

1.22 Proposition.

Sei $(\psi_t = \Phi(\varphi, F_t))$, so ist $S(\psi,\Sigma) \supseteq S(\varphi,\Lambda)$.

Bemerkung: Für (gleichmäßig beschränkte) Operatorhalbgruppen auf einem Banachraum $\mathbb{B}$ betrachtet man die „Summationsverfahren" $\Lambda_{(\tau)}^{(0)} = \varepsilon_\tau$,

$d\Lambda_{(\tau)}^{(1)}(s) := \frac{1}{\tau} 1_{[o,\tau]} ds$, bzw. $d\Lambda_{(\tau)}^{(2)}(s) := \tau e^{-\tau s} ds$ und erhält so die

Klassen von Operatorhalbgruppen vom Typ (C,o), falls $S(\varphi,\Lambda^{(0)}) = \mathbb{B}$;

vom Typ $(C,1)$ falls $S(\varphi,\Lambda^{(1)}) = \mathbb{B}$ bzw.

vom Typ $(A,1)$ falls $S(\varphi,\Lambda^{(2)}) = \mathbb{B}$.

(s. Hille-Phillips [14]).

1.23

Nun betrachten wir abermals Homomorphismen $\varphi : \mathbb{R}_+ \longrightarrow \mathcal{T}$ resp. $\Upsilon : \mathbb{Z}_+ \longrightarrow \mathcal{T}$
Sei $\tau > 0$, dann nennen wir τ "Zeiteinheit" und definieren vermöge Υ eine Abbildung $\dot{\Upsilon} : \mathbb{R}_+ \longrightarrow \mathcal{T}$ durch $\dot{\Upsilon}(t) := \Upsilon([t/\tau])$. $\dot{\Upsilon}$ nennen wir "diskrete Halbgruppe".

Nun seien $\varphi : \mathbb{R}_+ \longrightarrow \mathcal{T}, \Psi_n : \mathbb{Z}_+ \longrightarrow \mathcal{T}$, Homomorphismen. $\tau_n \geq o$ sei eine Folge von Zeiteinheiten mit $\tau_n \longrightarrow o$. $\dot{\Psi}_n$ seien die zugehörigen „diskreten Halbgruppen", deren „Generatoren" seien die Differenzenoperatoren $\frac{1}{\tau_n}(\dot{\Psi}_n(\tau_n) - I)$. Für Halbgruppen vom Typ C_0 folgt aus der verallgemeinerten Konvergenz der „Generatoren" von $\dot{\Psi}_n$ gegen den Generator von φ, die τ_{co}-Konvergenz $\dot{\Psi}_n \longrightarrow \varphi$ (s.z.B.T. Kato [20]). (Dabei bezeichnet wieder τ_{co} die Topologie der kompakt-gleichmäßigen Konvergenz).

Allgemein folgt aus dem Stetigkeitssatz I.1.13:

1.24 Proposition.

Sei $\varphi : \mathbb{R}_+ \longrightarrow \mathcal{T}$ ein stetiger Homomorphismus $\in \mathcal{M}_0(\mathbb{R}_+, \mathcal{T}; \mathbb{F})$, sei $\dot{\psi}_n$ eine Folge diskreter Halbgruppen mit Zeiteinheiten τ_n, so daß $\dot{\psi}_n \longrightarrow \varphi$ in τ_{co}. Sei $t \longrightarrow F_t \in M^1(\mathbb{R}_+)$ eine stetige Faltungshalbgruppe und sei $\mathbb{R}_+ \ni t \longrightarrow \dot{G}_{(t)}^{(n)}$ eine Folge diskreter Faltungshalbgruppen mit Zeiteinheiten σ_n, die gegen 0 konvergieren, so daß $(\dot{G}_{(t)}^{(n)}) \underset{n\to\infty}{\longrightarrow} (F_t)$ in τ_{co}.

Dann konvergieren die Mischungsintegrale $\int \dot{\Psi}_n(s) d\dot{G}^{(m)}_{(t)}(s) \xrightarrow[n,m\to\infty]{} \int \varphi(s) dF_t(s)$ in τ_{co}.

Berücksichtigt man, daß die $\dot{\Psi}_n$ konstant sind in $[k\tau_n,(k+1)\tau_n)$, nämlich gleich $\Psi_n(1)^k = \dot{\Psi}_n(\tau_n)^k$, und daß analog

$\dot{G}^{(n)}_{(t)} = \dot{G}^{(n)\,[t/\sigma_n]}_{(\sigma_n)}$, so erhält man mit

$$z_n := \Psi_n(1) = \dot{\Psi}_n(\tau_n),\quad H_n := \dot{G}^{(n)}_{(\sigma_n)}$$

$$\int \varphi(s) dF_t(s) = \lim_{n,m\to\infty} \sum_{k\in\mathbb{Z}_+} z_n^k \, H_m^{[t/\sigma_m]}([k\tau_n,(k+1)\tau_n)).$$

Wir geben später einige Anwendungen solcher Approximationsformeln an, im Zusammenhang mit Grenzwertsätzen der Wahrscheinlichkeitstheorie resp. Operatorentheorie, insbesondere im Zusammenhang mit stabilen Maßen.

1.26

Abschließend kehren wir in § 1 nochmals zu allgemeinen Mischungsintegralen der Form $\int_X \varphi d\lambda$ zurück. Dabei setzen wir aber voraus, daß $\mathcal{T}$ eine Halbgruppe von Kernoperatoren auf einem Raum $\mathbb{B} = C(Y)$, Y vollständig regulär, ist. I.e., es gibt für $t\in X$ einen submarkoffschen Übergangskern $N_t(\cdot,\cdot)$, so daß $\varphi(t)f(y) = \int_Y f(x) dN_t(y,dx)$.
$t \longrightarrow N_t$ sei stetig in dem Sinne, daß für $f\in C(Y)$, $y\in Y$, $X\ni t \longrightarrow$
$\longrightarrow \int f(x) N_t(y,dx) \in C(X)$ ist.

Zusätzlich setzen wir voraus, daß $\varphi\in\mathcal{M}_0(X,\mathcal{T};\mathbb{F})$. Es sei $\lambda\in M^1(X)$ und M sei der Kern $\int_X N_t(\ldots,.) d\lambda(t)$.

Wir stellen uns die Frage, welche Eigenschaften von (N_t) auf M übertragen werden.

1.27 Proposition .

a) Sei für $y \in Y$ und $t \in X$ das Maß $N_t(y,\cdot)$ diffus (=atomfrei), so ist auch $M(y,\cdot)$ diffus.

b) Sei $\sigma \in M^1(Y)$, $N_t(y,\cdot) \ll \sigma$ für alle $y \in Y$, $t \in X$, so ist $M(y,\cdot) \ll \sigma$.

c) Seien alle $N_t(y,\cdot)$ diskret und konzentriert auf einer abzählbaren Teilmenge $\Lambda \subseteq Y$, so ist auch $M(y,\cdot)$ diskret und konzentriert auf Λ.

Beweis: Sei $z \in Y$, $(f_\alpha) \subset C(Y)$ sei ein Netz mit $f_\alpha \downarrow 1_{\{z\}}$. Daher hat man $g_\alpha(t) = \int_Y f_\alpha(x) N_t(y,dx) \downarrow 0$ für alle t, und somit wegen $g_\alpha \in C(X)$ nach dem Satz von Dini: $\int g_\alpha d\lambda \longrightarrow 0$.

Analog beweist man b), indem man oberhalbstetige Funktionen betrachtet, die gegen die charakteristische Funktion einer σ-Nullmenge konvergieren; c) ist ebenso offensichtlich. (Die Aussage ist natürlich falsch, wenn man die Existenz von Λ nicht fordert!) □

1.28 Korollar. (s. auch Huff [17-19]):

Sei $X = \mathbb{R}_+$, Y sei eine (halb)topologische Halbgruppe mit Einheit e $(\mu_t)_{t \geq 0}$ sei eine stetige Halbgruppe in $M^1(Y)$ mit $\mu_0 = \varepsilon_e$, $t \longrightarrow \mu_t$ erfülle 1.8.4(i, ii). Es seien $\lambda \in M^1(\mathbb{R}_+)$, $\sigma \in M^1(Y)$, dann gilt:

Sind alle μ_t diffus [resp. $\ll \sigma$] für t>o, so ist auch $\int \mu_t d\lambda(t)$ diffus [$\ll \sigma$], falls $\lambda(\{0\}) = 0$. Allgemein erhält man:

Sei $\mu'_t := \mu_t|_{Y \setminus \{e\}}$ diffus [$\ll \sigma$] für $t > 0$, so ist $(\int \mu_t d\lambda(t))(\{0\}) = \int_{\mathbb{R}_+} \mu_t(\{e\}) d\lambda(t)$ und $(\int \mu_t d\lambda(t))|_{Y \setminus \{e\}}$ ist diffus [$\ll \sigma$].

[Man betrachtet die Faltungskerne $\int f(x) N_t(y,dx) := \int f(yx) d\mu_t(x)$.] □

§ 2 Affine Halbgruppen von Maßen auf $[0,\infty]$.

Wir betrachten folgende (topologischen, lokalkompakten) Halbgruppen:

$\overline{\mathbb{R}}_+ := [0,\infty] \supseteq \mathbb{R}_+ := [0,\infty) \supseteq \mathbb{R}_+^{\bullet} := \{0\} \cup (0,\infty) \supseteq \mathbb{R}_+^{\times} := (0,\infty)$, ebenso

$\overline{\mathbb{Z}}_+ := \mathbb{Z}_+ \cup \{\infty\} \supseteq \mathbb{Z}_+ \supseteq \mathbb{N}$.

Die Halbgruppenoperation ist stets „+" mit ∞ als Nullelement. Mit $\mathcal{U}(X)$, $X = \overline{\mathbb{R}}_+, \ldots$ werden stets die unendlich oft teilbaren Maße $\subseteq M^1(X)$ bezeichnet. Daher ist - da jedes Maß $\mu \in M^1(X)$ von der Gestalt $\mu = \mu|_{X \setminus \{\infty\}} + \mu\{\infty\}\varepsilon_\infty$ ist - jedes $\mu \in \mathcal{U}(X)$ in eindeutiger Weise in eine stetige Halbgruppe $(\mu_t, t \geq 0, \mu_1 = \mu)$ einbettbar. Für $X \neq \mathbb{R}_+^{\times}$, $\neq \mathbb{N}$ $\neq \mathbb{R}_+^{\bullet}$ existiert μ_0 und es ist $\mu_0 = \varepsilon_0$, (außer für $X = \overline{\mathbb{R}}_+$ oder $\overline{\mathbb{Z}}_+$, $\mu_t \equiv \varepsilon_\infty, t \geq 0$.)

Ebenso erhält man die Darstellung $\mu_t = e^{-t\mu_1(\infty)} \dot{\mu}_t + (1 - e^{-t\mu_1(\infty)})\varepsilon_\infty$, dabei ist $(\dot{\mu}_t, t \geq 0)$ eine stetige Halbgruppe $\subseteq M^1(X \setminus \{\infty\})$. Wegen $\mathcal{U}(\mathbb{N}) \subseteq \mathcal{U}(\mathbb{Z}_+) \subseteq \mathcal{U}(\mathbb{R}_+)$, $\mathcal{U}(\overline{\mathbb{Z}}_+) \subseteq \mathcal{U}(\overline{\mathbb{R}}_+)$, $\mathcal{U}(\mathbb{R}_+^{\times}) \subseteq \mathcal{U}(\mathbb{R}_+^{\bullet}) \subseteq \mathcal{U}(\overline{\mathbb{R}}_+)$ genügt es, die Maße aus $\mathcal{U}(\overline{\mathbb{R}}_+)$ genauer zu untersuchen.

2.1

Dazu führen wir die Laplacetransformierte ein: Für $\lambda \in \overline{\mathbb{R}}_+$ definiert man für ein Maß $\mu \in M^1(\overline{\mathbb{R}}_+)$ $\hat{\mu}(\lambda) := \int_{\overline{\mathbb{R}}_+} e^{-\lambda s} d\mu(s)$ (mit $e^{-\infty} := 0, e^{-\infty \cdot 0} := 1$).

Offenbar gilt: $1 \geq \hat{\mu}(\lambda) \geq \hat{\mu}(\infty) = 0$. Aus $\mu_n \to \mu$ schwach (in $M^1(\overline{\mathbb{R}}_+)$) folgt $\hat{\mu}_n \to \hat{\mu}$. Umgekehrt: $\hat{\mu}_n(\lambda) \longrightarrow \hat{\mu}(\lambda)$ für $\lambda > 0$ impliziert $\mu_n \longrightarrow \mu$ schwach; die Zuordnung $\mu \longrightarrow \hat{\mu}$ ist eineindeutig.

2.2

Nun betrachten wir wiederum stetige Halbgruppen $(\mu_t)_{t \geq 0} \subseteq \mathcal{U}(\overline{\mathbb{R}}_+) \subseteq M^1(\overline{\mathbb{R}}_+)$, die daher von der Gestalt $\mu_t = e^{-t\alpha}\dot{\mu}_t + (1 - e^{-t\alpha})\varepsilon_\infty$ sind, wobei $(\dot{\mu}_t)$ eine stetige Halbgruppe $\subseteq M^1(\mathbb{R}_+)$ ist.

Dann ist $\hat{\mu}_t(\lambda) = e^{-t\alpha}\hat{\mu}_t(\lambda)$, $t \geq 0$.

Sei A die erzeugende Distribution von μ_t $[f \in C^1(\mathbb{R}_+) \longrightarrow \frac{d^+}{dt}\mu_t(f)\Big|_{t=0}$

$=: A(f)]$, dann ist A von der Gestalt $A(f) = c\,\frac{d^+}{dt}f(o) + \int_{\mathbb{R}_+}(f(x)-f(o))d\eta(x)$

mit $c \geq o$ und mit einem auf $\mathbb{R}_+$ lokalbeschränkten Maß $\eta \geq 0$, so daß

$$\int_{(o,\infty)} \frac{x}{1+x}\,d\eta < \infty.$$

Die Laplacetransformierte von A hat die Gestalt

$$\hat{A}(\lambda) = -c\lambda + \int_{\mathbb{R}_+}(e^{-\lambda s} - 1)\,d\eta(s),$$

(s.z.B. W. Feller [8]).

Die Laplacetransformierten $\hat{\mu}$ von Maßen $\mu \in M^1(\mathbb{R}_+)$ sind die totalmonotonen Funktionen mit $\hat{\mu}(o) = 1$,

i.e. $(-1)^k \frac{d^k}{dx^k}\hat{\mu} \geq o$, ebenso sind die $\hat{A}$ gerade die total negativen

Funktionen (Bernsteinfunktionen) mit $\hat{A}(o) = 0$. Läßt man auch Maße in $M^1(\overline{\mathbb{R}_+})$ zu, so erhält man als Laplacetransformierte analog (stetige) total monotone Funktionen $\hat{\mu}: \mathbb{R}_+ \longrightarrow \mathbb{R}_+$ mit $1 \geq \hat{\mu} \geq o$ resp. total negative Funktionen $\hat{A}: \mathbb{R}_+^* \longrightarrow \overline{\mathbb{R}_+}$ mit $\hat{A}(o) \leq o$, $e^{t\hat{A}} = \hat{\mu}_t$. (Dabei ist $e^{-\infty} := o$ zu setzen).

2.3 Definition.

Wir identifizieren nun der Einfachheit halber $\mathcal{U}(\overline{\mathbb{R}_+})$ mit der Menge der Faltungshalbgruppen $\mathrm{Hom}(\mathbb{R}_+, M^1(\overline{\mathbb{R}_+}))$ resp. mit der Menge $\mathcal{B}$ der Bernsteinfunktionen $\hat{A}$ mit $\hat{A}(o) \leq o$.

Wir definieren nun in $\mathcal{U}(\overline{\mathbb{R}_+})$ [resp. $\mathrm{Hom}(\mathbb{R}_+, M^1(\overline{\mathbb{R}_+}))$ resp. $\mathcal{B}$] eine neue Komposition, die mit „$\circ$" bezeichnet wird:

Seien $\nu, \mu \in \mathcal{U}(\overline{\mathbb{R}_+})$ mit Halbgruppen $(\mu_t), \mu_1 = \mu$, $(\nu_t), \nu_1 = \nu$ und total negativen Funktionen $\hat{A} = \frac{d^+}{dt}\hat{\mu}_t\Big|_{t=0} = \log\hat{\mu}$, $\hat{B} = \frac{d^+}{dt}\hat{\nu}_t\Big|_{t=0}$. Dann sei

$(\mu\circledcirc\nu)_t := \int\mu_s d\nu_t(s)$ - in der Schreibweise von § 1 : $(\mu\circledcirc\nu)_t := \Phi((\mu_s),\nu_t)$ -. Dann ist $\lambda_1 := (\mu\circledcirc\nu)_1 \in \mathcal{U}(\overline{\mathbb{R}_+})$ mit Halbgruppe $(\lambda_t = \int\mu_s d\nu_t(s))$ und total negativer Funktion $\hat{C} = \hat{B}\circledcirc\hat{A}$ [die Komposition der Laplacetransformierten $\hat{C}(\lambda) = \hat{B}\circledcirc\hat{A}(\lambda) = \hat{B}(-\hat{A}(\lambda))$].

[Man prüft nämlich sofort nach, daß $\widehat{\mu\circledcirc\nu}(\alpha) = \int e^{-\alpha u} d(\int\mu_s d\nu_1(s))(u) = \iint e^{-\alpha u} d\mu_s(u) d\nu_1(s) = \int e^{-s\hat{A}(\alpha)} d\nu_1(s) = e^{\hat{B}(-\hat{A}(\alpha))}$.]]

2.4 Proposition.

$\mathcal{U}(\overline{\mathbb{R}_+})$ ist (schwach) abgeschlossen in $M^1(\overline{\mathbb{R}_+})$ und daher kompakt, aus I.1.13 folgt, daß $(\mathcal{U}(\overline{\mathbb{R}_+}),\circledcirc)$ eine kompakte topologische Halbgruppe ist.

ε_1 — i.e. die Halbgruppe $(\mu_t = \varepsilon_t)_{t\geq 0}$ mit $\hat{A} = -\mathrm{id}$ — ist das Einheitselement dieser Halbgruppe, idempotente Elemente sind ε_0 und ε_∞ sowie konvexe Kombinationen der beiden und man sieht leicht, daß jedes idempotente Element von dieser Gestalt ist. Es gilt nämlich der

2.5 Hilfssatz.

Sei F total negativ mit $F\circledcirc F = F$, so ist $F = -\mathrm{id}$ oder $F \equiv \text{const.}$, $-\infty \leq \text{const} \leq 0$. $F \equiv -\infty$ entspricht der Halbgruppe $(\mu_t \equiv \varepsilon_\infty)$, $F \equiv c > -\infty$ entspricht $(\mu_t = e^{-tc}\varepsilon_0 + (1-e^{-tc})\varepsilon_\infty)_{t\geq 0}$.

[Unmittelbar einzusehen: Sei $F(x) = F(-F(x)), x>0$.

a) Ist F strikt monoton in x_0, so muß $F(x_0) = -x_0$ sein,

b) $F(x_0) \neq x_0$, etwa $0 < x_0 < -F(x_0)$, dann muß, wegen der Monotonie von F auch $F(x) = F(x_0)$ sein für $x_0 < x < -F(x_0)$.

Nun ist aber F analytisch und daraus folgt $F \equiv \text{const.}$ $(=F(x_0))$ resp. $\mu_t = e^{-\text{const.}t}\varepsilon_0 + (1-e^{-\text{const.}t})\varepsilon_\infty$].

2.6 Proposition (s. K.H. Hofmann, P. S. Mostert [16]).

Nun sei X eine kompakte topologische Halbgruppe, $\varepsilon \in X$ sei idempotent und $\mathcal{A}(\varepsilon) := \{x \in X : \varepsilon x = x\varepsilon = \varepsilon\}$. Sei $y \in X$ fest, dann gibt es stets ein idempotentes $\varepsilon \in X$ mit $y \varepsilon \mathcal{A}(\varepsilon)$. (- Man wählt ein idempotentes Element in der kompakten Halbgruppe $\{y, y^2, \ldots\}^-$ -). Mit $H(\varepsilon)$ werde die Einheitengruppe bezeichnet, $H(\varepsilon) := \{x : \exists y \text{ mit } \varepsilon x \varepsilon = x,\ \varepsilon y \varepsilon = y$ und $xy = yx = \varepsilon\}$.

2.7. Folgerung.

Sei $\mu \in \mathcal{U}(\overline{\mathbb{R}_+})$, $\mu(\infty) = 0$, mit zugehöriger Halbgruppe $(\mu_t, \mu_0 = \varepsilon_0, \mu_1 = \mu)$ und totalmonotoner Funktion F, dann gelten:

a) $\mu \circledcirc \mu = \mu$ resp. $(\int \mu_s d\mu_t(s) = \mu_t)$ resp. $F \circledcirc F = F$ genau dann, wenn $F = -id$ resp. $(\mu_t = \varepsilon_t)$.

b) $\mathcal{A}(\varepsilon_1) = \{\varepsilon_1\}$ in $X := \{\mathcal{U}(\overline{\mathbb{R}_+}), \circledcirc\}$, i.e. $(\int \mu_s d\varepsilon_t(s) = \varepsilon_t)$ resp. $(\int \varepsilon_s d\mu_t(s) = \varepsilon_t)$ genau dann, wenn $\mu_t = \varepsilon_t$ für alle $t \geq 0$.

c) Sei $\mu \in \mathcal{U}(\overline{\mathbb{R}_+})$, $\mu \neq \varepsilon_1$, so ist jedes idempotente α mit $\mu \in \mathcal{A}(\alpha)$ von der Gestalt $(\alpha_t = e^{-tc}\varepsilon_0 + (1-e^{-tc})\varepsilon_\infty,\ t \geq 0)$ und $\{\varepsilon : \mu \in \mathcal{A}(\varepsilon)\} \neq \emptyset$.

d) Die Einheitengruppe $H(\varepsilon_1)$ besteht aus den Bildern von ε_1 unter den Automorphismen $\mathbb{R}_+ \ni t \longmapsto c \cdot t \in \mathbb{R}_+$, i.e. aus $\mu \circledcirc \nu = \varepsilon_1$ folgt $\mu = \varepsilon_c$, $\nu = \varepsilon_{1/c}$ für ein $c > 0$.

‖ Die Aussagen folgen unmittelbar, wenn man auf die Definition dieser Operation „$\circledcirc$" in $\mathcal{U}(\overline{\mathbb{R}_+})$ zurückgreift und beachtet, daß $(\mathcal{U}(\overline{\mathbb{R}_+}), \circledcirc)$ eine kompakte Halbgruppe ist, resp., wenn man die Laplacetransformátion der erzeugenden Distributionen einsetzt. Wir betrachten etwa d):

Seien $(\mu_t, \mu_1 = \mu)$, $(\nu_t, \nu_1 = \nu)$ die zugehörigen Halbgruppen mit total negativen Funktionen F und G.

So ist $\mu \circledcirc \nu = \varepsilon_1 \Leftrightarrow$ (i) $\int \mu_s d\nu_1(s) = \varepsilon_1 \Leftrightarrow$ (ii) $F \circledcirc G = -id$.

Aus (i) liest man unmittelbar die Behauptung ab, wenn man die Träger der Maße beachtet. Zieht man (ii) zum Beweis heran, setzt man voraus, daß

$F \neq -c\cdot id$: Sei H die $\circledcirc$-idempotente Funktion (Prop. 2.6), die zu F gehört und $(n_k) \subseteq \mathbb{N}$ sei eine Folge mit $F^{\circledcirc n_k} \longrightarrow H$. So ist $-id = F^{\circledcirc n_k} \circledcirc (G^{\circledcirc n_k})$. Sei $(k_l) \subseteq \mathbb{N}$ eine Folge, so daß $G^{\circledcirc n_{k_l}}$ konvergiert, etwa gegen L, so erhält man $-id = F^{\circledcirc n_{k_l}} \circledcirc (G^{\circledcirc n_{k_l}}) \xrightarrow[l\to\infty]{} H \circledcirc L$, also einen Widerspruch. □

2.7 Proposition.

Sei $(\mu_t)_{t\geq 0}$ eine stetige Halbgruppe in $\mathcal{U}(\overline{\mathbb{R}}_+)$, so existiert für $f \in C(\overline{\mathbb{R}}_+)$ $\lim_{t\to\infty} \int f d\mu_t = \int f d\varepsilon_\infty \begin{cases} = f(\infty) & \text{falls } \mu_t \not\equiv \varepsilon_0 \\ = f(0) & \text{falls } \mu_t \equiv \varepsilon_0 \end{cases}$

[Offensichtlich, da für $\mu_t \not\equiv \varepsilon_0$ die Restriktion $\mu_t\big|_{\mathbb{R}_+}$ $(t\to\infty)$ vage gegen 0 strebt].

2.8 Definition.

Wie schon früher erwähnt, bezeichnen wir stetige Halbgruppen $(F_t) \subseteq M^1(\mathbb{R}_+)$ als Subordinationshalbgruppen, deren erzeugende Distributionen $A := c\,\frac{d}{dx}\Big|_{x=0} + \int (\cdot - \cdot(0))\,d\eta$ als Subordinatoren. Abkürzend schreiben wir $(F_t) \sim A$, resp. $\sim (c,\eta)$, falls A die erzeugende Distribution von (F_t) ist.

2.9 Proposition.

Seien (F_t), (G_s) Subordinationshalbgruppen mit Subordinatoren $A \sim (c,\eta)$ $B \sim (d,\xi)$ und sei (H_r) die durch die Komposition $F_1 \circledcirc G_1$ gewonnene Halbgruppe ($H_r = \int F_t dG_r(t)$) mit Subordinator $C \sim (a,\zeta)$. Dann gilt (mit $F_t' := F_t\big|_{(0,\infty)}$) : $a = cd$, $\zeta = d\eta + \int_{\mathbb{R}_+^\times} F_t' d\xi(t)$.

[Unmittelbar einzusehen, da ja $C(f) = dA(f) + \int_{\mathbb{R}_+^\times} (F_t - \varepsilon_0)(f)\,d\xi(t)$.]

2.10 Proposition.

Sei $a>o$, dann bezeichne τ_a die Abbildung $x\to ax$, $\tau_a : \mathbb{R}\to\mathbb{R}$. τ_a ist in natürlicher Weise für Maße $\subseteq M(\mathbb{R})$ und Distributionen erklärt,

$\tau_a(F)(f) := \int f(x\cdot a)dF(x)$.

Sei $\Psi:\mathbb{R}_+ \longrightarrow \Upsilon$ - Υ wie in § 1 - ein Homomorphismus, $(F_t)\sim A\sim(c,\eta)$ eine Subordinationshalbgruppe und $\psi_1:t\to\Phi(\psi, F_t)$, $\psi^{(a)}:t\to\Psi(at)$ für $t>o$.

Sei $(G_t := \tau_{1/a}(F_t))$, so hat (G_t) den Subordinator $\tau_{1/a}(A)\sim(c/a,\tau_{1/a}(\eta))$ und es ist $\psi_1(t)=\Phi(\psi^{(a)},\tau_{1/a}(F_t))$.

[Beweis durch Einsetzen in die Integraldarstellung].

Korollar:

Insbesondere ist für $c>o$ stets $\psi_1(t)$ darstellbar, als $\Phi(\psi^{(c)},\tau_{1/c}(F_t))$, wobei die Subordinationshalbgruppe nun den Subordinator $(1,\tau_{1/c}(\eta))$ hat.

§ 3 Eine Ordnungsrelation, die durch Subordination definiert wird.

3.1 Definition.

$\mathcal{V}$ sei eine affine topologische Halbgruppe mit Einheit e, $\mathbb{E}\supseteq\mathbb{E}_{\mathcal{V}}\supseteq\mathcal{V}$, $\mathbb{F}\subseteq\mathbb{E}'$ seien ebenso wie in § 1 definiert. Insbesondere gelte also 1.17. X,Y seien topologische Halbgruppen mit Einheiten e resp. e', φ resp. Ψ seien Homomorphismen von X resp. Y $\longrightarrow\mathcal{V}$ die in $\mathcal{M}_0(X,\mathcal{V};\mathbb{F})$ resp. $\mathcal{M}_0(Y,\mathcal{V};\mathbb{F})$ liegen. Dann sagt man: Ψ ist φ untergeordnet (symbolisch: $\Psi\preceq_s\varphi$, falls ein Homomorphismus F: $Y\ni t\longrightarrow F_t\in M^1(X)$ existiert, der in $\mathcal{M}_0(Y,M^1(X);C\,(X))$ liegt, so daß $\Psi(t) = \int_X\varphi(s)dF_t(s)=\Phi(\varphi,F_t), t\in Y$. (Dabei setzen wir stets voraus, daß $t\longrightarrow F_t$ 1.8.4 erfüllt.)

In den folgenden §§ interessieren wir uns fast ausschließlich für die Fälle $X = Y = \mathbb{R}_+$ oder $\mathbb{Z}_+$.

Unser Ziel ist es, zu zeigen, daß $\preceq_s$ eine Präordnung ist und daß im Falle $X = \mathbb{R}_+$ $\varphi\preceq_s\Psi$ und $\Psi\preceq_s\varphi$ bedeuten, daß die Subordination trivial ist, (s.u.). Dies gelingt für den Spezialfall $\mathcal{V}= M^1(\mathfrak{G})$.

3.2 Hilfssatz.

Die Relation $\preceq_s$ ist transitiv.

[Seien X,Y,W topologische Halbgruppen, $\mathcal{V}$ eine affine topologische Halbgruppe. $\varphi:X\longrightarrow\mathcal{V}$, $\Psi:Y\longrightarrow\mathcal{V}$, $\xi:W\longrightarrow\mathcal{V}$, $F:Y\longrightarrow M^1(X)$, $G:W\longrightarrow M^1(Y)$ seien Halbgruppenhomomorphismen, die in den entsprechenden $\mathcal{M}_0(\cdot)$ liegen, so daß

$$\Psi(y) = \int_X\varphi(x)dF_y(x) = \Phi(\varphi,F_y), y\in Y$$

$$\xi(x) = \int_Y\Psi(y)dG_w(y) = \Phi(\Psi, G_w), w\in W.$$

Dann läßt sich ein Homomorphismus $H:W\longrightarrow M^1(X)$ angeben, nämlich $W\ni w\longrightarrow H_w\in M^1(X)$, $H_w = \int_Y F_y dG_w(y)$ (als schwaches Integral). Offenbar gilt $H\in\mathcal{M}_0(W,M^1(X); C(X))$ und es ist

$$\int_X\varphi(x)dH_w(x) = \int_Y\int_X\varphi(x)dF_y(x)dG_w(y) =$$

$$= \int_Y \psi(y) dG_w(y) = \xi(w),$$

i.e. aus $\psi \preceq_s \varphi$, $\xi \preceq_s \psi$ folgt $\xi \preceq_s \varphi$.]]

Für $X=Y=\mathbb{R}_+$ oder $\overline{\mathbb{R}_+}$ und $\mathcal{T}:=M^1(\overline{\mathbb{R}_+})$ sagt der Hilfssatz 3.2 insbes. aus, daß die in § 2 studierte Komposition „⊙" in $\mathcal{U}(\overline{\mathbb{R}_+})$ assoziativ ist.

Analoges gilt für $\mathbb{Z}_+$. Insbesondere hat man also:

a) Seien $\mathbb{R}_+ \ni t \longrightarrow F_t$, $\mathbb{R}_+ \ni t \longrightarrow G_t$ stetige Halbgruppen (Subordinationshalbgruppen) in $M^1(\mathbb{R}_+)$ [resp. $M^1(\overline{\mathbb{R}_+})$], weiter sei $\varphi : \mathbb{R}_+ \ni t \longrightarrow \varphi(t) \in \mathcal{T}$ ein Homomorphismus in $\mathcal{M}_o(\mathbb{R}_+, \mathcal{T}; \mathbb{F})$, $\psi(t) := \Phi(\varphi, F_t)$, $t \in \mathbb{R}_+$, $\xi(t) := \Phi(\psi, G_t)$, $t \in \mathbb{R}_+$, also $\xi \preceq_s \psi$, $\psi \preceq_s \varphi$. Setzt man $H: t \longrightarrow \Phi(F, G_t) =: H_t$, so erhält man $\xi(t) = \Phi(\varphi, H_t)$, $t \in \mathbb{R}_+$, i.e. $\xi \preceq_s \varphi$.

b) Seien $X = \mathbb{Z}_+ \ni t \longrightarrow \varphi(t) = T^t \in \mathcal{T}$,

$\mathbb{Z}_+ \ni t \longrightarrow F^t \in M^1(\mathbb{Z}_+)$, $t \to G^t \in M^1(\mathbb{Z}_+)$, Homomorphismen $(T \in \mathcal{T}, F, G \in M^1(\mathbb{Z}_+))$ $T^0 := I$, $F^0 := G^0 := \varepsilon_o$), weiter setze man

$\psi(t) := \sum_{k \geq 0} F^t(\{k\}) . T^k =: S^t$ mit $S = \psi(1)$,

$\xi(t) := \sum_{k \geq 0} G^t(\{k\}) S^k$, so ist

$\xi(t) = \sum_{k \geq 0} H^t(\{k\}) T^k$ mit $H = \sum_{k \geq 0} G(\{k\}) F^k \in M^1(\mathbb{Z}_+)$.

<u>Zusatz zu 3.2</u>

Sei $\tau : X \longrightarrow X$ ein stetiger Homomorphismus mit $\tau(e) = e$, $\varphi : X \longrightarrow \mathcal{T}$ wie vorhin und $\psi := \varphi \circ \tau$, dann ist $\psi \preceq_s \varphi$, nämlich $\psi(t) = \int_X \varphi(s) d\varepsilon_{\tau(t)}(s)$

Ist τ ein (topologischer) Automorphismus, so ist $\psi \preceq_s \varphi$ und $\varphi \preceq_s \psi$.

<u>Insbesondere ist $\preceq_s$ reflexiv!</u> Im Falle $X = \mathbb{R}_+$ ist für alle $c > 0$ durch $\tau_c : x \longrightarrow cx$ ein Automorphismus gegeben, im Fall $X = \mathbb{Z}_+$ gibt es keinen solchen außer dem trivialen.

Wir beschränken uns nun auf den Fall $X = \mathbb{R}_+$: Dann gibt es für jeden

Homomorphismus $\varphi: \mathbb{R}_+ \longrightarrow \mathcal{T}$ die trivialen Subordinationen: Für jedes $c>0$ sei $\varphi^c: \mathbb{R}_+ \ni t \longrightarrow \varphi(ct)$. Dann ist $\varphi \preceq_s \varphi^c$ und $\varphi^c \preceq_s \varphi$. Wir schreiben: $\varphi \approx_s \psi \Longleftrightarrow \varphi \preceq_s \psi$ und $\psi \preceq_s \varphi$.

Für einen Spezialfall zeigen wir, daß „$\approx_s$" eine Äquivalenzrelation ist und daß die Äquivalenzklassen sich durch die Automorphismen $\tau_c: x \longrightarrow cx$ beschreiben lassen:

3.3 Proposition.

$\mathcal{G}$ sei eine lokalkompakte Gruppe, $\mathcal{T} := M^1(\mathcal{G})$ wie in § 1. Sei (μ_t) eine stetige Halbgruppe von Wahrscheinlichkeitsmaßen mit erzeugender Distribution A, (F_s) sei eine Subordinationshalbgruppe, $(F_s) \sim F \sim (c,\eta)$. Dann gelten:

(i) Sei $(\lambda_t) = (\Phi((\mu_s), F_t)) = (\int \mu_s dF_t(s))$, so hat (λ_t) die erzeugende Distribution

$$f \longrightarrow C(f) = cA(f) + \int_{\mathbb{R}_+^\times} \left[\int f(x) d\mu_t(x) - f(e)\right] d\eta(t)$$

Seien η_A, η_C die Lêvy-Maße von A und C, so ist

$\eta_C = c\eta_A + \int_{\mathbb{R}_+^\times} \mu_t' d\eta(t)$, mit $\mu_t' := \mu_t|_{\mathcal{G}\setminus(e)}$.

(ii) Seien P_A, P_C, resp. Q_A, Q_C die primitiven resp. quadratischen Anteile von A und C, so gilt (bei fester Lêvy-Abbildung Γ)

$$Q_C = cQ_A,\ P_C = cP_A + \int_{\mathcal{G}\setminus(e)} \Gamma \cdot d\eta_C = cP_A + c\int_{\mathcal{G}\setminus(e)} \Gamma \cdot d\eta_A + \int_{\mathbb{R}_+^\times} \int_{\mathcal{G}^\times} \Gamma \cdot d\mu_t' \, d\eta(t)$$

[Dies gilt allgemeiner für Mischungen von Generatoren, s. [11] oder [12].].

3.4. Satz.

Seien $(\mu_t), (\nu_t)$ stetige Halbgruppen in $M^1(\mathcal{G})$ mit erzeugenden Distributionen A,B.

$(F_t) \sim F \sim (c,\eta), (G_t) \sim G \sim (d,\zeta)$ seien Subordinationshalbgruppen, so daß

$(\mu_t) = (\Phi((\nu_s),F_t))$, $(\nu_s) = (\Phi((\mu_t),G_s))$, so existiert ein $c>0$ mit $\mu_t = \nu_{ct}$.

Also: Aus $(\mu_t) \preceq_s (\nu_s)$ und $(\nu_s) \preceq_s (\mu_t)$ folgt, daß die Subordination trivial gewählt werden kann.

Beweis:

1. Sei $(H_s) = (\Phi((F_t),G_s)) = (\int F_t dG_s(t))$ so ist $\nu_s = \Phi((\nu_t),H_s)$.

Es wird zunächst gezeigt:

(i) Seien $A,B \neq 0$.

Aus $\nu_s = \int \nu_t dH_s(t)$, $H_s \sim (a,\xi)$ folgt $a = 1$, $\nu_t = \varepsilon_e$ für $t \in Tr(\xi)$.

[Sei $H_t^{(2)} := \int H_s dH_t(s)$, ..., $H_t^{(n)} = \int H_s^{(n-1)} dH_t(s)$, sei

$\nu_t^{(n)} = \Phi((\nu_s),H_t^{(n)})$, mit erzeugender Distribution $B^{(n)}$.

Dann gilt offenbar $B^{(n)} = B, \nu_t^{(n)} = \nu_t$, aber wegen 3.3

$$B^{(n+1)} = aB^{(n)} + \int(\nu_t^{(n)} - \varepsilon_e)d\xi = a^n B + (\sum_{k=0}^{n-1} a^k)\int(\nu_t - \varepsilon_e)\, d\eta(t),$$

somit $B = a^n B + (\sum_0^{n-1} a^k)\int (\nu_t - \varepsilon_e)\, d\eta(t)$.

Wählt man $f \in \mathcal{D}_+(G)$, $f(e) = 0$ so folgt $0 \le B(f) = a^n B(f) + (\sum_0^{n-1} a^k)\int(\nu_t - \varepsilon_e)(f)\, d\eta(t) = a^n B(f) + (\sum_0^{n-1} a^k)\int \nu_t(f)\, d\eta(t)$. Beide Summanden sind nicht negativ, also gilt: Ist $a>1$, so folgt $B(f) = 0$ und $\int \nu_t(f)\, d\eta(t) = 0$. Ist $B \neq 0$, nicht primitiv, dann gibt es ein $f \in \mathcal{D}_+$, $f(e) = 0$ mit $B(f) \neq 0$ und somit folgt also ein Widerspruch.

Ist $B \neq 0$, primitiv, so ist (ν_t) eine Halbgruppe von Punktmaßen, $(\nu_t = \varepsilon_{x_t})$ und $\int \nu_t(f) d\xi(t) = 0$, gilt genau dann für $f \in \mathcal{D}_+$, $f(e) = 0$, wenn $\nu_t = \varepsilon_{x_t} = \varepsilon_e$ für $t \in Tr(\xi)$. Dann ist aber $\int(\nu_t - \varepsilon_e)d\xi \equiv 0$, also $B = aB$ und es muß $a = 1$ sein.

Ist schließlich $B = 0$, so ist $(\nu_t \equiv \varepsilon_e)$ und damit auch für jede Subordination $\int \nu_t dF_s(t) \equiv \varepsilon_e$.

Im Falle $0<a<1$ erhält man (indem man den Grenzwert für $n \longrightarrow \infty$ betrachtet) $B = (\sum_0^\infty a^k) \int (\nu_t - \varepsilon_e)\, d\eta(t)$, also ist das Lêvymaß η_B von der Gestalt $\eta_B = \frac{1}{1-a} \int_{\mathbb{R}_+}^{x} \nu_t |_{G\setminus\{e\}}\, d\xi(t)$ und es ist wegen

$B(f) = \int_{G\setminus\{e\}} (f(x)-f(e)\, d\eta_B(x)$, $f \in \mathcal{D}(G)$, B vom Typ L_σ (s.[12],I.§5).
Dann ist $e \in Tr(\nu_t) = Tr(\eta_B)$ für $t \geq 0$ und daraus folgt leicht, daß

$\nu_\infty =: \lim_{t\to\infty} \nu_t$ in der vagen Topologie existiert. (ν_∞ ist gleich dem Haarmaß auf der Trägergruppe von ν_t, falls diese kompakt ist, ansonsten $= 0$.) Nun wähle man eine Teilfolge natürlicher Zahlen (n_k), sodaß $F^{\circledast n_k}$, $G^{\circledast n_k}$ und $H^{\circledast n_k}$ in $\mathcal{U}(\mathbb{R}_+)$ konvergieren und sodaß $H^\infty := \lim H^{\circledast n_k}$ $\circledast$-idempotent ist. F^∞ resp. G^∞ seien die Limiten von $F^{\circledast n_k}$ resp. $G^{\circledast n_k}$, mit (F_t^∞), (G_t^∞), (H_t^∞) werden die entsprechenden Halbgruppen bezeichnet.

H^∞ ist $\circledast$ idempotent, also nach 2.7 entweder $(H_t^\infty = \varepsilon_t)_{t\geq 0}$ oder $(H_t^\infty = e^{-\alpha t}\varepsilon_0 + (1-e^{-\alpha t})\varepsilon_\infty)_{t\geq 0}$ mit $\alpha \in (0,\infty)$. ($\alpha = \infty$ ist offenbar unmöglich).

Um zu zeigen, daß der zweite Fall nicht eintreten kann, unterscheiden wir zwei Fälle:

a) $\nu_\infty \in M^1(G)$. Also muß $\nu_t = \int_{\bar{\mathbb{R}}_+} \nu_s\, d\, H_t^\infty(s) = e^{-\alpha t}\varepsilon_e + (1-e^{-\alpha t})\nu_\infty$

sein, resp. (ν_t) Poissonsch mit $\eta_B = \text{const.}\ \omega_K |_{G\setminus\{e\}}$; dabei bezeichnet K die (kompakte) Trägergruppe von (ν_t).

Dann gilt stets für alle Faltungshalbgruppen $\subseteq M^1(\mathbb{R}_+)$, also insbesondere für (F_t)

$\int \nu_s\, dF_t\,(s) = \nu_{ct}$ mit $c = -\log \int_{[0,\infty)} e^{-cs}\, dF_t(s)$.

Also wirkt jede Falungshalbgruppe auf (ν_t) trivial.

b) $\nu_\infty = 0$. In diesem Fall führt die Annahme $\alpha > 0$ sofort zum Widerspruch zur Voraussetzung $(\nu_t) \subseteq M^1(G)$.

Es bleibt also der Fall

c) $(H_t^{\infty} = \mathcal{E}_t)_{t\geq 0}$ zu betrachten. Dann ist aber nach §2 $(F_t^{\infty}) \in H\,(\mathcal{E}_1)$ (Folgerung 2.7) und somit existiert ein $c>o$ mit $(F_t = \mathcal{E}_{ct})_{t\geq 0}$ und daher folgt sofort $\mu_t = \nu_{ct}$, $t\geq 0$.

(ii) Es sei also $a = 1$, $\nu_t = \mathcal{E}_e$ für $t\in Tr(\xi)$, also $(H_s) \sim H\sim(1,\xi)$. Nun ist aber auch $H\sim(cd, d\int_{\mathbb{R}_+^x} \Pi\, d\eta + \int_{\mathbb{R}_+^x} F_t'\, d\xi)$, also erhält man

$c>o$, $d = 1/c>o$ sowie $\xi = d\int_{\mathbb{R}_+^x} \mathcal{E}_t d\eta + \int_{\mathbb{R}_+^x} F_t'\, d\xi$, somit $Tr(\eta)\subseteq Tr(\xi)$.

Daraus folgt insbesondere: $\nu_t = \varepsilon_e$ für $t\in Tr(\eta)$, also $\int \nu_t' d\eta(t) = 0$.

Also $A = cB$ und damit $\mu_t = \nu_{ct}$. Analog zeigt man $\nu_t = \mu_{t/c}$, $t\geq 0$. □

Der Satz legt es nahe, nun zu untersuchen, wie sich Iterationen von Subordinationen verhalten. Dazu betrachten wir einen speziellen Typ von Homomorphismen, wir setzen nämlich voraus, daß $\Psi: \mathbb{R}_+ \longrightarrow \mathcal{V}$ so beschaffen ist, daß $\lim\limits_{t\to\infty} \Psi(t)\in\mathcal{V}$ existiert, also Ψ als Homomorphismus von $\overline{\mathbb{R}_+} \longrightarrow \mathcal{V}$ aufgefaßt werden kann:

<u>3.5 Satz.</u>

Sei $\Psi: \overline{\mathbb{R}_+} \longrightarrow \mathcal{V}$ ein Homomorphismus $(F_t)\sim C\sim(c,\eta)$ eine Subordinationshalbgruppe und $(F_t^{(n)})$ sei das n-fache $\circledast$-Produkt. Dann besitzt die Folge $\Phi(\Psi, F_t^{(n)})$ einen Häufungspunkt der Gestalt $(e^{-tc}\Psi(o) + (1-e^{-tc})\Psi(\infty))_{t\geq 0}$.

[Beweis folgt unmittelbar aus § 2.]

Überdies kann man sich leicht überlegen, daß sogar $\lim F_t^{(n)}$ und somit $\lim\limits_n \Phi(\Psi, F_t^{(n)})$ existieren: Je zwei Häufungspunkte von $F_t^{(n)}$ müssen ja äquivalent bezüglich $\preceq_s$ sein, daraus folgt durch leichte Rechnung die Behauptung.

Tatsächlich könnte man eine allgemeinere Form des Satzes 3.4 beweisen in dem man statt $M^1(\mathcal{G})$ positive Operatoren auf einem Funktionenverband betrachtet, wobei vorausgesetzt wird, daß der Definitionsbereich der Generatoren genügend reichhaltig ist. (Etwa Fellersche Halbgruppen auf einer Mannigkfaltigkeit, bei denen die C^{∞}-Funktionen zum Definitionsbereich des Generators gehören).

§ 4 Spezielle Halbgruppen.

4.1 Definition.

Sei $\mathcal{V}$ wie vorhin, $\mathbb{R}_+ \ni t \longrightarrow \varphi(t) \in \mathcal{V}$ sei ein Homomorphismus in $\mathcal{M}_0(\mathbb{R}_+, \mathcal{V}; \mathbb{F})$, so daß ein $z \in \mathbb{E}_{\mathcal{V}}$ existiert mit

(i) $\varphi(t) = \sum_{k \geq 0} \frac{t^k}{k!} z^k$,

(ii) für alle stetigen Halbnormen p existiert eine stetige Halbnorm q und ein $C = C(p,q,z) > 0$ mit $p(z^k) \leq C^k q(z), k \in \mathbb{N}$. Dann sagt man, φ besitzt einen beschränkten Generator.

(iii) φ heißt „von Poissonschem Typ", falls $z = \alpha(x - I), \alpha > 0, x \in \mathcal{V}$, also
$$\varphi(t) = e^{-t\alpha} \sum_{k \geq 0} \frac{t^k}{k!} x^k =: \exp t\alpha(x-I). \qquad (x^0 := I)$$

Unter den C_0-Operatorhalbgruppen sind die gleichmäßig stetigen gerade die Halbgruppen mit beschränkten Generatoren, unter den Faltungshalbgruppen von Wahrscheinlichkeitsmaßen sind die Poissonhalbgruppen die „Halbgruppen vom Poissonschen Typ." Sei Y eine vollständig reguläre topologische Halbgruppe mit Einheit e, $\mathbb{E} := \mathcal{L}(C(Y))$, so heißt eine Poissonhalbgruppe $(\mu_t = \exp \alpha t(\nu - \varepsilon_e), \nu \in M^1(Y), \alpha \geq 0, t \geq 0)$ elementar, wenn $\nu = \varepsilon_x, x \in Y$.

4.2 Definition.

Sei Y vollständig regulär, $\mathbb{E} = \mathcal{L}(C(Y))$, $\mathcal{V}$ sei eine affine Halbgruppe von Submarkoff-Operatoren auf C(Y), versehen mit der Topologie der punktweisen Konvergenz, i.e. $T_\alpha \to T \iff T_\alpha f(y) \longrightarrow Tf(y), f \in C(Y), y \in Y$.
Sei $\varphi: \mathbb{R}_+ \ni t \longrightarrow T_t \in \mathcal{V}$ ein stetiger Homomorphismus und sei $y_0 \in Y$.
φ resp. (T_t) heißt von lokalem Typ in y_0, falls für alle $f \in C(Y)$, die in einer Umgebung von y_0 verschwinden, $\frac{1}{t} T_t f(y_0) \longrightarrow 0$.

Betrachtet man das lineare Funktional $T \longrightarrow L(T) := Tf(y_0)$, so bedeutet die obige Bedingung, daß $t \longrightarrow L(T_t)$ in $t = 0$ differenzierbar ist, (i.e. $L\in D(\varphi)$, s. § 1.1.21) und es ist $D_\varphi(L) = 0$.

Die Faltungsoperatoren von Gaußhalbgruppen auf lokalkompakten Gruppen sind von lokalem Typ (E. Siebert [30], resp. H. Heyer [13]), Fellersche Halbgruppen auf einer Mannigfaltigkeit, so beschaffen, daß die C^∞-Funktionen zum Definitionsbereich des Generators gehören, sind von lokalem Typ, wenn der Generator in y_0 ein Differentialoperator ist.

4.3 Satz.

$\mathbb{R}_+ \ni t \longrightarrow F_t \in M^1(\mathbb{R}_+)$, $\mathbb{R}_+ \ni t \longrightarrow \varphi(t) \in \mathcal{T}$ seien stetige Homomorphismen.

a) Sei (F_t) eine Poissonhalbgruppe in $M^1(\mathbb{R}_+)$, $\varphi: \mathbb{R}_+ \longrightarrow \mathcal{T}$ sei ein stetiger Homomorphismus in $\mathcal{M}_0(\mathbb{R}_+, \mathcal{T}, \mathbf{F})$, so ist $(\int \varphi dF_t =: \Psi(t))$ von Poissonschem Typ.

b) Nun sei $\mathbf{E}_\gamma$ abgeschlossen in $\mathbf{E}$, $\mathbf{F} = \mathbf{E}'$. Dann gilt auch: Sei (F_t) eine beliebige stetige Faltungshalbgruppe $\subseteq M^1(\mathbb{R}_+)$, aber es sei φ von Poissonschem Typ [besitze einen beschränkten Generator], so hat auch Ψ diese Eigenschaft.

c) Sei G eine lokalkompakte Gruppe, (F_t) und (μ_t) seien stetige Faltungshalbgruppen in $M^1(\mathbb{R}_+)$ resp. $M^1(G)$ mit $\mu_0 = \varepsilon_e$. Dann gilt: Ist $(\lambda_t := \int \mu_s dF_t(s))$ von Poissonschem Typ, so ist (F_t) oder (μ_t) von diesem Typ.

Beweis:

a) und b) sind offensichtlich: Sei die erzeugende Distribution von (F_t) gegeben durch (c, η_F), so ist im Falle a) $c = o$, η_F beschränkt und man erhält $\Psi(t) = \exp t\|\eta_F\|(\int \varphi \frac{d\eta_F}{\|\eta_F\|} - \varphi(o))$.

Im Falle b) - mit $\varphi(t) := \exp t z$ - ist für jede stetige Halbnorm p

$$p(\tfrac{1}{t}(\varphi(t) - I)) \leq \frac{1}{t} \sum_{k\geq 1} \frac{t^k}{k!} p(z^k) \leq \frac{1}{t}(e^{at}-1) \text{ mit passendem } a>o \text{ (s. 4.1.(ii)).}$$

Wegen $\int \frac{x}{1+x}\, d\eta_F < \infty$ ist daher $\limsup_{t\to 0} p(\frac{1}{t}(\Psi(t)-I)) \leq \limsup_{t\to 0} c\cdot p(\frac{1}{t}\varphi(t)-I)$

$+ \limsup_{t\to 0} \int \frac{1}{t}\,(e^{at}-1)\,d\eta_F(t)$.

Aus der Abgeschlossenheit von $\mathbb{E}_\gamma$ folgt nun die Existenz des Integrals $\int_{\mathbb{R}_+^x} (\varphi(t)-I)\,d\eta_F(t) \in \mathbb{E}_\gamma$ und man erhält leicht mit

$z_1 := cz + \int(\varphi(t)\ -I)\,d\eta_F(t)$, daß $\Psi(t) = \exp t\, z_1$.

c) (s. [11]. Der Vollständigkeit halber geben wir eine Beweisskizze an):

Sei $(\int \mu_s dF_t(s) =: \lambda_t)$ Poissonsch, dann ist also die erzeugende Distribution von (λ_t): $cA + \int(\mu_s - \varepsilon_e)\,d\eta(s)$ beschränkt.

Dabei ist A die erzeugende Distribution von (μ_s) und die erzeugende Distribution von (F_t) ist durch (c,η) gegeben. Betrachtet man die kanonische Zerlegung von A, so folgen: Ist $c>0$, so verschwinden der Gaußanteil und der primitive Anteil von A und das Lévymaß von A ist beschränkt, i.e. A erzeugt eine Poissonhalbgruppe.

Ist $c = 0$, so ist das Maß

$\int \mu_t'\, d\eta(t)$ beschränkt, und somit $\int(1-\mu_t(\{e\}))\,d\eta(t) < \infty$. Nun wählt man $x_t \in G$, so daß $\mu_t(\{x_t\}) = \max_{x\in G} (\mu_t(\{x\})) =: u(\mu_t)$.

Wenn η unbeschränkt, also (F_t) nicht Poissonsch ist, dann folgt aus der Konvergenz des Integrals $\int(1-\mu_t(\{e\}))\,d\eta(t)$, daß für eine Folge $t_n \downarrow 0$ $\mu_{t_n}(\{e\}) \longrightarrow 1$.

Nun ist aber $\|\mu_t - \varepsilon_{x_t}\| \leq 2(1-\mu_t(\{x_t\}))$, daraus folgt leicht

$\|\mu_t - \varepsilon_e\| \longrightarrow 0$, also ist (μ_t) Poissonsch. □

Es ist anzunehmen, daß c) auch allgemein gilt, doch ist mir kein Beweis bekannt. Wir kommen später auf ähnliche Fragestellungen noch zurück (bei Halbgruppen invertierbarer Maße).
Im Moment betrachten wir noch Halbgruppen von lokalem Typ und elementare Poissonhalbgruppen und zeigen, daß diese bezüglich $\lesssim_s$ maximal sind:

4.4 Satz.

Sei Y vollständig regulär, $y_0 \in Y$, (T_t) sei eine Halbgruppe von submarkoffschen Operatoren auf C(Y), die in einer Umgebung von y_0 von lokalem Typ ist. Dann ist für jede Darstellung $T_t = \int S_s dF_t(s)$, [(S_s) submarkoffsch, (F_t) stetige Halbgruppe in $M^1(\mathbb{R}_+)$] auch (S_s) von lokalem Typ in einer Umgebung von y_0 oder $T_t f(y_0) \equiv 0$ für $f \in C(Y)$, $f \equiv 0$ in einer Umgebung $U(y_0)$.

Beweis:

Sei die erzeugende Distribution von (F_t) durch (c,η) gegeben. Dann gilt nach 4.2 für jede stetige beschränkte Funktion f, die in einer Umgebung von y_0 verschwindet, daß $t \longrightarrow T_t f(y)$ differenzierbar ist in t = o mit Ableitung 0 für alle y in einer Umgebung $U=U(y_0)$.

Also gilt für alle $f \geq 0$, $f \equiv 0$ in einer Umgebung von y_0:

$$\frac{1}{t}\int_{\mathbb{R}_+} S_s f(y) dF_t(s) \longrightarrow 0, \quad y \in U(y_0).$$

Man setzt $g: s \longrightarrow S_s f(y)$ und wählt für $o < \varepsilon$ ein $h^\varepsilon \in C^1(\mathbb{R}_+)$, mit $h^\varepsilon \equiv 0$ in $[o,\varepsilon)$, $\equiv 1$ in $[2\varepsilon,\infty)$, $0 \leq h^\varepsilon \leq 1$ sonst.

Wegen (i) $0 \leq \frac{1}{t}\int h^\varepsilon g dF_t \xrightarrow[t \downarrow 0]{} \int h^\varepsilon g d\eta \geq 0$

und (ii) $\frac{1}{t}\int (1-h^\varepsilon) g dF_t \geq 0$, sowie

(i) + (ii) $\frac{1}{t}\int h^\varepsilon g dF_t + \frac{1}{t}\int (1-h^\varepsilon) g dF_t = \frac{1}{t}\int g dF_t \xrightarrow[t \downarrow 0]{} 0$ folgt

$\lim_{t\downarrow 0} \frac{1}{t} \int h^{\varepsilon} g dF_t = \int h^{\varepsilon} g d\eta = 0$ und

$\lim_{t\downarrow 0} \frac{1}{t} \int (1-h^{\varepsilon}) g dF_t = 0.$

Schließlich ist $\lim_{\varepsilon\downarrow 0} h^{\varepsilon} g = g$, also folgt

$\int_{\mathbb{R}^x} g\, d\eta = 0$, i.e. $\int S_s f(y) d\eta(s) = 0$ und somit $S_s f(y) = 0$ für s Tr( ), y U(y_0).

Daraus folgt aber $S_s f(y_0) = 0$ für $s \in \langle Tr(\eta)\rangle$ (der von $Tr(\eta)$ erzeugten Halbgruppe). [Es ist nämlich für $s,t \in Tr(\eta)$: $S_{s+t} f(y_0) = S_s g(y_0)$ mit $g := S_t f \equiv 0$ in einer Umgebung von y_0. Somit ist $S_{s+t} f(y_0) = 0 = S_s g(y_0)$].

Damit ist gezeigt: Ist c = 0, so ist $0 \equiv \int S_s f(y_0) dF_t(s) = T_t f(y_0), t \geq 0$.

Nun stellt man (F_t) in kanonischer Form dar,

$F_t = \varepsilon_{ct} \bar{F}_t$ mit $(\bar{F}_t) \sim (0,\eta)$, $Tr(\bar{F}_t) = \langle Tr(\eta)\rangle^-$. Dann ist für alle $f \in \mathcal{C}(Y), f \equiv 0$ in $U(y_0)$:

$\frac{1}{t} T_t f(y_0) = \frac{1}{t} \iint S_{s+s'} f(y_0) d\bar{F}_t(s) d\varepsilon_{ct}(s') \xrightarrow[t\downarrow 0]{}$ (wegen $S_{s+s'} = S_{s'} S_s$)

$\lim_{t\to 0} \left[\frac{1}{t} \int S_s d\bar{F}_t (s) + \frac{1}{t} \int S_{s'} d\varepsilon_{ct}(s') \right] . f(y_0) = \int S_s f(y_0) d\eta + \frac{d^+}{dt} S_{ct} f(y_0) \Big|_{t=0}$

Der erste Summand verschwindet, wie oben gezeigt. Also ist

$0 = \frac{d^+}{dt} T_t \Big|_{t=0} f(y_0) = c \frac{d^+}{dt} S_t \Big|_{t=0} f(y_0)$, i.e. (S_t) ist von lokalem Charakter □

Für Halbgruppen von Gaußmaßen ist der Satz wohlbekannt: Für Abelsche Gruppen wurde er von S. Bochner [2] bewiesen, für kompakte Gruppen von H. Carnal [4], für beliebige lokalkompakte Gruppen s. [11], unabhängig von einer dem Verfasser damals nicht bekannten Arbeit von . W. Woll [37], die das selbe Resultat enthält.

Nun erwähnen wir noch, daß die elementaren Poissonhalbgruppen ebenfalls maximal bezüglich $\preceq_s$ sind:

5 Satz.

Sei X eine vollständig reguläre topologische Halbgruppe, $x_0 \in X$.

Sei $\mu_t = \exp t\alpha(\varepsilon_{x_0} - \varepsilon_e)$. Wenn $x_0^2 \neq e, \neq x_0, \alpha > 0$, dann folgt aus der Darstellung $\mu_t = \int \nu_s dF_t(s)$, daß $\mu_t = \nu_{ct}$ für ein $c \geq 0$.

Beweis: (s. auch [11]).

Sei $f \in C(X)$, $f \geq 0$, $f(e) = f(x_0) = 0$.

So ist $\frac{1}{t} \int_X f d\mu_t \xrightarrow[t \downarrow 0]{} 0$, also $\frac{1}{t} \int_{\mathbb{R}_+} \int_X f(x) d\nu_s(x) dF_t(s) \xrightarrow[t \downarrow 0]{} 0$.

Wiederum ist $g = \int_X f d\nu_s \geq 0$ und wie in 4.4 zeigt man, daß das Lévy-Maß η von (F_t) so beschaffen sein muß, daß $\int_X f d\nu_s = o$ für alle $s \in Tr(\eta)$. Daraus folgt aber: Ist $Tr(\eta) \neq \emptyset$, so ist $Tr(\nu_s) \subseteq \{e, x_0\}$ für alle $s \in Tr(\eta)$, dies führt zu einem Widerspruch zur Voraussetzung.

Ist aber $Tr(\eta) = \emptyset$, also $\eta = 0$, so ist $F_t = \varepsilon_{ct}$ für ein $c \geq 0$ und es folgt die Behauptung des Satzes. □

Wir schließen den § 4 ab mit der Betrachtung einiger spezieller Zusammenhänge zwischen Halbgruppen $\mathbb{R}_+ \ni t \longrightarrow \varphi(t)$ und untergeordneten Halbgruppen $\int \varphi(s) dF_t(s) =: \Psi(t)$.

Zunächst ein Beitrag zu dem "Markoff-Gruppen-Problem" (s. [11, 5]). Wir interessieren uns für Homomorphismen $\varphi: \mathbb{R}_+ \longrightarrow \mathcal{T}$, so beschaffen, daß jedes $\varphi(t)$ ein Inverses besitzt (das in $\mathbf{E}_{\mathcal{T}}$, aber nicht notwendig in $\mathcal{T}$ zu liegen hat). Dazu beschränken wir uns auf folgenden Spezialfall: Sei $\mathcal{T} := M^1(G)$, wobei G eine lokalkompakte Gruppe ist.

4.6 Definition.

$(\mu_t) \subseteq M^1(G)$ heißt vom Typ "P+L", falls die erzeugende Distribution A von (μ_t) von der Gestalt $A(f) = P(f) + \int (f(x) - f(o)) d\eta_A$ mit primitivem P und beschränktem η_A ist. Ist $P = 0$, liegt eine Poissonhalbgruppe vor. Ist $\eta_A = 0$, so liegt eine Punktmaßhalbgruppe vor. In vielen Fällen (z.B. für Halbgruppen normaler Maße) läßt sich zeigen, daß die Halbgruppen invertierbarer Maße gerade die Halbgruppen vom Typ "P+L" sind. Es ist ein ungelöstes Problem, ob dies allgemein gilt. (Natürlich ist jedes Maß einer Halbgruppe vom Typ "P+L" invertierbar).

4.7 Satz.

(ν_t) sei eine stetige Halbgruppe in $M^1(G)$, (F_t) eine Subordinationshalbgruppe, A resp. $F \sim (c, \eta_F)$ seien die erzeugenden Distributionen. Es sei $\mu_t = \int_{\mathbb{R}_+} \nu_s dF_t(s)$ mit erzeugender Distribution B. Dann gelten

(i) Sind A und F vom Typ "P+L", so ist auch B von diesem Typ.

(ii) Ist B vom Typ "P+L", $c \neq 0$ und A nicht Poissonsch, so sind A und F vom Typ "P+L".
Ist B vom Typ "P+L", $c = 0$, A nicht Poissonsch, so ist F Poissonsch. (Dann braucht A nicht vom Typ "P+L" zu sein).

(iii) Der Fall "B vom Typ "P+L", A Poissonsch" ist in 4.3.b enthalten.

(iv) Ist A oder F nicht vom Typ "P+L" und ist die andere Distribution nicht Poissonsch, so ist B nicht vom Typ "P+L".

Beweis:

(i) Es ist $B(f) = cA(f) + \int (\nu_t(f) - f(e)) d\eta_F(t)$. Sind η_A und η_F beschränkt und verschwindet der Gaußanteil von A, so ist offenbar B vom Typ "P+L".

(ii) Ist B vom Typ "P+L" und seien P_A, Q_A der primitive resp. der Gaußanteil von A, dann ist $cQ_A = 0$, $c\eta_A$ beschränkt, $\int \nu_t' d\eta_F(t)$ beschränkt, also ist A vom Typ "P+L", falls $c \neq 0$.

Analog folgt aus der Endlichkeit von $\int \nu_t' d\eta_F(t)$, daß -s. 4.3.c)-, (ν_t) Poissonsch oder η_F beschränkt ist. Nach Voraussetzung ist A nicht Poissonsch, also ist η_F beschränkt und somit (F_t) ebenfalls vom Typ "P+L". Die übrigen Aussagen sind offensichtlich.

(iv) Nun sei etwa A nicht vom Typ "P+L", also $Q_A \neq 0$ oder $\eta_A(G^\times) = \infty$.

F sei vom Typ "P+L", aber nicht Poissonsch, also c>o, und $\eta_F(\mathbb{R}_+^x)<\infty$. Dann ist $B = cA + \int(\nu_s - \varepsilon_e)d\eta_F(s)$. Ist c>0, so ist $c\eta_A(\mathcal{G}^x) = \infty$ oder $cQ_A \neq 0$, also B nicht vom Typ "P+L".

Ist dagegen (F_s) nicht vom Typ "P+L" (also η_F unendlich) und A nicht Poissonsch, so ist $Bf = cA(f) + \iint(f(x)-f(e))d\nu_t(x)d\eta_F(t)$, dann ist nach 4.3 c) wieder $\int \nu_t' d\eta_F$ nicht endlich und somit ist das Lévymaß von B nicht endlich. □

Ein Analogon zu 4.4 und 4.5 läßt sich auch für Halbgruppen vom Typ P+L beweisen: Sei (ν_t) vom Typ P+L mit erzeugender Distribution $A=P+Q+\int\cdot d\eta_A$, $Q=0, \eta_A$ beschränkt. Es sei $Tr(\eta_A)=\{x_0\}$ mit $x_0^2 \neq e$, dann ist (ν_t) maximal bezüglich $\preccurlyeq_s$.

Nun untersuchen wir die Träger untergeordneter Halbgruppen: Sei wiederum $\mathbb{E} := \mathcal{L}(C(Y))$, Y vollständig regulär, $\mathcal{T}$ die Menge der Submarkoffschen Operatoren mit Kerndarstellung $Tf(y) = \int fN(y,dx)$.

$\varphi: \mathbb{R}_+ \ni t \longrightarrow T_t := N_t(\cdot\cdot,\cdot) \in \mathcal{T}$ sei ein stetiger Homomorphismus in $\mathcal{M}_0(\mathbb{R}_+, \mathcal{T}; \mathbb{E}')$ und (F_t) sei eine Subordinationshalbgruppe mit erzeugender Distribution $F\sim(c,\eta_F)$. Weiter sei $(M_t := \int N_s dF_t(s))_{t\geq 0}$.

Wir stellen (F_t) dar in der Form $(F_t = \varepsilon_{ct}\bar{F}_t)$ mit $(\bar{F}_t)\sim(0,\eta_F)$.

4.8 Satz.

Für $y \in Y$ ist
$$Tr(M_t(y,\cdot)) = \left[\bigcup_{s\in\langle Tr(\eta_F)\rangle} Tr(N_{ct+s}(y,\cdot))\right]^-$$

Insbesondere für c = 0 ist der Träger der untergeordneten Halbgruppe - bei festem y - unabhängig von t>o.

Ist Y eine Halbgruppe mit Einheit e und N_t vom Faltungstyp, d.h. $\int f(x)\, N_t(y,dx) = \int f(yx)N_t(e,dx) = \int f(yx)d\nu_t(x)$ mit $\nu_t := N_t(e,\cdot)$, dann ist im Falle c = 0 $Tr(\int \nu_s dF_t(s)) =: L_t$ eine abgeschlossene Unterhalbgruppe von Y und es ist $L_t = L_{t'}$ für alle t,t'>0. (Dabei wird, wie vereinbart, die "Prohorov Bedingung" 1.8.4 vorausgesetzt, i.e. $t \longrightarrow N_t[\nu_t] \in \mathcal{M}_0$).

Beweis: Der erste Teil der Behauptung gilt für eine beliebige stetige Abbildung $\mathbb{R}_+ \ni t \longrightarrow F_t \in M^1(Y)$, wenn man bedenkt, daß

$Tr(F_t) = ct + \langle Tr(\eta_F)\rangle^-$.

Es ist noch zu zeigen, daß im zweiten Fall $L := \left[\bigcup_{s\in\langle Tr(\eta_F)\rangle} Tr(\nu_s)\right]^-$

eine Halbgruppe von Y bildet. Dazu betrachte man $x, y \in \bigcup_{s\in\langle Tr(\eta_F)\rangle} Tr(\nu_s)$,

dann gibt es $s = s(x), t = t(y) \in \langle Tr(\eta_F)\rangle \subseteq Tr(F_t)$ mit

$x \in Tr(\nu_s)$, $y \in Tr(\nu_t)$. Daher ist auch $xy \in Tr(\nu_t \nu_s) = Tr(\nu_{t+s}) \subseteq L$, also ist L eine Halbgruppe $\subseteq Y$. □

Nun betrachten wir abschließend nochmals allgemein Homomorphismen $(\varphi : \mathbb{R}_+ \longrightarrow \mathcal{T}) \in \mathcal{M}_0(\mathbb{R}_+, \mathcal{T}; \mathbf{F})$, es sei $\mathcal{T}_0 := \langle \varphi(\mathbb{R}_+)\rangle^-$ die kleinste von $\varphi(\mathbb{R}_+)$ erzeugte affine abgeschlossene Unterhalbgruppe. Es sei nun $\alpha : \mathcal{T}_0 \longrightarrow (\mathbb{C}, \cdot)$ ein stetiger Homomorphismus mit $|\alpha(\cdot)| \leq 1$.

4.9 Proposition.

Sei α wie vorhin definiert, sei (F_t) eine Subordinationshalbgruppe mit erzeugender Distribution $A \sim (c, N)$. Man definiert $\psi(t) := \int \varphi(s)dF_t(s)$. $\hat{F}_t(\lambda) = \int e^{-\lambda s}dF_t(s)$ sei die Laplacetransformierte definiert auf $\{\lambda : \Re e\lambda > o\}$. Analog sei $\hat{A}$ definiert. Dann gilt: $\alpha \circ \varphi(t)$, $\alpha \circ \psi(t)$ sind stetige Homomorphismen von $\mathbb{R}_+ \longrightarrow \mathbb{C}$ und daher von der Gestalt $\alpha\circ\varphi(t) = \exp(-t\, u)$, $\alpha\circ\psi(t) = \exp(-t\, v)$, $\mathrm{Re}\, u \geq 0$, $\mathrm{Re}\, v \geq 0$. Daher ist $\alpha\circ\psi(t) = \hat{F}_t(u) = \exp(t\,\hat{A}(u))$. [[Offensichtlich]].

Korollar 1.

Sei Y eine Halbgruppe, $\gamma: Y \longrightarrow \mathbb{C}$, $|\gamma| \leq 1$ ein stetiger Semicharakter, so gilt für jede stetige Halbgruppe $(\nu_t) \subseteq M^1(Y)$ und jede Subordinationshalbgruppe $(F_t) \sim A$, $\mu_t = \int \nu_s dF_t(s)$:

$\mathbb{R}_+ \ni t \longrightarrow \int_Y \gamma d\nu_t \in \mathbb{C}$, $t \longrightarrow \int_Y \gamma d\mu_t \in \mathbb{C}$ sind stetige Homomorphismen und daher von der Gestalt $\int_Y \gamma d\nu_t = e^{-tu}$, $\int_Y \gamma d\mu_t = e^{-tv}$,

Re $u \geq 0$, Re $v \geq 0$.

$e^{-tv} = \int \gamma d\mu_t = e^{t\hat{A}(u)}$. Anders ausgedrückt

$$\int_Y \gamma d\left(\int_{\mathbb{R}_+} \nu_s dF_t(s)\right) = \exp t\, \hat{A}\left(-\frac{d^+}{ds} \int \gamma d\nu_s \Big|_{s=0}\right).$$

Korollar 2.

Sei $\mathbf{B}$ ein Hilbertraum und $(\varphi(t))$ eine Halbgruppe von normalen Kontraktionen mit Spektraldarstellung nach einer Spektralschar $(E_\lambda) \subseteq \mathcal{L}(\mathbf{B})$

$$\varphi(t) = \int_{\mathbb{R}} f(\lambda)^t e^{itg(\lambda)} dE_\lambda = \int_{\mathbb{R}} e^{t(\log f(\lambda) + ig(\lambda))} dE_\lambda, \quad f \geq 0,\ g \text{ reell},$$

so ist

$$\Psi(t) = \int \varphi(s) dF_t(s) = \int_{\mathbb{R}} \exp t \left[\hat{A}(-\log f(\lambda) - ig(\lambda))\right] dE_\lambda.$$

Ist insbesondere $g = 0$, d.h. ist $(\varphi(t))$ eine Halbgruppe positiv definiter Kontraktionen und wählt man $(F_t := \pi_t^{(\alpha)})_{t \geq 0}$ mit den Laplace-transformierten $\hat{\pi}_t^{(\alpha)}(\lambda) := e^{-\lambda^\alpha}$, $\hat{A}(\lambda) = -\lambda^\alpha$, Re $\lambda > 0$, so erhält man klarerweise die Darstellung der gebrochenen Potenz (für $0 < \alpha < 1$)

$$\int \varphi(s) d\pi_t^{(\alpha)}(s) = \int_{\mathbb{R}_+} \exp(-t(-\log f(\lambda))^\alpha) dE_\lambda .$$

(s. auch den folgenden § 5).

§ 5 Invariante Wellengleichungen, (Bochner-)-stabile Verteilungen und Verteilungen von lokalem Typ der Ordnung α, $0 \le \alpha \le 1$.

5.1

C_u sei der Raum der gleichmäßig stetigen Funktionen $\mathbb{R}_+ \longrightarrow \mathbb{R}$, C_u^k der Teilraum der k-fach stetig differenzierbaren Funktionen (wobei in 0 von rechts zu differenzieren ist). Mit $D: f \in C_u^1 \longrightarrow D(f)$ werde die (rechte) Ableitung in 0 bezeichnet. $\mathbb{D}$ sei der Operator $f \longrightarrow \frac{d}{dx} f$ (mit $\mathbb{D} f(o) := D(f)$).

Entsprechend sei $\mathbb{D}^k$, $k \in \mathbb{Z}_+$ definiert.

$\mathbb{D}$ ist der Generator der Halbgruppe der Translationen

$(R_{\varepsilon_t}: f \longrightarrow f(\cdot + t))_{t \geq 0}$.

Die Laplacetransformierten haben die Gestalt $\hat{\varepsilon}_t(\lambda) = e^{-t\lambda}$, $\hat{D}(\lambda) = -\lambda$. Nun seien für $0 < \alpha < 1$ $(\pi_t^{(\alpha)})_{t \geq 0}$ die Halbgruppen $\subseteq M^1(\mathbb{R}_+)$ mit den Laplacetransformierten $\hat{\pi}_t^{(\alpha)}(\lambda) = e^{-t\lambda^\alpha}$ (also die einseitigen stabilen Verteilungen). Die erzeugenden Distributionen haben dann die Gestalt $C_u^1 \ni f \longrightarrow C_\alpha \int_{\mathbb{R}_+^x} (f(x) - f(o)) \frac{1}{x^{1+\alpha}} dx =: D_\alpha(f)$ mit Laplacetransformierten $\hat{D}_\alpha(\lambda) = -\lambda^\alpha$, mit einer Konstanten $C_\alpha > o$.

Zu jedem Homomorphismus $\varphi: \mathbb{R}_+ \longrightarrow \mathcal{T}$ definiert man nun den untergeordneten Homomorphismus $\mathbb{R}_+ \ni t \longrightarrow \varphi^{(\alpha)}(t) := \Phi(\varphi, \pi_t^{(\alpha)}) = \int \varphi(s) d\pi_t^{(\alpha)}(s) \in \mathcal{T}$.

Wir beschränken uns im folgenden Text auf C_o-Kontraktionshalbgruppen $\varphi: t \longrightarrow T_t$ auf einem Banachraum $\mathbb{B}$. Sei A der Generator von (T_t) - wir schreiben dann $(e^{tA} := T_t)$ -, weiter sei $(T_t^{(\alpha)} := \int_{\mathbb{R}_+} T_\tau \, d\pi_t^{(\alpha)}(\tau))_{t \geq 0}$, die untergeordnete Halbgruppe mit Generator A_α, dann ist bekanntlich $A_\alpha = -(-A)^\alpha$, wobei $(-A)^\alpha$ die α-te Potenz von -A ist (s.z.B. Komatsu [23] oder U. Westphal [35] und die dort zitierte Literatur).

Nun erweitert man das Konzept der verallgemeinerten Differentiation, man setzt $\mathbb{D}_{(k)} := (-\mathbb{D})^k$, $\mathbb{D}_k := (-1)^k \mathbb{D}_{(k)}$, $k \in \mathbb{Z}_+$, also insbesondere $\mathbb{D}_{(0)} = I = -\mathbb{D}_0$, $\mathbb{D}_{(1)} = -\mathbb{D} = -\mathbb{D}_1$. Für $\alpha \in (0,1)$ sei $\mathbb{D}_\alpha$ derFaltungsoperator $\mathbb{D}_\alpha = R_{D_\alpha} =: -\mathbb{D}_{(\alpha)}$, allgemein für $\alpha \in \mathbb{R}_+$ setze man $\mathbb{D}_{(\alpha)} := \mathbb{D}_{([\alpha])} \mathbb{D}_{(\alpha - [\alpha])}$ und $\mathbb{D}_\alpha := \mathbb{D}_{[\alpha]} \mathbb{D}_{\alpha - [\alpha]} = (-1)^{[\alpha]+1} \mathbb{D}_{([\alpha])} \mathbb{D}_{(\alpha - [\alpha])}$. Dann bilden die $(\mathbb{D}_{(\alpha)})_{\alpha \geqslant 0}$ eine Halbgruppe (von unbeschränkten) Operatoren, es gilt nämlich $\mathbb{D}_{(\alpha+\beta)} = \mathbb{D}_{(\alpha)} \mathbb{D}_{(\beta)}, \alpha, \beta \geqslant 0$.
Schließlich definiert man zu jeder Halbgruppe $(T_t = e^{tA})$ resp. zu deren Generator A $A_{(\alpha)} := (-A)^{[\alpha]} (-A_{\alpha - [\alpha]})$, $A_{(0)} = I$, $A_\alpha := (-1)^{[\alpha]+1} A_{(\alpha)} A_{(\alpha - [\alpha])}$ mit Definitionsbereich $D_{A^{[\alpha]}} \supseteq D_{A_\alpha} \supseteq D_{A^{[\alpha]+1}}$.

Die Operatoren $\mathbb{D}_\alpha$, $\mathbb{D}_{(\alpha)}$ resp. die zugehörigen Distributionen sind also so gewählt, daß die Laplacetransformierten die Gestalt $\lambda \longrightarrow D_\alpha(\lambda) = -\lambda^\alpha$, $\alpha \in \mathbb{R}_+$, haben.
Die Operatoren $\mathbb{D}_\alpha$, $\alpha \in [0,1]$ sind selbst Halbgruppengeneratoren und man kann daher $(\mathbb{D}_\alpha)_\beta$ für $\beta \in \mathbb{R}_+$ bilden: Aus der Gestalt der Laplacetransformierten folgert man $(\mathbb{D}_\alpha)_\beta = \mathbb{D}_{\alpha\beta}$, $1 \geqslant \alpha \geqslant 0, \beta \geqslant 0$ und damit ganz allgemein $A_{\alpha\beta} = (A_\alpha)_\beta$.

Schließlich weiß man, daß $\alpha \longrightarrow A_\alpha$ stetig ist im Sinne der verallgemeinerten Konvergenz von Generatoren. Dies folgt einmal aus der Konvergenz der Laplacetransformierten von D_α resp. $\pi_t^{(\alpha)}$. Wir können dies aber auch direkt aus unseren früheren Überlegungen folgern:

<u>5.2 Proposition.</u>

a) Für $1 > \alpha > 0$ ist $\alpha \longrightarrow D_{(\alpha)}(f)$ stetig für $f \in C_u^1$. Sei $\Lambda := \{\hat{f}, f \in L_1(\mathbb{R}_+)\}$, so ist $\mathbb{R}_+ \ni \alpha \longrightarrow D_{(\alpha)}(f)$ stetig für $f \in \Lambda$, insbesondere ist $D_{(1)} = -D$, $D_{(0)} = \varepsilon_0$.

b) Die Halbgruppen $(\pi_t^{(\alpha)}) \xrightarrow[\alpha \to \beta]{} (\pi_t^{(\beta)})$ konvergieren schwach, kompakt-gleichmäßig in t für $0<\alpha,\beta\leq 1$, $(\pi_t^{(\alpha)}) \xrightarrow[\alpha\to 0]{} (e^{-t}\varepsilon_0)$ vage.

c) Für jede Kontraktionshalbgruppe $(T_t = e^{tA})$ mit $(T_t^{(\alpha)}) = (e^{tA_\alpha}) = (\Phi((T_s),\ \pi_t^{(\alpha)}))$ erhält man somit, daß $(0,1]\ni\alpha \longrightarrow T_t^{(\alpha)}$ stetig ist etwa bezüglich der schwachen Operatortopologie, kompakt-gleichmäßig in t. Ist für $x\in\mathbb{B}$, $y\in\mathbb{B}'$ $t \longrightarrow \langle T_t x,y\rangle \in C_0(\mathbb{R}_+)$ $[C(\overline{\mathbb{R}}_+)]$ so gilt überdies $\langle T_t^{(\alpha)}x,y\rangle \xrightarrow[\alpha\to 0]{} e^{-t}\langle x,y\rangle$ $\left[\longrightarrow e^{-t}\langle x,y\rangle + (1-e^{-t})\lim_{s\to\infty}\langle T_s\, x,y\rangle\right]$.

⟦ Die Beweise folgen unmittelbar aus den in § 1 angegebenen Stetigkeitsaussagen ⟧.

5.3 Bemerkung.

Faßt man $(\pi_t^{(\alpha)})$, $0\leq\alpha\leq 1$, auf als Elemente der in § 2 studierten Halbgruppe $\mathcal{U}(\overline{\mathbb{R}}_+)$, identifiziert $(\pi_t^{(\alpha)})$ mit D_α und setzt $D_0 := \varepsilon_0$, $\tilde{\pi}_t^{(0)} = e^{-t}\varepsilon_0 + (1-e^{-t})\varepsilon_\infty$, so erhält man:

$\alpha \xrightarrow{\Psi} \pi_t^{(\alpha)}\left[\alpha \xrightarrow{\varphi} (-1)^{[\alpha]+1} D_\alpha\right]$ ist ein stetiger Homomorphismus der multiplikativen Halbgruppe $([0,1],\cdot) \longrightarrow (\mathcal{U}(\overline{\mathbb{R}}_+),\circledast)$ [resp. ein stetiger Homomorphismus der additiven Halbgruppe $(\mathbb{R}_+,+)$ in die Halbgruppe der Distributionen bezüglich der Faltung].

Es ist $\Psi(0) = (e^{-t}\cdot\varepsilon_0 + (1-e^{-t})\varepsilon_\infty)_{t\geq 0}$, $\Psi(1) = (\varepsilon_t)_{t\geq 0}$.

5.4. Proposition.

Sei $(T_t = e^{tA})$ eine Kontraktionshalbgruppe, $x\in\mathbb{B}$, $y\in\mathbb{B}'$, $x\in D_{A_\beta}$, weiter sei $g(s) := \langle e^{sA}x,y\rangle$. Dann gelten:

(i) $\mathbb{D}_\beta g(s) = \langle A_\beta e^{sA}x,y\rangle$, $\beta\geq 0$, $s\in\mathbb{R}_+$ insbesondere

(i') $\mathbb{D}_\beta g(0) = D_\beta(g) = \langle A_\beta\, x,y\rangle$.

(ii) Sei $g^{(\alpha)}(s) := \langle e^{sA_\alpha}x,y\rangle$, $0<\alpha<1$, so ist für $x\in D_{A_\alpha}\cap D_{A_{\alpha\beta}}$ und $\beta\in\mathbb{R}_+$: $\mathbb{D}_\beta g^{(\alpha)}(s) = \langle A_{\alpha\beta}e^{sA_\alpha}x,y\rangle$, insbesondere

(ii') $D_\beta(g^{(\alpha)}) = \langle A_{\alpha\beta}x,y\rangle$. Setzt man speziell $\beta = 1/\alpha$, so folgt

(ii") $D_{1/\alpha}(g^{(\alpha)}) = \langle A\, x,y\rangle$.

[Folgt unmittelbar aus der Definition der gebrochenen Potenzen von Erzeugern und den Eigenschaften der verallgemeinerten Differential=operatoren $\mathbb{D}_\alpha$].

Man kann tatsächlich zeigen, daß die Aussagen in 5.4 sogar bezüglich der starken Operatorentopologie richtig sind. Faßt man also die D_β und $\mathbb{D}_\beta$ als Operatoren auf den $\mathbb{B}$- wertigen C^1-Funktionen auf, so erhält man

(i) $(\mathbb{D}_\beta e^{\cdot A}x)(s) = A_\beta e^{sA}x,$

(i') $D_\beta(e^{\cdot A}x) = A_\beta x$

(ii) $(\mathbb{D}_\beta e^{\cdot A_\alpha}x)(s) = A_{\alpha\beta}\, e^{sA_\alpha}x$

(ii') $D_\beta(e^{\cdot A_\alpha}x) = A_{\alpha\beta}\, x,$

(ii") $D_{1/\alpha}(e^{\cdot A_\alpha}x) = Ax$

<u>5.5 Definition.</u>

Sei $(T_t = e^{tA})$ eine Kontraktionshalbgruppe. Sei B ein Generator einer weiteren Kontraktionshalbgruppe $(S_t = e^{tB})$, dann heißen Gleichungen des Typs

$$\mathbb{D}_\beta(T_\cdot x)(s) = BT_s x = T_s Bx \text{ (s.o.) resp.}$$

$$D_\beta(T_\cdot x) = B\,x,\; x \in D_B$$

(verallgemeinerte) Wellengleichungen und (T_t) heißt Lösung dieser Wellengleichung.

(Im Falle $\beta = 2$, $\mathbb{B} = C_0(\mathbb{R})$, $B = \Delta$ erhält man die Wellengleichung $-\frac{d^2}{dt^2} = \Delta$).

Sei $\mathcal{G}$ eine lokalkompakte Gruppe, $\mathbb{B} = \mathbb{C}_0(\mathcal{G})$, sei R_B der Faltungsoperator, der zu einer erzeugenden Distribution B mit erzeugter Halbgruppe (μ_t) gehört - wir werden R_B mit dem Generator $(R_B|\mathcal{D}(G))^-$ identifizieren - dann sagen wir: Eine stetige Halbgruppe (μ_t) mit erzeugender Distribution A erfüllt eine verallgemeinerte invariante Wellengleichung, falls

$$\mathbb{D}_\beta(\mathbb{R}_+ \ni t \longrightarrow \int f d\mu_t) = B(R\mu_s f) = \mu_s(R_B f) \text{ resp. } D_\beta(\mathbb{R}_+ \ni t \longrightarrow \int f d\mu_t) = B(f)$$

für $f\in\mathcal{D}(G)$ gilt.

(Für Subordination von Gaußmaßen auf Lie-Gruppen s. E. Stein[32]: $B=\Delta, \beta=2$)

5.6 Proposition.

a) Sei $\beta>1$. Jede verallgemeinerte Wellengleichung hat eine eindeutig bestimmte Lösung, nämlich ($T_t = S_t^{(1/\beta)} = e^{tB_{1/\beta}}$.)

b) Jede invariante Wellengleichung
$\mathbb{D}_\beta\,(t \longrightarrow \int_G f d\mu_t)(s) = B(R\mu_s f) = \mu_s(R_B f)$ hat eine eindeutig bestimmte Lösung, nämlich $(\int_{\mathbb{R}_+} \nu_s d\pi_t^{(1/\beta)} = \mu_t)$, wobei (ν_s) die von B erzeugte Halbgruppe ist, also $\mu_t = \nu_t^{(1/\beta)}$. Die erzeugende Distribution A von (μ_t) hat die Gestalt $A = C_{1/\beta}\int_{\mathbb{R}_+^x} (\nu_t - \varepsilon_e)\,\frac{dt}{t^{1+1/\beta}}$

Beweis:

a) Aus der Gleichung
$\mathbb{D}_\beta(T_\cdot x)(s) = BT_s x = T_s Bx, s\geq 0,\ x\in D_B$,
folgt für $s = 0$
$D_\beta(T_\cdot x) = A_\beta x\left[=-(-A)^\beta x\right] = Bx,\ x\in D_B$, also $A_\beta \supseteq B$. Da A_β und B maximal dissipativ sind, folgt $A_\beta = B$ und es ist bekanntlich dadurch A eindeutig bestimmt, nämlich $A = B_{1/\beta}$. Die Aussage b) folgt ganz analog. Dabei ist zu beachten, daß die Definition einer "invarianten Wellengleichung" etwas von der allgemeinen Form abweicht, da wir die Betrachtung auf $f\in\mathcal{D}(G)$ beschränkt hatten. Da aber $\mathcal{D}(G)$ für die erzeugenden Distributionen A stetiger Halbgruppen von Wahrscheinlichkeitsmaßen, (resp. ihre Faltungsoperatoren R_A) reich genug ist, („reich genug" soll heißen: $(R_A, \mathcal{D}(G))^-$ ist der volle Generator), folgt die Behauptung. □

5.7 Proposition.

$\mathbb{B}$ sei ein Hilbertraum, die Generatoren A,B seien wie in 5.6a Lösungen der Wellengleichung. A,B seien selbstadjungiert (somit negativ semidefinit), wie vorher seien $(S_t := e^{tA})_{t\geq 0}$, $(T_t := e^{tB})_{t\geq 0}$ definiert.

Es sei (E_λ) die Spektralschar von $B, B = -\int_{\mathbb{R}_+} \lambda dE_\lambda$, (F_λ) die Spektralschar von A, so ist wegen $A_\beta = -\int \lambda^\beta dF_\lambda = -\int \lambda dE_\lambda = B$, $E_\lambda = F_{\lambda^{1/\beta}}$, ebenso $S_t = \int e^{-\lambda t} dF_\lambda = \int \exp(-\lambda^{1/\beta} t)\, dE_\lambda$.

[Wurde allgemein für Subordinationshalbgruppen in 4.9 gezeigt]

5.8 Definition.

Sei wieder $\mathcal{G}$ eine lokalkompakte Gruppe, $(\mu_t)_{t\geq 0}$ sei eine symmetrische Gaußhalbgruppe und es sei mit $\alpha: 0<\alpha\leq 1$, $\mu_t^{(\alpha)} := \int \mu_s \, d\pi_t^{(\alpha)}(s)$, dann heiße $(\mu_t^{(\alpha)})$ eine Halbgruppe Bochner-stabiler Maße. Anders ausgedrückt: (ν_t) heißt Bochner-stabil, wenn es einen symmetrischen Gaußgenerator B und ein $\alpha: 0<\alpha\leq 1$ gibt, so daß (ν_t) der Wellengleichung $\mathbb{D}_{1/\alpha}(t \longrightarrow \nu_t(f)) = B(f), f \in \mathcal{D}(\mathcal{G})$ genügt.

Bemerkung: Für $\mathcal{G} = \mathbb{R}$ sind alle symmetrisch stabilen Verteilungen von diesem Typ (s. W. Feller [8,9]). Bisher wurden in der Literatur zumeist die sogenannten operatorstabilen Verteilungen studiert, Halbgruppen (ν_t) von W-Maßen (zumeist auf $G = \mathbb{R}^k$), bei denen kurz gesagt, die "Wurzeln" ν_t aus ν_1 durch einen Homomorphismus von $\mathbb{R}$ in die Automorphismen $\varphi: t \in \mathbb{R} \longrightarrow \mathrm{Aut}(G)$ gewonnen werden. Tatsächlich besteht ein Zusammenhang zwischen Subordinationen, Bochner-stabilen- und operatorstabilen Verteilungen, wie die Lévy-Hinčin-Darstellungen der letzteren zeigen (s. z.B. K. Schmidt [27,29]), im Rahmen dieser Arbeit können wir darauf aber nur am Rande eingehen, und zwar im Zusammenhang mit Verteilungen von lokalem Typ, s.u.

Das Interesse an den Bochner-stabilen Maßen ist leicht erklärbar, wenn man bedenkt, daß invariante Wellengleichungen unter Gruppenhomomorphismen invariant bleiben. Sei somit $\varphi: \mathcal{G} \longrightarrow \mathcal{G}_1$ ein Homomorphismus lokalkompakter Gruppen, seien $\mathcal{B}(\mathcal{G})$ resp $\mathcal{B}(\mathcal{G}_1)$ die Klassen der B-stabilen Verteilungen, so ist offenbar $\varphi(\mathcal{B}(\mathcal{G})) \subseteq \mathcal{B}(\mathcal{G}_1)$. Betrachtet man dagegen die (symmetrischen) stabilen Verteilungen auf $\mathcal{G} = \mathbb{R}$,

φ: $\mathbb{R} \longrightarrow \mathbb{R}/\mathbb{Z}$ =: T sei der kanonische Homomorphismus, dann ist das φ-Bild stabiler Verteilungen wieder B-stabil auf T, aber nicht operatorstabil. (In $M^1(T)$ können nur triviale operatorstabile Verteilungen existieren, da Aut(T) diskret ist und somit keine Einparameterhalbgruppen enthält).

Es folgt nun eine besondere Darstellung der gebrochenen Potenzen als Limes verallgemeinerter Differenzenoperatoren; dies ermöglicht es, die Struktur von A und A_α in Beziehung zu setzen:

5.9 Hilfssatz.

(U. Westphal [36]). Sei $(T_t = e^{tA})_{t\geq 0}$ eine C_o-Kontraktionshalbgruppe, dann ist $Ax = \lim\limits_{n\to\infty} n(T_{1/n} - I)x =: \lim A^{(n)}x$ für $x\in D_A$. Sei $\alpha > 0$, Dann kann man A_α als Limes der verallgemeinerten Differenzenoperatoren

$A_\alpha x = \lim\limits_{n\to\infty} A_\alpha^{(n)} x$ darstellen, wobei man für $n\in\mathbb{N}$ definiert.

$$A_\alpha^{(n)} := -\sum_{j=0}^{\infty} n^\alpha \binom{j-\alpha-1}{j} T_{j/n} = -\left[-n(T_{1/n} - I)\right]^\alpha = (A^{(n)})_\alpha .$$

[Für $0<\alpha\leq 1$ sind $A_\alpha^{(n)}$ und $e^{tA_\alpha^{(n)}}$ offenbar durch Subordinationen darstellbar, aus unseren früheren Überlegungen erhalten wir, daß $(e^{tA_\alpha^{(n)}}) \xrightarrow[n\to\infty]{\tau_{co}} (e^{tA_\alpha})$, nicht aber unmittelbar die obige Aussage]

Nun kehren wir zurück zu Maßhalbgruppen und erhalten mit den vorher vereinbarten Bezeichnungen:

5.10 Satz.

Sei (μ_t) eine stetige Faltungshalbgruppe mit erzeugender Distribution A, sei $0<\alpha\leq 1$, A_α sei die erzeugende Distribution von $(\mu_t^{(\alpha)})$, so ist für $f\in\mathcal{D}(G)$

$$A_\alpha(f) = \lim_{n\to\infty} n^\alpha \sum_{j=0}^{\infty} -\binom{j-\alpha-1}{j}\mu_{j/n} = \lim_{n\to\infty} n^\alpha\left[\sum_{j=1}^{\infty} -\binom{j-\alpha-1}{j}\mu_{j/n} - \varepsilon_e\right](f).$$

5.11 Folgerung.

Sei A eine Gaußsche Distribution, $o<\alpha\leq 1$, A_α sei wie vorhin die erzeugende Distribution der untergeordneten Halbgruppe $(\mu_t^{(\alpha)} =: \nu_t)$. Dann gilt:

Für alle $f\in\mathcal{D}(G)$ mit $0\leq f$, und $f \equiv 0$ in einer Umgebung von e, ist

$$n^{1/\alpha}\left(\sum_{j=1}^{\infty} -\binom{j-1/\alpha-1}{j} \nu_{j/n}\right)(f) \xrightarrow[n\to\infty]{} 0,$$

umso mehr

$$n^{1/\alpha}\nu_{1/n}(f) \xrightarrow[n\to\infty]{} 0$$

[Analog $\frac{1}{t^{1/\alpha}} \sum_{j=1}^{\infty} -\binom{j-1/\alpha-1}{j} \nu_{jt}(f) \xrightarrow[t\downarrow 0]{} 0$

resp. $\frac{1}{t^{1/\alpha}} \nu_t(f) \xrightarrow[t\downarrow 0]{} 0$].

5.12 Definition.

Eine Halbgruppe (ν_t) heißt von lokalem Typ der Ordnung α, $o<\alpha\leq 1$, falls für alle $f\in\mathcal{D}(G)_+$, $f \equiv 0$ in einer Umgebung von e,

$$\frac{1}{t^{1/\alpha}} \sum_{j=1}^{\infty} -\binom{j-1/\alpha-1}{j} \nu_{tj}(f) \xrightarrow[t\to 0]{} 0,$$

(ν_t) heißt von lokalem Typ der Ordnung α im engeren Sinne, wenn

$$\frac{1}{t^{1/\alpha}} \nu_t(f) \xrightarrow[t\to 0]{} 0 .$$

5.13

Wir erhalten somit:

Die Bochner-stabilen Verteilungen sind (einbettbar in) Maßhalbgruppen vom lokalen Typ (im engeren Sinne) der Ordnung α, $o<\alpha\leq 1$.

Für $\alpha = 1$ erhält man die Halbgruppen von lokalem Typ, also die Gaußhalbgruppen [13]. Umgekehrt gibt es aber Halbgruppen von lokalem Typ, die nicht Gaußsch und nicht stabil (aber auch nicht symmetrisch) sind:

[Man setzt etwa $\mathcal{G} := \mathbb{R}$, $A := \frac{1}{2}\frac{d^2}{dx^2} + a\frac{d}{dx}$, $0<\alpha<1$. Es sei $\mu_t := N(at,t)$ und $\mu_t^{(\alpha)} := \int_{\mathbb{R}_+} \mu_s d\pi_t^{(\alpha)}(s)$ mit der Fouriertransformierten $\int_{\mathbb{R}_+} e^{ixy} d\mu_t^{(\alpha)}(x) = \text{const.}\exp(t[iay - \frac{y^2}{2}]^\alpha)$,

dann ist $(\mu_t^{(\alpha)})$ nicht Gaußsch und auch nicht stabil (im gewöhnlichen Sinne), wohl aber von lokalem Typ!]

5.14 Vermutung.

Die bisher in der Literatur betrachteten (operator)stabilen Verteilungen lassen sich durch Verteilungen lokalen Typs beschreiben. (Dabei ist notfalls der Begriff "Verteilung lokalen Typs" geeignet zu modifizieren). Die Vermutung stützt sich darauf, daß aus der Definition der operatorstabilen Verteilungen (s. K. Schmidt [29], K.R. Parthasarathy, K. Schmidt [27] und die dort zitierte Literatur) die Existenz eines stetigen Homomorphismus $\Psi: \mathbb{R}_+ \longrightarrow \mathrm{Aut}(G)$ folgt, so daß, abgesehen von einem Translationsanteil, $\mu_t = \Psi(t)(\mu_1)$.
Also: Die Wurzeln μ_t werden durch einen Automorphismus aus μ_1 gewonnen. Die explizite Gestalt der Lévy-Hinčin-Formel als Mischungsintegral über die Bahnen von $\Psi(\cdot)$ läßt einen Zusammenhang zu Wellengleichungen erkennen.

5.15 Zusammenfassend:

Die Gaußverteilungen sind definiert als Lösungen einer Wärmeleitungsgleichung $\frac{d}{dt}\cdot = \Delta\cdot$, die Bochner-stabilen Verteilungen als Lösun-

gen einer Wellengleichung, (wobei nun Δ als symmetrisch voraus gesetzt wurde) die Verteilungen lokalen Typs umfassen diese Klassen.
Um nicht symmetrische Bochner-stabile Verteilungen zu definieren, müßte man statt der Subordinationen mit $\pi_t^{(\alpha)}$ allgemeiner mit Riesz'schen Potentialen arbeiten (s. W. Feller [9]). Also, ausgehend von symmetrischen Gaußhalbgruppen müßte man (wie für $G = \mathbb{R}$) mittels eines allgemeineren Subordinationsbegriffs (M. Riesz Potentiale), der die gebrochenen Potenzen und verallgemeinerte Differentation umfaßt, nicht symmetrische Verteilungen konstruieren. Analog könnte man die Theorie der gebrochenen Potenzen von Gruppengeneratoren (U. Westphal [35], Teil II) zur Definition allgemeinerer Bochner-stabilen Verteilungen heranziehen.
Dies führt aber von dem Ziel, das wir uns in dieser Arbeit gesteckt hatten fort, die Erörterung dieser Fragen müssen einer anderen Arbeit vorbehalten werden.

Abschließend bringen wir als Anwendungsbeispiel für die Konvergenzsätze des § 1 (Proposition 1.24 etc.) einige Grenzwertsätze für Verteilungen resp. für Operatorhalbgruppen:

5.16

Sei $(T_t = e^{tA})_{t\geq 0}$ eine C_0-Kontraktionshalbgruppe, sei B_n eine Folge von Kontraktionen und $\alpha_n \geq 0$ mit $(\exp t\alpha_n(B_n - I))_{t\geq 0} \longrightarrow (e^{tA})$, sei C_n eine Folge von Kontraktionen, $\tau_n \downarrow 0$ mit $(C_n^{[t\tau_n]}) \longrightarrow (T_t)$.
Seien (G_n), (F_n) Folgen von Wahrscheinlichkeitsmaßen auf $\mathbb{R}_+$, $a_n, b_n \geq 0$ mit $(\exp t a_n(F_n - \varepsilon_0))_{t\geq 0} \xrightarrow[n\to\infty]{} (\pi_t^{(\alpha)})_{t\geq 0}$,
resp. $(G_n^{[tb_n]})_{t\geq 0} \longrightarrow (\tilde{\pi}_t^{(\alpha)})_{t\geq 0}$.

Dann gelten:

a) $\left(\int_{\mathbb{R}_+} \exp s\alpha_n(B_n - I)\, d(\exp t a_m(F_m - \varepsilon_0)(s))\right)_{t\geq 0} \xrightarrow{n,m\to\infty} \left(T_t^{(\alpha)} := \int T_s\, d\pi_t^{(\alpha)}(s)\right)_{t\geq 0}$.

b) $\left(\int_{\mathbb{R}_+} \exp s\alpha_n(B_n - I)\, dG_m^{[tb_m]}(s)\right)_{t\geq 0} \xrightarrow{n,m\to\infty} (T_t^{(\alpha)})_{t\geq 0}$

c) $\left(\int_{\mathbb{R}_+} C_n^{[s\tau_n]}\, d\exp t\alpha_m (F_m - \varepsilon_0)(s)\right)_{t\geq 0} \longrightarrow (T_t^{(\alpha)})_{t\geq 0}$

d) $\left(\int_{\mathbb{R}_+} C_n^{[s\tau_n]}\, dG_m^{[tb_m]}(s)\right)_{t\geq 0} \longrightarrow (T_t^{(\alpha)})_{t\geq 0}.$

Eine Reihe von Grenzwertsätzen lassen sich nun ableiten, in dem man für B_n, C_n, G_n, F_n speziell einsetzt, z.B.

$C_n, B_n := T_{1/n}, \alpha_n = n, \tau_n = \frac{1}{n}$, bzw. $C_n, B_n := \lambda_n \int_{\mathbb{R}_+} e^{-\lambda_n s}\, T_s ds,\ \alpha_n = \lambda_n \to \infty$

$G_n, F_n := \pi_{1/n}^{(\alpha)}$.

Setzt man für $(T_t)_{t\geq 0}$ etwa Maßhalbgruppen resp. deren Generatoren ein, so erhält man spezielle Grenzwertsätze für Wahrscheinlichkeitsmaße, z.B:

5.17

a) $\left(\int_{\mathbb{R}_+} \exp sn\,(\varepsilon_{1/n} - \varepsilon_0)\, d\pi_t^{(\alpha)}(s)\right)_{t\geq 0} \longrightarrow \left(\int \varepsilon_s\, d\pi_t^{(\alpha)}(s) = \pi_t^{(\alpha)}\right)_{t\geq 0}$

i.e. für $f \in C(\mathbb{R}_+)$ ist

$$\int_{\mathbb{R}_+} f(x)\, d\pi_t^{(\alpha)}(x) = \lim_{n\to\infty} \sum_{k=0}^{\infty} \frac{n^k}{k!} f\left(\frac{k}{n}\right) \int_{\mathbb{R}_+} s^k \cdot e^{-ns}\, d\pi_t^{(\alpha)}(s) .$$

b) $\left(\int_{\mathbb{R}_+} \rho_\lambda^{[s/\lambda]}\, d\pi_t^{(\alpha)}(s)\right)_{t\geq 0} \xrightarrow[\lambda\to\infty]{} (\pi_t^{(\alpha)})_{t\geq 0}$

dabei ist ρ_λ die Exponentialverteilung mit der Dichte

$\lambda \cdot e^{-\lambda x} \cdot 1_{[0,\infty)}(x)$ (und somit $(\rho_\lambda)^{[s/\lambda]} \xrightarrow[\lambda\to\infty]{} (\varepsilon_s)$.

i.e. für $f \in C(\mathbb{R}_+)$ ist

$$\int f d\pi_t^{(\alpha)} = \lim_{\lambda\to\infty} \sum_{k=0}^{\infty} \left(\int_{\mathbb{R}_+} f(x)\, d\rho_\lambda^k(x)\right) \cdot \pi_t^{(\alpha)}\left\{\left[\frac{k}{\lambda}, \frac{k+1}{\lambda}\right)\right\} .$$

5.18

Seien (Ω, Σ, P) ein Wahrscheinlichkeitsraum, seien $(f_i)_{i=1}^{\infty}$ eine Folge unabhängiger reeller identisch verteilter zufälliger Variabler mit

$E(f_i) = 0$, $V(f_i) = 1$ und Verteilung $f_i(P) =: \mu$. Weiter sei $\tau_a : x \to xa$.
Sei $0<\alpha\leq 1$ und sei $(\sigma_t^{(\alpha)})_{t\geq 0}$ die Halbgruppe der symmetrischen stabilen Verteilungen auf $\mathbb{R}_+$ mit Index 2α. Dann gilt der zentrale Grenzwertsatz, i.e.

$[\tau_{1/n}(\mu)]^{[t/n]} \longrightarrow N(0,t)$ und daher für $f \in C(\mathbb{R})$

$$\text{z.B.}\ \sum_{k=0}^{\infty} \left[\int_{\mathbb{R}} f(x)d(\tau_{\frac{1}{n}}(\mu))^n(x)\right] \pi_t^{(\alpha)}\{[\tfrac{k}{n}, \tfrac{k+1}{n})\} \longrightarrow \int f(x)d\sigma_t^{(\alpha)}(x) .$$

Weitere Anwendungsmöglichkeiten zeichnen sich ab, wenn man Evolutionsfamilien betrachtet, i.e. Familien von Kontraktionen $\{U_{s,t}, 0\leq s\leq t\}$, die stetig von s und t abhängig sind mit

(i) $U_{s,s} = I$, $s\geq 0$, (ii) $U_{s,t}U_{t,v} = U_{s,v}$, $0\leq s\leq t\leq v$.

Als Subordination bieten sich Mischungen der Form $\int_{\mathbb{R}_+} U_{0,s}d\lambda(s)$ an.

Wir interessieren uns für folgenden Fall:

Seien $0\leq t_1^{(n)} \leq t_2^{(n)} \leq \ldots \leq t_{k_n}^{(n)} \leq \ldots \uparrow \infty$ Unterteilungen von $\mathbb{R}_+$

mit $\max\limits_{i\leq j\leq k_n(T)} |t_i^{(n)} - t_{i+1}^{(n)}| \xrightarrow[n\to\infty]{} 0$, $T>0$.

Dabei sei für jedes $T>0$ $k_n(T)$ so gewählt, daß $t_{k_n(T)} \leq T < t_{k_n(T)+1}$

Sei $(U_k^{(n)})_{k\in\mathbb{N}}$ eine Folge von Kontraktionen

mit $\max\limits_{1\leq i\leq k_n(T)} \|(U_k^{(n)} - I)x\| \longrightarrow 0$, $x\in B$, $T>0$.

Es sei $T_{s,t}^{(n)} := U_j^{(n)} \ldots U_k^{(n)}$, falls $t\in[t_k^{(n)}, t_{k+1}^{(n)})$, $s\in[t_j^{(n)}, t_{j+1}^{(n)})$.

Also ist $t \to T_{s,t}^{(n)}$ eine stückweise stetige Abbildung, die für $0\leq s\leq t\leq r$

$T_{s,t}^{(n)}\ T_{t,r}^{(n)} = T_{s,r}^{(n)}$, erfüllt. (I.e. $(T_{s,t}')$ ist eine (unstetige) Evolutionsfamilie.)

Nun setzten wir zusätzlich voraus, daß die $(T_{s,t}^{(n)})$ so gewählt sind, daß eine Evolutionsfamilie $(T_{s,t})$ existiert mit $T_{s,t}^{(n)} \xrightarrow[n\to\infty]{} T_{s,t}$ in der starken Operatorentopologie, kompakt-gleichmäßig in s,t.

Nun kann man wieder Mischungen erklären,

$\int_{\mathbb{R}_+} T_{o,t}\, d\lambda(t)$ resp. $\int_{\mathbb{R}_+} T_{o,t}^{(n)}\, d\lambda(t)$, $\lambda \in M^1(\mathbb{R}_+)$ und die Konvergenzsätze aus § 1 lassen sich übertragen. Wir beschränken uns auf ein Anwendungsbeispiel:

5.19

$\mathfrak{G}$ sei eine lokalkompakte Gruppe, $(\mu_{n,k})_{1\leq k\leq k_n}$ sei ein infinitesimales Dreieckssystem von Wahrscheinlichkeitsmaßen, so daß

$\mu_n := \mu_{n,1} * \ldots * \mu_{n,k_n}$ gegen ein Wahrscheinlichkeitsmaß μ konvergieren. Zusätzlich setzen wir voraus, daß das Dreieckssystem so ergänzt wird zu einem Schema $(\mu_{n,k})_{k,n\geq o}$, so daß zu jedem $t>o$ eine Folge $k_n(t)\uparrow\infty$ natürlicher Zahlen existiert, so daß

$\prod_1^{k_n(t)} \mu_{n,k} \longrightarrow \mu_{0,t} \in M^1(\mathfrak{G})$. Analog für $s<t$ $\prod_{k_n(s)+1}^{k_n(t)} \mu_{n,k} \longrightarrow \mu_{s,t}$,

so daß $(\mu_{s,t})$ eine Evolutionsfamilie ist (s. z.B. H. Heyer [13], Strook, Varadhan [33], J. Feinsilver [7]).

Strook und Varadhan geben Bedingungen an, unter denen $(\mu_{s,t})$ Gaußsch (in einem verallgemeinerten Sinne) ist.

Dann folgt z.B. aus den vorhergehenden Überlegungen, daß

$$\int \mu_{0,t}\, d\lambda(t) = \lim_{n\to\infty} \sum_{k=1}^{\infty} \lambda(\{t : k_n(t)=k\}) \prod_{j=1}^{k} \mu_{n,j}$$

insbesondere etwa

$$\int \mu_{0,t}\, d\pi_s^{(\alpha)}(t) = \lim_{n\to\infty} \pi_s^{(\alpha)}(\{k_n(t)=k\}) \prod_{j=1}^{k} \mu_{n,j}\, .$$

Setzt man nun zusätzlich voraus, daß

$\mu_{s,t} = \mu_{0,t-s}$ ist, i.e. daß $t \longrightarrow \mu_{0,t}$ eine (Gaußsche) Faltungshalb gruppe bildet, so ist damit ein Grenzwertsatz für stabile Verteilungen gegeben.

5.20 Beispiel.

Spezialisiert man und setzt $G := \mathbb{R}$, $(\mu_{n,k})$ sei eine Doppelfolge von symmetrischen Wahrscheinlichkeitsmaßen, die Bedingungen vom Lindeberg-Feller Typ erfüllen, so daß für jedes $t > 0$ eine Folge $k_n(t) \uparrow \infty$ existiert mit

$$\prod_{j=k_n(s)+1}^{k_n(t)} \mu_{n,j} \longrightarrow N(0,t-s),\quad 0 \leq s < t.$$

Dann erhält man die symmetrisch-stabilen Verteilungen $\sigma_t^{(\alpha)}$ (definiert durch die Fouriertransformierten $\hat{\sigma}_t^{(\alpha)}(y) = C(\alpha)\exp - t\left(\frac{y^2}{2}\right)^{\alpha}$, $0 < \alpha \leq 1$)

als Grenzwert von konvexen Kombinationen

$$\sigma_t^{(\alpha)} = \int_{\mathbb{R}_+} N(0,s)\,d\pi_t^{(\alpha)}(s) = \lim_{n\to\infty} \sum_{k=1}^{\infty} \pi_t^{(\alpha)}\{t : k_n(t) = k\}\, \prod_{j=1}^{k} \mu_{n,j}.$$

Wichtigste Bezeichnungen und Symbole

(soweit es sich nicht um Standardbezeichnungen handelt und sie etwa dem Buch von H. Heyer [13] entnommen werden können).

$T(P)$	Bild eines Maßes P unter der meßbaren Abbildung T
$\mathbb{R} / \mathbb{R}_+ / \mathbb{Z} / \mathbb{Z}_+ / \mathbb{N} / \overline{\mathbb{R}_+}$:=	reelle Gerade/ $t \in \mathbb{R} : t \geq o$/ ganze Zahlen/ $k \in \mathbb{Z} : k \geq o$/ natürliche Zahlen/ $\mathbb{R}_+ \cup \{\infty\}$
$M(X)/M_+(X)/M^1(X)$	beschränkte Maße/ nicht negative beschränkte Maße/Wahrscheinlichkeitsmaße . (Maße sind stets als straff zu verstehen).
μ, ν, ε_x	Maße resp. Punktmaß in $x \in X$
R_μ	Faltungsoperator, $R\mu\, f(x) = \int f(xy)\, d\mu(y)$
$\mu\, \nu$	Faltung von μ und ν
$C(X)/C_0(X)$	Banachraum der beschränkten, stetigen Funktionen/ die in ∞ verschwinden
$\mathcal{D}(G)$	Testfunktionen (im Sinne von Bruhat)
$(\mu_t)_{t \geq 0}$	Faltungshalbgruppe
(T_t)	C_0- Kontraktionshalbgruppe
$\mathcal{V}$	affine Halbgruppe
$\preceq_s$	Ordnungsrelation für Homomorphismen $X \longrightarrow \mathcal{V}$ s. §3
$\mathcal{M}(X, \mathbb{E}; \mathbb{F}), \mathcal{M}_0(\ldots), \mathcal{M}_1(\ldots) \mathcal{M}_2(\ldots)$	s. Definition 1.3, 1.4, 1.9.
τ_{co}	Topologie in $\mathcal{M}_2(\ldots)$, s. § 1
$\Phi(\varphi, \mu)$	$:= \int \varphi\, d\mu$
$\mathbb{B} / \mathcal{L}(\mathbb{B})$	Banachraum/ beschränkte Operatoren auf $\mathbb{B}$
$\mathcal{U}(\overline{\mathbb{R}_+})$, $\odot$	unendlich teilbare Maße auf $[o, \infty]$ versehen mit der durch Subordination definierten Komposition $\odot$,s.§2
η_F / Γ.	Lévy-Maß / Lévy Abbildung.

Literatur

1 Berens, H. - Westphal, U.: A Cauchy problem for a generalized wave equation. Acta Sc. Math. 29, 93-106 (1968).

2 Bochner, S.: Harmonic analysis an the theory of probability. Univ. of California Press, Berkeley-Los Angeles 1955.

3 Bourbaki, N.: Eléments de Mathématique XXV: Livre VI: Intégration. Chapitre 6. Actual. Scient. Ind. 1281. Paris: Hermann 1959.

4 Carnal, H.: Unendlich oft teilbare Wahrscheinlichkeitsverteilungen auf kompakten Gruppen. Math. Ann. 153, 351-383 (1964).

5 Cohen,J.W.: Derived Markov Chains I-III, Indag. Math. 24, 55-70, 71-81, 82-92, (1962).

6 Collins, H. S.: Remarks on affine semigroups. Pacific. J. Math.12, 449-455 (1962).

7 Feinsilver, Ph.: Processes with independent increments on a Lie group. To appear in Trans. Amer. Math. Soc. (1978).

8 Feller, W.: An introduction to probability theory. Volume II. New York - London - Sydney: John Wiley & Sons 1966.

9 Feller, W.: On a generalization of the Marcel Riesz potentials and the semigroups generated by them. Comm. Sem. Math. Lund. 73-81 (1952).

10 Glicksberg, I.: Weak compactness and separate continuity. Pacific. J. Math. 11, 205-214, (1961).

11 Hazod, W.: Subordination von Gauß- und Poissonmaßen. Z. Wahrscheinlichkeitstheorie und verw. Gebiete 35, 45-55 (1976).

12 Hazod, W.: Stetige Halbgruppen von Wahrscheinlichkeitsmaßen und erzeugende Distributionen. Lecture Notes in Math. Vol. 595. Berlin-Heidelberg-New York: Springer 1977.

13 Heyer, H.: Probability measures on locally compact groups. Ergebnisse der Mathematik und ihrer Grenzgebiete 94, Berlin-Heidelberg-New York: Springer 1977.

14 Hille, E., Phillips, R.S.: Functional analysis and semigroups. Amer. Math. Soc. Colloquium Publications, Vol. 31. Revised edition. Providence, R. I, Amer. Math. Soc. 1957.

15 Hirsch, F.: Intégrales de résolvantes et calcul symbolique. Ann. Inst. Fourier, Grenoble 22, 4 239-264 (1972).

16 Hofmann, K.H., Mostert, P.S.: Elements of compact semigroups. Columbus, Ohio: Charles E. Merrill Books, Inc. 1966.

17 Huff, B.W.: The strict subordination of differential processes. Sankhyā Indian J. Statistics Ser. A. 31, 403-412 (1969).

18 Huff, B.W.: Superpositions and the continuity of infinitely divisible distributions. Ann. Math. Stat. 42, 410-412 (1971).

19 Huff. B.W.: Comments on the continuity of distribution functions obtained by superposition. Proc. Amer. Math. Soc. 27, 141-146 (1971).

20 Kato, T.: Perturbation theory for linear operators. Berlin-Heidelberg-New York, Springer 1966.

21 Kendall, D.G.: On Markov groups, Proc. 5th Berkeley-Symposiom on Mathematical Statistics and Probability Vol. II, 2. 165-173, (1967).

22 Khurana, S.S.: Weak integration of vector-valued functions.Journal Indian Math. Soc. 39, 155-166 (1975).

23 Komatsu, H.: Fractional powers of operators. Pacific J. Math. Vol. 19, 285-346, 1966.

24 Nelson, E.: A functional calculus using singular Laplace integrals. Trans Amer. Math. Soc. 88, 400-413 (1958).

25 Nollau, V.: Über gebrochene Potenzen infinitesimaler Generatoren Markovscher Übergangswahrscheinlichkeiten I, Mathematische Nachrichten, Bd. 65, 235-246, (1975).

26 Nollau, V.: Über gebrochene Potenzen infinitesimaler Generatoren Markoffscher Übergangswahrscheinlichkeiten.II. Math. Nachr. 72, 99-107, (1976).

27 Parthasarathy, K. R., Schmidt, K.: Stable positive definite functions. Trans. Amer. Math. Soc. 203, 161-174 (1975).

28 Phillips, R.S.: On the generation of semigroups of linear operators. Pacific J. Math. 2, 343-369 (1952).

29 Schmidt, K.: Stable probability Measures on $\mathbb{R}^d$, Z. Wahrscheinlichkeitstheorie verw. Gebiete 33, 19-31 (1975).

30 Siebert, E.: Über die Erzeugung von Faltungshalbgruppen auf beliebigen lokalkompakten Gruppen. Math. Z. 131, 313-333 (1973).

31 Siebert, E.: Absolut-Stetigkeit und Träger von Gauß-Verteilungen auf lokalkompakten Gruppen. Math. Ann. 210, 129-147 (1974).

32 Stein, E.M.: Topics in harmonic analysis related to the Littlewood-Paley theory. Ann. of Math. Studies No. 63, Princeton University Press 1970.

33 Stroock, D. W.:Varadhan, S.R.S.: Limit theorems for random walks on Lie groups. Sankhyā Ser. A. 35, 277-294 (1973).

34 Tortrat, A.: Mélange de lois et lois indéfiniment divisibles. Proc. Fourth Conf. Probability Theory, Sept. 1971, Braşov, Romania, Ed. Acad. R. S. R., 1973.

35 Westphal, U.: Ein Kalkül für gebrochene Potenzen infinitesimaler Erzeuger von Halbgruppen und Gruppen von Operatoren. Teil I,II Compositio Math. 22,67-103, 104-136, (1970).

36 Westphal, U.: An approach to fractional powers of operators via fractional differences. Proc. London Math.Soc. (3) 29, 557-576 (1974).

37 Woll jr., J.W.: Homogeneous stochastic processes, Pacific. J. Math. 9, 293-325 (1959).

Wilfried Hazod
Universität Dortmund
Abteilung Mathematik
Postfach 500 500
D-4600 Dortmund 50

MARCHES ALEATOIRES SUR LES ESPACES HOMOGÈNES

H. Hennion

Soit G un groupe topologique localement compact à base dénombrable et H un sous-groupe fermé de G, on désigne par M l'espace homogène G/H et par π la surjection canonique de G sur M : G opère à gauche sur M par la formule $g\pi(k) = \pi(gk)$. Soit maintenant μ une probabilité adaptée sur G (i.e. telle que le plus petit sous-groupe fermé contenant le support de π soit égal à G) et $(X_n)_{n=1}^{\infty}$ une suite de variables aléatoires indépendantes à valeurs dans G, de loi μ, définies sur un espace de probabilité (Ω,F,P), on appelle marche aléatoire (m.a.) de loi μ sur M la chaine de Markov $Y = (Y_n^x)_{x\in M, n\in \mathbb{N}}$,

$$Y_n^x = X_n \dots X_1 x \quad \text{par} \quad n \geq 1 \;, \quad Y_0^x = x \;,$$

obtenue par l'action sur M de la m.a. gauche de loi μ sur G; sa probabilité de transition et son potentiel s'écrivent pour f borélienne positive sur M:

$$Pf(x) = \int f(gx)\mu(dg) = \int f\circ\pi(gk)\mu(dg)$$

$$Uf(x) = \sum_{n=0}^{\infty} P^n f(x) = \int f\circ\pi(gk)\Big(\sum_{n=0}^{\infty}\mu^n\Big)(dg) \quad \text{si } x = \pi(k) \;.$$

Si H est distingué dans G, la chaine Y n'est autre que la m.a. de loi $\pi(\mu)$ sur le groupe G/H, mais, dans le cas général, le comportement d'une m.a. sur un espace homogène diffère sensiblement de celui d'une m.a. sur un groupe. Ainsi, dans le cas d'une m.a. sur un groupe, on a la dichotomie suivante: soit la m.a. visite p.s. une infinité de fois tout ouvert non vide et le potentiel de ces ouverts est infini, soit la m.a. visite p.s. un nombre fini de fois tout compact et le potentiel de ces compacts est borné; comme le montre l' exemple qui suit; ceci ne subsiste pas en général dans les espaces homogènes.

Soit L le groupe des transformations affines de $\mathbb{R}$, i.e. le produit $\mathbb{R}_+^* \times \mathbb{R}$ muni de la loi $(a,b)(a',b') = (aa',b+ab')$, et $H = \mathbb{R}_+^* \times \{o\}$ le sous groupe des homothèties, $M = G/H$ s'identifie à $\mathbb{R}$ et L agit sur $\mathbb{R}$ par la formule $(a,b)x=ax+b$.

Donnons nous μ adaptée sur L telle que support $(\mu) \subset]0,\frac{1}{2}] \times [-1,+1]$ et posons $X_n = (A_n, B_n)$, nous avons donc $0 < A_n \leq \frac{1}{2}$ et $|B_n| \leq 1$.
La position à l'instant n de la m.a. de loi μ sur $\mathbb{R}$, issue de x , est:

$$Y_n^x = (A_n, B_n) \dots (A_1, B_1)x = A_n \dots A_1 x + B_n + A_n B_n + \dots + A_n \dots A_2 B_1 ,$$

et

$$|Y_n^x| \leq \frac{|x|}{2n} + 1 + \frac{1}{2} + \dots + \frac{1}{2n-1} \leq \frac{|x|}{2^n} + 2 ,$$

de sorte que

$$U\, 1_{[-2,+2]}(0) = +\infty$$

tandis que

$$U\, 1_k(0) = 0$$

si $K \cap [-2,+2] = \emptyset$,

Dans ce paragraphe nous énonçons un théorème de dichotomie pour les m.a. sur les espaces homogènes en faisant l'hypothèse que M possède une mesure σ-finie m relativement invariante par G , i.e. telle que, pour tout $g \in G$, $g(m) = \chi(g^{-1}).m$, où χ est un caractère continue de G .

Rappelons que μ est dite étalée si il existe n_0 tel que μ^{*n_0} ne soit pas étrangère à une mesure de Haar de G et notons

$$h_A(x) = P([\sum_{n=1}^{\infty} 1_A(Y_n^x) = +\infty])$$

pour A borelien de M .

Théorème 1. [4]

Supposons μ étalée et $\int \chi(g^{-1})\mu(dg) = 1$ alors, pour la m.a. de loi μ sur M :

-soit Y est m-récurrente au sens de Harris, i.e. $m(A) > 0$ implique $h_A \equiv 1$ et $m(A) = 0$ implique $h_A = 0$,

-soit le potentiel de tout compact est borné.

Lorsque G est discret, la mesure de comptage sur M est G-invariante et les hypothèses de ce théorème sont toujours vérifiées, dans ce cas, la preuve en est notablement plus simple et mérite d'être traitée en détail ici.

Proposition.

Si G est discret alors, pour la m.a. de loi μ sur M,

-soit tout point est récurrent et il n'y a qu'une seule classe de récurrence,

-soit tous les points sont transitoires.

Démonstration.

Soit $\check{\mu}$ l'image de μ par l'application $g \to g^{-1}$ et $\check{P}$ la m.a. de loi $\check{\mu}$ sur M, on a clairement pur tout $x,y \in M$:

$$P^n(x,y) = \mu^x(\{g:gx=y\}) = \check{\mu}^n(\{g:gy=x\}) = \check{P}^n(y,x)$$

et

$$U(x,y) = \check{U}(y,x) ,$$

il en résulte, en particulier, que $x \in M$ est P-récurrent si et seulement si il est $\check{P}$-récurrent.

Donnons nous $x \in M$; x P-récurrent, désignons par C_x sa classe de récurrence et notons T_μ le semi-groupe engendré par le support de μ :

i) si $g \in T_\mu$ on a $U(x,g\ x)) > 0$, par suite $g(x)$ est P-récurrent et appartient à C_x, ce que nous résumons par $T_\mu x \subset C_x$,

ii) si $g \in T_\mu^{-1}$ on a $0 < U(g(x),x) = \check{U}(x,g(x))$, puisque x est $\check{P}$-récurrent $g(x)$ est $\check{P}$-récurrent donc P-récurrent et dans C_x, ceci s'écrit $T_\mu^{-1} x \subset C_x$,

iii) enfin, μ étant adaptée, il vient

$$M = Gx \subset C_x .$$

q.e.d.

Le théorème 1 se complète d'une condition de transience:

Théorème 2. [4]

Supposons μ étalée, si $\int \chi(g^{-1})\mu(dg) < 1$ ou si, hypothèse en général plus faible, $\int \log \chi(g^{-1})\mu(dg) < 0$, la m.a. de loi μ sur M satisfait à la deuxième alternative du théorème 1.

Ce théorème s'applique au cas, précédemment cité, de l'action sur $\mathbb{R}$ du groupe L : la mesure de Lebesgue sur $\mathbb{R}$ est relativement invariante par L de facteur $\chi, \chi(a,b) = a$, si μ est une probabilité adaptée étalée sur G, la m.a. de loi μ sur $\mathbb{R}$ est transitoire dès que $\int \log a . \mu(da.db) > 0$.

L'hypothèse d'existence d'une mesure σ-finie relativement invariante sur M est satisfaite avec χ non nécessairement trivial si G est un groupe de Lie résoluble connexe et simplement connexe. Elle est satisfaite avec $\chi \equiv 1$ lorsque H est compact ou de Lie semi-simple ou lorsque G et H sont unimodulaires ; ce dernier cas contient outre celui des groupes discrets, déjà signalé, celui des groupes de Lie de type rigide.

Le cas où l'espace homogène M possède une mesure σ-finie invariante est particulièrement intéressant puisque toute m.a. étalée sur M satisfait alors aux conclusions du théorème 1, et l'on peut envisager une classification des espaces homogènes en espaces homogènes récurrents , i.e. portant une m.a. étalée récurrente et non-récurrents ou transi toires. Avant d'énoncer quelques résultats obtenus dans cette direction, notons que B. Roynette a prouvé que, si G est de Lie connexe, H moyennable, M = G/H compact et si toute probabilité étalée sur G satisfait aux conclusions du théorème 1, alors M porte une mesure finie G-invariante.

a) Si M porte une mesure σ-finie invariante et est non moyennable alors M est transitoire . (Y. Derriennic et Y. Guivarc'h [2], C. Berg [1])

b) Soit G nilpotent à génération finie, $G_1 = [G,G]$ le sous-groupe engendré par les commutateurs de G et H_1 le sous-groupe distingué $H.G_1$, alors $M = G/H$ est récurrent si et seulement si G/H_1 est un groupe récurrent et H_1/H est fini. (H. Hennion [5])

c) Si G de Lie nilpotent simplement connexe et H connexe non-distingué, alors $M = G/H$ est transitoire (R.Schott [6])

d) Soit $G = So(d) \times \mathbb{R}^d$ le groupe des déplacements de $\mathbb{R}^d$ et H connexe, désignons par $\pi_2(H)$ la projection de H sur $\mathbb{R}^d$ et notons p la dimension du sous espace vectoriel de $\mathbb{R}^d$ engendré par les directions asymptotiques de $\pi_2(H)$, alors M est récurrent si et seulement si $d-p \leq 2$ (L. Gallardo et V. Ries [3]) .

Les points b-c-d conduisent à formuler la conjecture suivante: si G est de Lie de type rigide, M est récurrent si et seulement si il est à croissance polynomiale de degré inférieur ou égal à 2, i.e. si pour tout $x \in M$ et tout V voisinage compact de $e \in G$,

$$\overline{\lim_n} \; n^{-2} \, m \, (V^n x) < + \infty \, .$$

REFERENCES

(1) C. Berg et J.P.R. Christensen , Sur la norme des opérateurs de convolution. Inventiones math. 23, 1974.

(2) Y. Derriennic et Y. Guivarc'h, Théorème de renouvellement pour les groupes non moyennables. C.R.A.S. t 277, Octobre 1973.

(3) L. Gallardo et V. Ries, Marches aleatoires sur les espaces homogènes des groupes de déplacement. Nancy 1977.

(4) H. Hennion et B. Roynette, Un théorème de dichotomie pour les marches aleatoires sur les espaces homogènes. C.R.A.S, t 285 Septembre 1977. (démonstrations complètes à paraitre)

(5) H. Hennion, Marches aléatoires sur les espaces homogènes des groupes nilpotents à génération finie. Zeitschr. für Wahrsch. Th., 34, 1976.

(6) R. Schott, Marches aléatoires sur un espace homogène de groupe de Lie nilpotent simplement convexe. Thèse de 3e cycle, Nancy, 1976.

Université de Rennes I
U.E.R. Mathématiques et Informatique
Avenue du Général Leclerc
35042 RENNES CEDEX
(FRANCE)

Über die Meßbarkeit der Mengen der zulässigen und singulären Translationen von Maßen: Der Lebesguesche Zerlegungssatz für Kerne

Arnold Janssen

Abstract: A general Lebesgue-decomposition-theorem for kernels is proved. A corollary shows that several sets of measures are measurable. In particular it is shown that the set of admissible translates of a measure μ is a $F_{\sigma\delta}$-set and that the set of singular translates is a G_δ-set, if μ is a regular and τ-smooth finite measure on a topological semigroup.

1. Einführung: Translationsfamilien von Wahrscheinlichkeitsmaßen haben in der Statistik eine große Bedeutung. Skorohod betrachtete in den 60er Jahren als einer der ersten zulässige Translationen von Maßen über Hilberträumen. Allgemein liege folgende Situation vor: Sei μ ein endliches Borelmaß über einer topologischen Halbgruppe H. μ_x bezeichne das Bildmaß von μ unter der Rechtstranslation von $x \in H$.

Die Mengen $r(\mu) := \{x \in H: \mu_x \ll \mu\}$ und

$sr(\mu) := \{x \in H: \mu_x \perp \mu\}$

heißen die Mengen der zulässigen bzw. singulären Rechtstranslationen von μ. In verschiedenen Arbeiten, z. B. Hudson [3], Brokett [1], wird versucht, die Menge $r(\mu)$ "zu messen", obwohl die Meßbarkeit nur für Spezialfälle geklärt ist; z.B. wenn H ein separabler Hilbertraum ist, Skorohod [6], bzw. wenn H ein polnischer Vektorraum ist, Zinn [8], 1976.

In dieser Arbeit wird nun gezeigt, daß für beliebige topologische Halbgruppen H und τ-glatte, reguläre[1]) Maße μ die Menge $r(\mu)$ ($sr(\mu)$) eine $F_{\sigma\delta}$ (G_δ)-Menge ist (Satz 5). Aus einem Resultat von Ressel [5] folgt direkt, daß sich die Faltung zweier τ-glatter Maße über H erklären läßt. Bezüglich der Faltung bildet die Menge $M_\tau(H)$ der τ-glatten, endlichen Borelmaße über H, versehen mit der schwachen Topologie, eine topologische Halbgruppe. Weiter benötigt man:

2. den Lebesgueschen Zerlegungssatz für Kerne.

Da es zu jedem σ-endlichen Maß ein äquivalentes Wahrscheinlichkeitsmaß gibt, genügt es endliche Maße zu betrachten. Deshalb seien alle vorkommenden Maße endlich. Sei X ein topologischer Raum (nicht notwendig Hausdorff), versehen mit der Borelschen σ-Algebra $\mathcal{B}(X)$.
Die Begriffsbildung der regulären[1]) bzw. τ-glatten Borelmaße wird von Topsøe [7] P.13 übernommen. Die Menge der endlichen Borelmaße M(X)

[1]) regulär bzgl. abgeschlossener Mengen von innen.

über $(X, \mathfrak{B}(X))$ wird mit der Borelschen σ-Algebra $\mathfrak{B}(M(X))$ versehen, die von den offenen Mengen - bzgl. der schwachen Topologie - erzeugt wird. Die schwache Topologie ist die gröbste Topologie über $M(X)$, so daß für alle $f:X \to \mathbb{R}$, die beschränkt und halbstetig von unten sind, $\mu \to \int f d\mu$ halbstetig von unten ist. Daneben trägt $M(X)$ die sogenannte Stern σ-Algebra $\mathfrak{B}(X)^*$, die gröbste σ-Algebra, so daß für alle $A \in \mathfrak{B}(X)$ $\mu \to \mu(A)$ meßbar ist. Es gilt $\mathfrak{B}(X)^* \subset \mathfrak{B}(M(X))$.
Ein Kern ist eine Abbildung $y \longmapsto \mu(y,\cdot)$ von einem Meßraum (Y,Σ) in $M(X)$, so daß $y \longrightarrow \mu(y,A)$ für alle meßbaren Mengen A meßbar wird.
Daher gilt:

$$\begin{array}{l} Y \longrightarrow M(X) \\ y \longmapsto \mu(y,\cdot) \end{array} \quad \text{ist genau dann ein Kern,}$$

wenn die Abbildung $(\Sigma, \mathfrak{B}(X)^*)$ meßbar ist.

Bevor Meßbarkeitsfragen erörtert werden, wird die Lebesguesche Zerlegung von Kernen bzgl. eines anderen Maßes ρ untersucht. Für festes $y \in Y$ sei $\mu(y,\cdot) = \mu_1(y,\cdot) + \mu_2(y,\cdot)$ die Lebesguesche Zerlegung von $\mu(y,\cdot)$ bzgl. ρ mit $\mu_1(y,\cdot) \ll \rho$ und $\mu_2(y,\cdot) \perp \rho$.

Satz 1

Sei ρ ein reguläres Maß im Sinne von Topsøe [7], $\mu = \mu_1 + \mu_2$, $\mu_1 \ll \rho$, $\mu_2 \perp \rho$. Dann gilt:

a) Die Abbildungen $\begin{array}{l} M(X) \longrightarrow M(X) \\ \mu \longmapsto \mu_{1(2)} \end{array}$ sind $(\mathfrak{B}(M(X)), \mathfrak{B}(X)^*)$-meßbar

b) Sei $\begin{array}{l} \mu: Y \longrightarrow M(X) \\ y \longmapsto \mu(y,\cdot) \end{array}$ ein $(\Sigma, \mathfrak{B}(M(X)))$-meßbarer Kern.

Dann sind $y \longmapsto \mu_i(y,\cdot)$ Kerne $(i = 1,2)$.

Satz 2

Sei ρ ein reguläres Maß, $\mu \in M(X)$ und $\rho = \rho_1(\mu) + \rho_2(\mu)$ mit $\rho_1(\mu) \ll \mu$, $\rho_2(\mu) \perp \mu$.
Dann gilt:

a) Die Abbildungen $\begin{array}{l} M(X) \longrightarrow M(X) \\ \mu \longmapsto \rho_{1(2)}(\mu) \end{array}$ sind $(\mathfrak{B}(M(X)), \mathfrak{B}(X)^*)$-meßbar.

b) Sei $\begin{array}{l} \mu: Y \longrightarrow M(X) \\ y \longmapsto \mu(y,\cdot) \end{array}$ ein $(\Sigma, \mathfrak{B}(M(X)))$-meßbarer Kern

und $\rho_1(y,\cdot) := \rho_1(\mu(y,\cdot))$
$\rho_2(y,\cdot) := \rho_2(\mu(y,\cdot))$.

Dann sind $y \longmapsto \rho_i(y,\cdot)$ Kerne $(i = 1,2)$.

Die Voraussetzungen für Satz 1(2) b) sind z. B. dann erfüllt, wenn die Abbildung $y \longmapsto \mu(y,\cdot)$ (von einem topologischen Raum Y in M(X)) stetig ist.

Korollar 3 Sei ρ ein reguläres Maß. Dann gilt:
$\{\mu \in M(X) : \mu \ll \rho\}$,
$\{\mu \in M(X) : \rho \ll \mu\}$ und
$\{\mu \in M(X) : \rho \approx \mu\}$ sind $F_{\delta\sigma}$-Mengen, ($\rho \approx \mu$, falls $\rho \ll \mu$ und $\mu \ll \rho$).
$\{\mu \in M(X) : \mu \perp \rho\}$ ist eine G_δ-Menge
bzgl. der schwachen Topologie.

Bemerkung 4
a) Für abzählbar erzeugte σ-Algebren sind die Sätze bekannt. Vgl. z.B. Dubins und Freedman [2], Lange [4].
b) Wenn ρ nicht regulär ist, gibt es Beispiele, die zeigen, daß die Lebesguesche Zerlegung von Kernen im allgemeinen nicht wieder Kerne liefert, auch dann nicht, wenn die Zuordnung (von einem topologischen Raum Y in M(X)) $y \longmapsto \mu(y,\cdot)$ stetig ist. Die Meßbarkeitseigenschaften aus Korollar 3 gehen ebenfalls verloren.

3. Zulässige und singuläre Translationen von τ-glatten und regulären Maßen.

Die obigen Resultate finden eine Anwendung auf Translationsfamilien. Sei H eine beliebige topologische Halbgruppe. $M_\tau(H)$ bezeichne die τ-glatten Borelmaße.
Ressel [5] T.1 zeigt, daß sich das Produktmaß $\mu \otimes \nu$ zweier τ-glatter Maße zu einem τ-glatten Maß $\mu \hat{\otimes} \nu$ über der Borelschen σ-Algebra, die von der Produkttopologie erzeugt wird, fortsetzen läßt. Diese Fortsetzung ist stetig (als Abb. von $M_\tau(H) \times M_\tau(H)$ in $M_\tau(H \times H)$).
Sei $\varphi: H \times H \longrightarrow H$.
$(x,y) \longmapsto x \cdot y$ Durch das Bildmaß $(\mu \hat{\otimes} \nu)^\varphi =: \mu * \nu$ läßt sich die Faltung τ-glatter Borelmaße über einer topologischen Halbgruppe erklären. Da die Faltung τ-glatter Maße ein τ-glattes Maß liefert und diese Operation stetig ist, wird $(M_\tau(H), *)$ eine topologische Halbgruppe (bezüglich der schwachen Topologie).

Sei μ ein reguläres und τ-glattes Borelmaß über $(H, \mathfrak{B}(H))$. Da die Translation
$x \longmapsto \mu_x = \mu * \delta_x$ stetig ist, lassen sich die Zerlegungssätze

auf diesen Kern anwenden und es gilt:

Satz 5

Sei μ ein τ-glattes und reguläres Maß über $(H,\mathcal{B}(H))$.

Dann gilt: $\{x : \mu_x \ll \mu\}$,

$\{x : \mu \ll \mu_x\}$ und

$\{x : \mu \approx \mu_x\}$ sind $F_{\sigma\delta}$-Mengen,

$\{x : \mu \perp \mu_x\}$ ist eine G_δ-Menge.

Eine analoge Aussage gilt für Linkstranslationen.

Die Beweise werden später an anderer Stelle veröffentlicht.

Literatur

[1] Brockett, P.L.: Admissible Transformations of measures. Semigroup Forum 12 (1976) 21-33

[2] Dubins, L. and Freedman, D.: Measurable sets of measures Pacific J. Math. 14 (1964) 1211-1222

[3] Hudson, W.: Admissible Translates for Probability Distributions Ann. of Prob. 4 (1976) 505-508

[4] Lange, K.: Decomposition of substochastic transition functions Proc. A.M.S. 37 (1973) 575-580

[5] Ressel, P.: Some continuity and measurability results on spaces of measures, Math. Scand. 40 (1977) 69-78

[6] Skorohod, A.V.: Integration in Hilbert space Springer 1974

[7] Topsøe, F.: Topology and measure, Lecture Notes in Math. 133 (1970)

[8] Zinn, J.: Admissible translates of stable measures Studia Math. 54 (1975/76) 245-257.

Arnold Janssen

Universität Dortmund

Abteilung Mathematik

Postfach 500 500

D-4600 Dortmund 50

A NOTE ON RANDOM WALKS ON COMPACT GROUPS

By V. Losert and H. Rindler

Let G be a separable compact topological group. A sequence (x_n) is called <u>uniformly distributed</u> in G if for any Haar-measurable subset Y of G, such that the boundary of Y has measure 0:

$$\lim_{n\to\infty} 1/N \sum_{n\leq N} C_Y(x_n) = \lambda(Y)$$

(λ = normalized Haar-measure, C_Y the characteristic function of Y). If μ is a probability measure on G, consider the infinite product space G^N with the product measure μ^N.
If $\omega = (x_n) \in G^N$ put $Y_n(\omega) = x_1 x_2 \ldots x_n$, then $Y_n(\omega)$ is a random walk on G (we consider multiplicative groups). Let $T: G^N \to G^N$ be the one-sided shift $(T(x_n) = (x_{n+1}))$. An element $\omega \in G^N$ is called generic for (T,μ), if $(T^n\omega)$ is uniformly distributed with respect to μ^N.
The following theorem is a consequence of Satz 2 in Sigmund [9].

<u>Theorem 1</u>: If μ is concentrated on a countable union of metrizable subsets of G and supp μ is not contained in a proper coset, then for μ^N-almost all sequences ω, the sequence $Y_n(\omega)$ is uniformly distributed in G.

<u>Proof</u>: Let $\omega = (x_n)$ be a generic point for (T,μ). It follows that for any $m \in \mathbb{N}$ the sequence $x_1 x_2 \ldots x_m, x_2 x_3 \ldots x_{m+1}, \ldots, x_n x_{n+1} \ldots x_{n+m-1}, \ldots$ is u.d. in G with respect to μ^m (convolution power). It follows from Van der Corput's difference theorem in the generalized version

of Cigler [1], and from the theorem of Ito-Kawada that the sequence $x_1, x_1x_2, \ldots, x_1x_2\ldots x_n, \ldots$ is u.d. in G with respect to λ. Therefore it suffices to show that μ^N-almost all $\omega \in G^N$ are generic for (T,μ). But this follows from the same argument as in the proof of Proposition 3 of [6].

Theorem 2: If G is not metrizable and $\mu = \lambda$ then the set of all sequences $\omega \in G^N$ such that $Y_n(\omega)$ is u.d. in G has outer measure 1 and inner measure 0.

Proof: By Lemma 1 in [4] a subset of G^N has outer measure one iff its image in $(G/K)^N$ has outer measure one for any metrizable quotient group G/K. Let M be the set of points $\omega \in G^N$ such that $Y_n(\omega)$ is u.d. in G. Let U be a symmetric neighbourhood of the unit element in G such that K is not contained in U^2. If $\omega_1 = (z_n) \in$ $\in (G/K)^N$ is arbitrary, one can choose a sequence (x_n') in $G \setminus K$ such that $\pi(x_n') = z_1 \ldots z_n$ for all n ($\pi: G \to G/K$ denotes the canonical projection). Put $x_n = x_{n-1}'^{-1} x_n'$ $(x_0 = e)$.
Then we have $\pi(x_n) = z_n$ and $\omega = (x_n)$ does not belong to M. This shows that the image of $G^N \setminus M$ is all of $(G/K)^N$. Consequently M has inner measure zero. Conversely take a sequence $\omega_1 = (z_n) \in$ $\in (G/K)^N$ such that $Y_n(\omega_1)$ is u.d. in G/K. By Theorem 2 in [4] there exists a u.d. sequence (x_n') in G such that $\pi(x_n') = z_1 \ldots z_n$ for all n. Put again $x_n = x_{n-1}'^{-1} x_n'$. Then $\pi(x_n) = z_n$ and $\omega = (x_n) \in M$. This shows that the image of M consists of all sequences $\omega_1 \in (G/K)^N$ such that $Y_n(\omega_1)$ is u.d. in G/K. By Theorem 1 this set has measure one in G/K and consequently M has outer measure one.

If μ is an arbitrary probability measure on G, it follows from a well-known zero-one law that the set of all ω for which $Y_n(n)$ is u.d. in G has measure zero or one provided that it is measurable. If it is not measurable, its outer measure is one, its inner measure is zero (with respect to μ^N). If G is metrizable the set has always measure one. In the case of a separable, non metrizable group all cases are possible. Two cases have already been treated, the next theorem will give an example for the third case.

<u>Theorem 3</u>: There exists a probability measure μ on $\{0,1\}^c$ (where c denotes the cardinality of the continuum) such that supp μ is not contained in a proper coset, and for μ^N - almost all sequences ω the sequence $Y_n(\omega)$ is <u>not</u> uniformly distributed.

<u>Remark</u>: A similar result for the general case of uniformly distributed sequences on G was proved in [6] Example 4. It is perhaps interesting to note that the construction which was used there cannot be applied to the case of random walks.

<u>Proof of Theorem 3</u>: We consider the interval $[0,1]$ with ordinary Lebesgue measure ν. Let $\mathfrak{A}$ be the Boolean algebra of Lebesgue measurable subsets of $[0,1]$ modulo sets of measure zero. $A \Delta B$ shall denote the symmetric difference of elements $A,B \in \mathfrak{A}$. If addition is defined by Δ, then $\mathfrak{A}$ is a vector space over $\mathbf{Z}_2 = \{0,1\}$. Let I be a family in $\mathfrak{A}$ such that $I \cup \{[0,1]\}$ is a basis of $\mathfrak{A}$. Since card($\mathfrak{A}$) = c, we have also card(I) = c. Let X be the Stone

representation space of $\mathfrak{A}$. Each element $A \in \mathfrak{A}$ defines a continuous $\{0,1\}$-valued function f_A on X. This gives rise to a continuous map $j: X \to \{0,1\}^I$ by $j(x)_A = f_A(x)$ (for $A \in I$). The measure ν on $\{0,1\}$ induces a measure ν_1 on X whose support is X. Put $\mu = j(\nu_1)$. The support of μ is $j(X)$. We claim that $j(X)$ is not contained in a proper closed coset of $\{0,1\}^I$. For this it is sufficient to show that for any finite subset $\{A_1,\dots,A_n\}$ of I the set $\{(f_{A_i}(x))_{i=1}^n : x \in X\}$ is not contained in a proper coset of $\{0,1\}^n$. Assume the contrary. Then there exist $(\alpha_i)_{i=1}^n \in \{0,1\}^n$ and $c \in \{0,1\}$ such that $\Sigma\, \alpha_i f_{A_i}(x) = c$ (in $\mathbb{Z}_2$) for all $x \in X$. Addition of the f_{A_i} in $\mathbb{Z}_2$ corresponds to vector addition in $\mathfrak{A}$ and the equation $\Sigma\, \alpha_i f_{A_i} = c\, f_{[0,1]}$ contradicts the linear independence of $I \cup \{[0,1]\}$. Similarly it follows from the fact that $I \cup \{[0,1]\}$ is a basis that j is injective.

Now we claim that no point $\omega \in j(X)^N$ can define a u.d. sequence $Y_n(\omega)$. Assume that $(\sum_1^n j(x_i))_{n=1}^{\infty}$ is u.d. in $\{0,1\}^I$ (the group is written additively). If $A \in \mathfrak{A}$, $A \neq \emptyset$, then $f_A = \sum_{i=1}^n d_i f_{A_i}$ for some $d_i \in \mathbb{Z}_2$, $A_i \in \mathfrak{A}$. The sequence $(j(x_n))$ cannot be contained in a proper closed subgroup of $\{0,1\}^I$ consequently there exists $n \in \mathbb{N}$ such that $f_A(x_n) = 1$, and this means that x_n belongs to the clopen subset corresponding to A. It would follow that (x_n) is dense in X. But this is impossible since X is not separable (an easy consequence of the fact that ν is non-atomary).

If $(Y_n(\omega))$ is u.d. in $\{0,1\}^I$, it follows that $\omega \in (\{0,1\}^I)^N \setminus j(X)^N$. But since μ is concentrated on $j(X)$, the measure μ^N is concentrated on $j(X)^N$. This finishes the proof of Theorem 3.

In the proof of Theorem 1 we made use of the connection between generic points of certain measures and u.d. sequences of the form $Y_n(\omega)$. Now we want to state some general results about the existence of generic points.

<u>Theorem 4</u>: If is separable, the there exists a generic element for (T,λ).

<u>Proof</u>: We use a similar construction as in [8]. Let F be a countable dense subset of G. If $U_1,\dots,U_k$ are continuous, irreducible representations of G, let $P_{U_1},\dots,P_{U_k}$ be the orthogonal projection onto the subspace on which the representation $U_1 \otimes \dots \otimes U_k$ of G^k acts trivially. An element $\omega = (z_n)$ is generic for (T,λ) iff $\lim_{N\to\infty} \| 1/N \sum_{n\le N} U_1(z_n) \otimes \dots \otimes U_k(z_{n+k-1}) - P_{U_1 \dots U_k} \| = 0$ for all $U_1,\dots,U_k$ as above.

If $B = \{x_1,\dots,x_m\}$ is a finite subset of F, $m > k$, put

$$(B_j\ U_1,\dots,U_k) = \| 1/(m-k) \sum_{n=1}^{m-k} U_1(z_n) \otimes \dots \otimes U_k(z_{n+k-1}) - P_{U_1 \dots U_k} \| .$$

If $D = \{y_1,\dots,y_l\}$ is another subset we define the product BD of the blocks as in [8]. It follows easily from the equation

$$P_{U_1 \dots U_k} = P_{U_1 \dots U_k} \cdot U_1(a_1) \otimes \dots \otimes U_k(a_k) = U_1(a_1) \otimes \dots \otimes U_k(a_k) \cdot$$

$\cdot P_{U_1 \dots U_k}$ that $(BD; U_1,\dots,U_k) \in (B; U_1,\dots,U_k)$ and $(BD; U_1,\dots,U_k) \le$ $\le (D; U_1,\dots,U_k)$. Now we can use the same construction as in [8] - successive addition and multiplication of blocks - to get a generic point for (T,λ). T_ω^n is even well distributed, i.e. ω is strictly generic.

<u>Remark</u>: If G is metrizable, then μ^N almost all elements are generic for (T,μ). If G is separable and not metrizable, the same statements as in Theorem 1-3 hold for the set of generic points for (T,μ) (with similar proofs), i.e. the set can have either measure zero or one or its outer measure is one and its inner measure zero. If μ is an arbitrary probability measure on G one can show under assumption of the continuum hypothesis that generic points do exist (the proof uses a similar construction as in [5] Theorem 1). But this needs no longer be true for measures on arbitrary separable, compact topological spaces (see [4] Proposition).

We want to state a consequence of Theorem 1. Consider the group R/Z and a measurable subset Y: $0 < \lambda(Y) < 1$. For $\omega = (x_n)$ put $Y_n(\omega) = \sum_{n \leq N} C_Y(x_n)$.

Then it was proved by Hlawka in [2] that this sequence of integers is uniformly distributed modulo m for all $m = 1,2,\ldots$ for almost all sequences in $(R/Z)^N$. Identifying Y with {1} and the complement of Y with {0} in the Bohr-compactification of the integers we obtain that almost every sequence is even u.d. in the Bohr-compactification, (first stated by Maxones, Muthsam et al. in [7]).

The following interesting result was obtained by Veech [10]. If $(x_n) = (na)$ where a is an irrational number with bounded partial quotients and Y a proper subinterval then the corresponding sequence is uniformly distributed mod 2. (If the length of the interval is a multiple of a it may happen that the limit measure assigns different values to 0 and 1). The converse of this result was also proved in that paper.

The example concerning the Bohr-compactifiaction of the integers shows that nonmetrizable groups may appear in a natural way. We give an application of Theorem 4. The Bohr-compactification of the free group of countably many generators (x_k) may be considered as the maximal compact separable group (for metrizable groups no such universal metrizable object exists). The proof shows that a generic element $\omega = (y_n)$ can be constructed with y_n belonging to the free group. Thus y_n can be written in the form $p_n((x_k))$, where p_n are functions of the generators x_k of the group involving only group multiplication and inversion. It is easy to show that the p_n are universal in the following sense: If G is any compact group and (a_k) any sequence in G generating a dense subgroup, then $p_n((a_k))$ is a generic element.

REFERENCES

[1] Cigler, J.: Über eine Verallgemeinerung des Hauptsatzes der Gleichverteilung, J. reine und angewandte Math. 216 (1961) 141 - 147.

[2] Hlawka, E.: Ein metrisches Gegenstück zu einem Satz von W.A. Veech, Monatshefte f. Math. 76, (1972) 436 - 447.

[3] Kuipers, L., Niederreiter,H.: Uniform distribution of sequences, John Wiley and Sons, New York (1974).

[4] Losert, V.: Uniformly distributed sequences on compact, separable, non metrizable groups, Acta Sci. Math. 40 (1978) 107 - 110.

[5] Losert, V.: On the existence of uniformly distributed sequences in compact topological spaces I, TAMS to appear.

[6] Losert, V.: On the existence of uniformly distributed sequences in compact topological spaces II, erscheint in Monatshefte f. Math.

[7] Maxones, W., Muthsam, H., Rindler, H.: Bemerkungen zu einem Satz von E. Hlawka, Anz. Österr. Ak. Wiss. 1976, No. 7, 82 - 83.

[8] Rindler, H.: Uniform distribution on locally compact groups, Proc. A.M.S. 57 (1976) 130 - 132.

[9] Sigmund,K.: Über Verteilungsmaße von Maßfolgen auf kompakten Gruppen, Compositio Math. 21, (1969) 299 - 311.

[10] Veech, W.A.: Strict ergodicity in zero dimensional dynamical systems and the Kronecker-Weyl theorem mod 2, TAMS 140, (1969) 1 - 33.

[11] Veech, W.A.: Some questions of uniform distribution, Annals of Math. (2) 94 (1971) 125 - 138.

Institut für Mathematik der Universität Wien
A-1090 Wien, Strudlhofgasse 4, Austria

A CLASS OF PROBABILITY MEASURES ON GROUPS ARISING FROM SOME PROBLEMS IN ERGODIC THEORY

Viktor Losert and Klaus Schmidt

1. Introduction

Let A be a locally compact second countable abelian group with dual $\hat{A}$. For every probability measure σ on A we define a subgroup E_σ of A as follows:

$$E_\sigma = \{a \in A: \lim_{n\to\infty} \sup_{\{\chi\in\hat{A}: \int|\chi(a)-1|\, d\sigma(a) < \frac{1}{n}\}} |1-\chi(a)| = 0\} \tag{1.1}$$

It is clear that E_σ is a Borel set for every σ. Following [3,Chapter 3] we call a probability measure σ on A *full* if $E_\sigma = A$. In the following we shall write P(A) for the convolution semigroup of probability measures on A, and we denote by $F(A) \subset P(A)$ the subset of full probability measures on A. From [3, Chapter 3] we have:

Proposition 1.1. A probability measure σ on A is full if and only if its Fourier transform $\hat{\sigma}$ satisfies

(1) $\hat{\sigma}(\chi) \neq 1$ whenever $\chi \neq 1$, $\chi \in A$, (1.2)

(2) $\limsup_{\chi\to\infty} |\hat{\sigma}(\chi)| < 1$. (1.3)

Corollary 1.2. F(A) is dense in P(A) in the weak*-topology. If A is connected, F(A) is a norm-open ideal in P(A).

Proof: The first assertion is obvious. If A is connected, $\hat{A}$ has no compact subgroups, and condition (1.3) implies that $|\hat{\sigma}(\chi)| < 1$ for every $\chi \neq 1$ in $\hat{A}$. It follows immediately that F(A) is norm-open, and that it is an ideal in P(A).

There are many interesting open questions associated with the Borel subgroups E_σ, $\sigma \in P(A)$, which are worth investigating in their own right, but one of the main motivations for the study of these subgroups and of full measures in particular comes from some problems in ergodic theory, several of which we shall describe below (Chapters 2-4). In all these problems one encounters Borel subgroups B of a locally compact second countable abelian group A which have the following property:

$$\sigma(B) < 1 \quad \text{for every } \sigma \in F(A). \tag{1.4}$$

It is clear that (1.4) implies

$$\lambda_A(B) = 0, \tag{1.5}$$

where λ_A is the Haar measure on A, and the question arises whether (1.4) is strictly stronger than (1.5). In Chapter 5 we shall prove that, for uncountable A, (1.4) is indeed stronger than (1.5). In fact we shall show that every uncountable locally compact second countable abelian group A has a Borel subgroup B satisfying (1.5) and a $\sigma \in F(A)$ with $\sigma(B) = 1$, and whose Fourier transform tends to zero at infinity. In Chapter 6 we shall prove that every uncountable locally compact second countable abelian group A contains an uncountable dense subgroup B which is a Borel set and which satisfies (1.4). Chapter 7 contains some concluding remarks on a deeper property of the groups arising in the Chapter 2 and 4.

2. Full measures and cocycles

Let G be a locally compact second countable group which acts non-singularly and ergodically on a standard measure space (X,S,μ), and let A be a locally compact second countable abelian group with dual $\hat{A}$.

A Borel map $h: G \times X \to A$ is called a (1-) *cocycle* for G if, for every $g_1, g_2 \in G$,

$$h(g_1, g_2 x) - h(g_1 g_2, x) + h(g_2, x) = 0$$

for μ-a.e.x. A cocycle $h: G \times X \to A$ is said to be a *coboundary* if there exists a Borel map $\phi: X \to A$ such that, for every $g \in G$,

$$h(g,x) = \phi(gx) - \phi(x)$$

for μ-a.e.x. For any (fixed) cocycle $h: G \times X \to A$, and for every $\chi \in \hat{A}$, we can define a cocycle $\chi.h: G \times X \to T$ (= the circle group) in the obvious way. Let

$$B(h) = \{\chi \in \hat{A} : \chi.h \text{ is a coboundary}\}. \qquad (2.1)$$

It is not difficult to verify that B(h) is a Borel subgroup of $\hat{A}$ for every cocycle $h: G \times X \to A$, and we have the following result ([3, Chapter 4]):

Theorem 2.1. Let $h: G \times X \to A$ be a cocycle, and let $B(h) \subset \hat{A}$ be the Borel subgroup defined by (2.1). If $\sigma \in P(\hat{A})$ satisfies

$$\sigma(B(h)) = 1,$$

then we have

$$E_\sigma \subset B(h).$$

Moreover, if there exists a $\sigma \in F(\hat{A})$ with $\sigma(B(h)) = 1$, then h is a coboundary.

Corollary 2.2. If h is not a coboundary, we have

$$\sigma(B(h)) < 1$$

for every $\sigma \in F(\hat{A})$.

For a detailed discussion of the rôle of cocycles and coboundaries in ergodic theory we refer to [1], [3], [4] and [6].

3. Full measures and ergodic flows

Let $(V_t : t \in R)$ be a nonsingular ergodic flow on a standard measure space (X, S, μ) which is not type I (i.e. $\mu(\{V_t x : t \in R\}) = 0$ for every $x \in X$). Then it is known that there exists a $t_o \in R$ for which V_{t_o} is an ergodic automorphism of (X, S, μ). The authors are grateful to W. Parry for a crucial idea in the following proof of this fact, which also throws light on the possible size of the set $\{t \in R : V_t$ is non-ergodic$\}$. We define a cocycle $h_o : R \times X \to R$ for the flow (V_t) by setting

$$h_o(t,x) = t \tag{3.1}$$

for every $t \in R$, $x \in X$. Next we define, for every integer $k \geq 1$, a group $B_k \subset R$ by

$$B_k = \{s \in R : \exp(2\pi i s k h_o) \text{ is a coboundary}\}. \tag{3.2}$$

From Chapter 2 we know that B_k is a Borel subgroup of R, and it is easy to see that

$$B_k = \frac{1}{k}.B_1 = \{\frac{s}{k} : s \in B_1\}. \tag{3.3}$$

for every $k \geq 1$. In particular we have

$$B_1 \subset B_{2!} \subset B_{3!} \subset \ldots \subset B_{k!} \subset \ldots$$

and

$$B_o = \bigcup_{k\geq 1} B_{k!} = \bigcup_{k\geq 1} B_k \tag{3.4}$$

is a Borel subgroup of R.

<u>Lemma 3.1</u>. $B_o = \{0\} \cup \{s \in R : s \neq 0 \text{ and } V_{\frac{1}{s}} \text{ is nonergodic}\}$.

Proof: Let $t_o \in R$, $t_o \neq 0$, and assume that V_{t_o} is nonergodic. We put $S_{t_o} = \{Y \in S: V_{t_o} Y = Y\}$ and denote the restriction of μ to S_{t_o} by μ_{t_o}. On the measure space (S_{t_o}, μ_{t_o}) the flow (V_t) induces an ergodic circle action. In particular we can find a probability measure μ'_{t_o} on S_{t_o} which is equivalent to μ_{t_o}, and which is invariant under the flow induced by (V_t) on S_{t_o}. The space $L^2(S_{t_o}, \mu'_{t_o})$ is thus spanned by the set of S_{t_o}-measurable eigenfunctions of (V_t). In particular we can find a complex valued S_{t_o}-measurable function ϕ_{t_o} of modulus one and an integer $k_{t_o} \geq 1$ such that

$$\phi_{t_o}(V_s x) = \exp\,(2\pi i k_{t_o} s/t_o).\phi_{t_o}(x)$$

μ-a.e. on X, for every $s \in R$. But this is exactly the same as saying that $t_o^{-1} \in B_{k_{t_o}} \subset B_0$. Conversely, if $s_o \in B_0$, we have $s_o \in B_{k_o}$ for some $k_o \geq 1$, which in turn implies the existence of a Borel map $\phi: X \to T$ with

$$\phi(V_t x)\overline{\phi(x)} = \exp\,(2_\pi i k_o s_o t)$$

μ-a.e., for every $t \in R$. Hence ϕ is non-constant, but $\phi.V_{\frac{1}{s_o}} = \phi$. This shows that $V_{\frac{1}{s_o}}$ is non-ergodic.

Theorem 3.2. Let $(V_t : t \in R)$ be an ergodic flow on a standard measure space (X, S, μ), which is not type I. Then the set

$$B_o = \{0\} \cup \{s \in R: s \neq 0 \text{ and } V_{\frac{1}{s_o}} \text{ is non-ergodic}\}$$

is a Borel subgroup of R with the following property: If σ is a full probability measure on R, then

$$\sigma(B_o) < 1.$$

Proof: If $\sigma \in F(R)$ satisfies $\sigma(B_o) = 1$, we define a sequence $(\sigma_k:$

$k \geq 1$) in $P(R)$ by

$$\sigma_k(C) = \sigma(C \cap B_{k!}).\sigma(B_{k!})^{-1}$$

for every Borel set $C \subset R$ (if the first few terms of this sequence are not defined, we just ignore them - (3.3) shows that (3.4) makes sense for all sufficiently large k). It is easy to see that (σ_k) converges to σ in norm. By Corollary 1.2, there exists a $k_o \geq 1$ such that σ_{k_o} is full. Theorem 2.1 now yields $B_1 = B_{k_o} = R$, and hence there exists a Borel map $\phi : X \to R$ with

$$h_o(t,x) = t = \phi(V_t x) - \phi(x) \tag{3.5}$$

μ-a.e., for every $t \in R$. (3.5) contradicts our assumption about (V_t), since it is easily seen to imply that the flow is of type I. This proves that $\sigma(B_o) < 1$ for every $\sigma \in F(R)$.

Remark 3.3. If $(V_t : t \in R)$ is an ergodic type I flow on (X,S,μ), which is aperiodic, we have $B_o = R$. If (V_t) is periodic with period $\alpha > 0$, we get $B_o = \alpha^{-1}.Q$ (Q denotes the rational numbers).

4. Full measures and eigenfunctions

Let G be a locally compact second countable abelian group which acts non-singularly and ergodically on a standard measure space (X,S,μ), and assume that the action is not type I (i.e. $\mu(\{gx : g \in G\}) = 0$ for every $x \in X$). We define a cocycle $h_o : G \times X \to A$ by

$$h_o(g,x) = g \tag{4.1}$$

for every g,x, and define $B(h_o)$ by (2.1). An element $\gamma \in \hat{G}$ is called an *eigenvalue* of G if there exists a nonzero Borel map $\phi : X \to C$ (the complex numbers) with $\phi(gx) = \gamma(g).\phi(x)$ μ-a.e., for every $g \in G$. We write E(G) for the group of eigenvalues of G and observe that

$$E(G) = B(h_o). \tag{4.2}$$

In particular, E(G) is a Borel subgroup of G.

Theorem 4.1. Let G be a locally compact second countable abelian group which acts non-singularly and ergodically on a standard measure space $(X,\mathcal{S},\mu)$, and let $E(G) \subseteq \hat{G}$ denote its group of eigenvalues. Then E(G) is a Borel subgroup of $\hat{G}$. If the action of G is not type I, then we have, for every full probability measure σ on $\hat{G}$,

$$\sigma(E(G)) < 1.$$

Proof: If there exists a $\sigma \in F(G)$ with $\sigma(E(G)) = 1$, we conclude from (4.2) and from Theorem 2.1 that h_o is a coboundary. As before, one can then show that the action of G is type I, which is absurd. This contradiction proves our assertion.

Remark 4.2. From Theorem 2.1 we get the following: If $\sigma \in P(\hat{G})$ satisfies $\sigma(E(G)) = 1$, then $E(G) \supset E_\sigma$. If the action of G is type I, E(G) is of the form $\widehat{G/K}$, where K is a closed subgroup of G.

Remark 4.3. [3, Chapter 8] gives an explicit construction of an uncountable subgroup B of T, which is the group of eigenvalues of some integer action. The same method can be used to find an uncountable subgroup of R which is the group of eigenvalues of a suitable ergodic action of R.

5. The existence of full probability measures on subgroups of Haar measure zero.

Theorem 5.1. Let A be an uncountable locally compact second countable abelian group. Then there exists a Borel subgroup B of A and a

$\sigma \in P(A)$ such that

$$\lambda_A(B) = 0, \text{ where } \lambda_A \text{ is the Haar measure of } A, \tag{5.1}$$

$$\sigma(B) = 1, \tag{5.2}$$

$$|\hat{\sigma}(\chi)| < 1 \text{ for every } \chi \in \hat{A} \text{ with } \chi \neq 1, \tag{5.3}$$

$$\lim_{\chi\to\infty} \hat{\sigma}(\chi) = 0 \tag{5.4}$$

For the proof of Theorem 5.1 we need several lemmas.

Lemma 5.2. Let $\sigma \in P(A)$ satisfy (5.4), and let $f \in L^1(A,\sigma)$. Then we have

$$\lim_{\chi\to\infty} \int f(a)\chi(a)\, d\sigma(a) = 0. \tag{5.5}$$

Proof: It is easy to see that the set of trigonometric polynomials lies dense in $L^1(A,\sigma)$. Since (5.5) is obviously satisfied for every trigonometric polynomial f, (5.5) must hold for every $f \in L^1(A,\sigma)$.

Lemma 5.3. Let C be a closed subgroup of A and suppose Theorem 5.1 holds for the quotient group A/C. Then it also holds for A.

Proof: We put $C^\perp = \{\chi \in \hat{A}: \chi = 1 \text{ on } C\}$ and identify $\hat{C}$ with $\hat{A}/C^\perp$ in the obvious way. $\pi: A \to A/C$ and $\hat{\pi}: \hat{A} \to \hat{A}/C^\perp = \hat{C}$ are the projection maps, and we choose Borel cross-sections $\gamma: A/C \to A$ and $\hat{\gamma}: \hat{C} \to \hat{A}$ such that $\gamma(0) = 0$, $\hat{\gamma}(1) = 1$, and such that $\hat{\gamma}(\hat{K})$ has compact closure in $\hat{A}$ whenever $\hat{K}$ is compact in $\hat{C}$. ν will denote a probability measure in the Haar measure class on C. In particular ν will satisfy (5.3) and (5.4) with ν replacing σ and with C replacing A. Assume now Theorem 5.1. holds on A/C, and take a Borel group $B \subset A/C$ and a measure $\sigma \in P(A/C)$ satisfying (5.1) - (5.4) with A/C replacing A. Put $B_1 = \pi^{-1}(B)$ and define $\sigma_1 \in P(A)$ by $\int f\, d\sigma_1 = \int_{A/C} \int_C f(\gamma(x)+c)\, d\nu(c).\, d\sigma(x)$ for every bounded continuous complex valued f on A. Clearly we have $\lambda_A(B_1) = 0$ and $\sigma_1(B_1) = 1$. To prove (5.3) we note that every $\chi \in \hat{A}$ can be written uniquely as $\chi = \chi_1 \cdot \chi_2$ with $\chi_1 \in C^{\perp}$ and $\chi_2 \in \hat{\gamma}(\hat{C})$. If $|\hat{\sigma}_1(\chi)| = 1$, we get $1 = |\int_{A/C} \int_C \chi(\gamma(x) + c)\, d\nu(c)\, d\sigma(x)| = |\int_{A/C} \chi_1(\gamma(x))\chi_2(\gamma(x))\, d\sigma(x)| \cdot |\int_C \chi_2(c) d\nu(c)|$. The properties of ν imply that $\chi_2 = 1$ on C, and the choice of $\hat{\gamma}$ shows that $\chi_2 = 1$. It follows that $|\int_{A/C} \chi_1(\gamma(x))\, d\sigma(x)| = 1$, and hence that $\chi_1 = 1$, on account of (5.3). This shows that $\chi = 1$, and proves (5.3) for σ_1. Turning now to (5.4), we assume there exists a sequence $(\chi^{(k)} : k \geq 1)$ in $\hat{A}$ with $\lim_k \chi^{(k)} = \infty$ and with $\lim_k \inf |\hat{\sigma}_1(\chi^{(k)})| > 0$. Writing each $\chi^{(k)}$ as a product $\chi^{(k)} = \chi_1^{(k)} \cdot \chi_2^{(k)}$ with $\chi_1^{(k)} \in C^{\perp}$ and $\chi_2^{(k)} \in \hat{\gamma}(\hat{C})$, we get

$$\lim_k \inf |\hat{\sigma}_1(\chi^{(k)})| = \lim_k \inf |\int_{A/C} \chi_1^{(k)}(\gamma(x))\chi_2^{(k)}(\gamma(x))\, d\sigma(x)$$

$$\cdot |\int_C \chi_2^{(k)}(c)\, d\nu(c)| > 0.$$

The properties of ν imply that $(\chi_2^{(k)} : k \geq 1)$ has a convergent subsequence. Without loss in generality we assume that $(\chi_2^{(k)})$ itself converges, $\lim_k \chi_2^{(k)} = \chi_2 \in \hat{A}$, say. We conclude immediately that

$$\lim_k \inf |\int_{A/C} \chi_1^{(k)} (\gamma(x))\chi_2(\gamma(x))\, d\sigma(x)| > 0.$$

Considering each $\chi_1^{(k)}$ now as an element of $\widehat{A/C}$, we get

$$\lim_k \inf |\int_{A/C} \chi_1^{(k)}(x).f(x) d\sigma(x)| > 0,$$

where $f(x) = \chi_2(\gamma(x))$, contradicting Lemma 5.2. We have proved that $\lim_k \hat{\sigma}_1(\chi^{(k)}) = 0$ whenever $\lim_k \chi^{(k)} = \infty$, so σ_1 satisfies (5.4).

Lemma 5.4. Suppose Theorem 5.1 holds for an open subgroup A_o of A. Then it also holds for A.

Proof: A/A_o is countable, and we find a countable subgroup D of A such that $A_o + D = A$. Let $D = (d_1, d_2, \ldots)$ be an enumeration of D, and let ν be a probability measure on A with $\nu(D) = 1$, which has positive mass at every point of D. If B is a subgroup of A_o and $\sigma \in P(A_o)$ satisfies (5.1) - (5.4) with A_o replacing A, we put $B_1 = B + D$ and $\sigma_1 = \sigma * \nu$. Again one checks easily that B_1 and σ_1 satisfy (5.1) - (5.4).

Lemma 5.5. Theorem 5.1 holds for A = T and A = R.

Proof: This is Theorem 4 in [5].

For the following we assume that $A = \prod_{j=1}^{\infty} Z_{n_j}$, an infinite product of finite cyclic groups.

Lemma 5.6. Let $0 < \varepsilon < 1$ and let N be a positive integer. Then there exists an integer $m \geq 1$, a Borel set $M \subset A_m = \prod_{j=1}^{m} Z_{n_j}$, and a $\sigma \in P(A_m)$ such that the following conditions are satisfied:

$$\sigma(M) = 1, \tag{5.6}$$

$$|\hat{\sigma}(\chi)| < \varepsilon \text{ for every } \chi \in \hat{A}_m, \chi \neq 1, \tag{5.7}$$

$$\{x_1 + x_2 + \ldots + x_N : x_i \in M \cup -M \cup \{0\} \text{ for } i=1,\ldots,N\} \neq A_m. \tag{5.8}$$

Proof: For every $1 \leq l \leq k$ we denote by $\pi_{kl}: A_k \to A_l$ the projection of A_k onto A_l. Using induction, we shall construct subsets $M_k \subset A_k$ and measures $\sigma_k \in P(A_k)$ such that, for every $k \geq 1$, $1 \leq l \leq k$,

$$\sigma_k(M_k) = 1, \tag{5.9}$$

$$\sigma_k \pi_{kl}^{-1} = \sigma_l, \tag{5.10}$$

$$\pi_{kl}(M_k) = M_l, \tag{5.11}$$

and

$$|\hat{\sigma}_k(\chi)| < \varepsilon \quad \text{for every } \chi \in \hat{A}_k, \quad \chi \neq 1. \tag{5.12}$$

Finally we shall prove the existence of $m \geq 1$ for which M_m satisfies (5.8). To start the induction process, we choose k_0 to be the smallest integer satisfying $k_0 \geq 2$ and $\prod_{j=1}^{k_0} n_j > N/\varepsilon$. We define $M_l = A_l$ and $\sigma_l = \lambda_{A_l}$ for every $l = 1,\dots,k_0$, where λ_{A_l} denotes the Haar measure on A_l. To outline the continuation of the induction process we assume for the moment that $k \geq k_0$ is fixed and that M_l, σ_l, $1 \leq l \leq k$ have been defined in accordance with (5.9) - (5.12). We write the elements of $A_{k+1} = A_k \times Z_{n_{k+1}}$ as pairs (x,y) with $x \in A_k$ and $y \in Z_{n_{k+1}}$. For every $x \in A_k$, the values $\sigma_{k+1}(x,y)$, $y \in Z_{n_{k+1}}$, will be defined according to one of the following rules:

$$\sigma_{k+1}(x,y) = (n_k+1)^{-1}.\sigma_k(x) \quad \text{for every } y \in Z_{n_{k+1}}. \tag{5.13}$$

or

$$\sigma_{k+1}(x,0) = \sigma_k(x), \text{ and } \sigma_{k+1}(x,y) = 0 \text{ for } y \neq 0. \tag{5.14}$$

If we put

$$M_{k+1} = \{(x,y) \in A_{k+1} : \sigma_{k+1}(x,y) \neq 0\}, \tag{5.15}$$

M_{k+1} and σ_{k+1} will satisfy (5.9) - (5.11). (5.12) will hold if

$$\Sigma\{\sigma_k(x) : x \in A_k \text{ and (5.14) is applied to } x\} < \varepsilon. \tag{5.16}$$

In order to achieve (5.8) we shall have to choose carefully between (5.13) and (5.14) at each step. Let, for every $j \geq 1$, $z^{(j)}$ be a fixed nonzero element in Z_{n_j}, and write $z_k = (z^{(1)},\dots,z^{(k)}) \in A_k$ for every $k \geq 1$. We now turn to a detailed description of how to continue the induction process after having defined $M_{k_0} = A_{k_0}$ and $\sigma_{k_0} = \lambda_{A_{k_0}}$.

We put

$$S_{k_0} = \{\{x_1,\dots,x_p\} \subset A_{k_0} : 1 \le p \le N, \text{ and there exist integers } m_i \ge 1 \text{ with } \sum_{i=1}^{p} m_i = N \text{ and } \sum_{i=1}^{p} m_i x_i = z_k\}, \tag{5.17}$$

and select an element $s_{k_0} \in S_{k_0}$. For every $x \in s_{k_0}$ we apply (5.14), and we apply (5.13) to all $x \notin s_{k_0}$, $x \in A_{k_0}$. This will give us the measure σ_{k_0+1}, and (5.15) yields M_{k_0+1}. Since (5.16) is satisfied, σ_{k_0+1} fulfils (5.12). If we put

$$S_{k_0+1} = \{\{x_1,\dots,x_p\} \subset M_{k_0+1} \cup -M_{k_0+1} \cup \{0\}: 1 \le p \le n, \text{ and there exist integers } m_i \ge 1 \text{ with } \sum_{i=1}^{p} m_i = N \text{ and } \sum_{i=1}^{p} m_i x_i = z_{k_0+1}\}, \tag{5.18}$$

we clearly have $s_{k_0} \notin \pi_{k_0+1,k_0} S_{k_0+1} = \{\pi_{k_0+1,k_0} s : s \in S_{k_0+1}\}$, since $z^{(k_0+1)} \neq 0$. Let now $k \ge k_0$, and assume we have defined M_l, σ_l and S_l for $k_0 \le l \le k$, satisfying (5.9) - (5.12), where S_l is given by (5.18) with l replacing k_0+1, and such that

$$S_{k_0} \not\supseteq \pi_{k_0+1,k_0} S_{k_0+1} \not\supseteq \cdots \not\supseteq \pi_{k,k_0} S_k .$$

If $\pi_{k,k_0} S_k = \emptyset$, (5.8) is satisfied with $m = k$. Otherwise we choose an element $s_k \in \pi_{k,k_0} S_k$ and apply (5.14) to every $x \in A_k$ with $\pi_{k,k_0}(x) \in s_k$, and (5.13) to all other elements of A_k. This defines σ_{k+1} and M_{k+1} through (5.15). (5.16) is easily verified, and our construction implies that once again $\pi_{k+1,k_0} S_{k+1} \not\subseteq \pi_{k,k_0} S_k$. Since S_{k_0} is a finite set, this procedure has to stop after finitely many steps, i.e. there must exist an $m \ge k_0$ with $\pi_{m,k_0} S_m = \emptyset$, and such an m will satisfy (5.8).

<u>Lemma 5.7</u>. Theorem 5.1 holds for $A = \prod_{j=1}^{\infty} Z_{n_j}$.

Proof: Take $\varepsilon_i = i^{-1}$, $N_i = i$. By Lemma 5.6 there exist positive integers m_i, subsets M_i of $\prod_{j=m_{i-1}}^{m_i-1} Z_{n_j} = A^{(i)}$ ($m_0 = 1$), and measures $\sigma_i \in P(A^{(i)})$ such that the conditions (5.6) - (5.8) are satisfied for every $i \geq 1$, with σ_i, M_i, m_i and $A^{(i)}$ replacing σ, M, m and A_m. We put $M = \prod_{i=1}^{\infty} M_i, \sigma = \bigotimes_{i=1}^{\infty} \sigma_i$, and let B be the subgroup generated by M. Clearly B is a Borel set in A, and one checks easily from (5.8) that $B \neq A$, and that (5.2) - (5.4) are satisfied. (5.3) shows that B is dense in A, and this implies (5.1).

<u>Lemma 5.8</u>. Theorem 5.1 holds for $A = \Delta_p$, the group of p-adic integers.

Proof: The construction is similar to that of Lemma 5.6 and 5.7, and we shall only indicate what has to be changed. We write a typical element of a A as a $\sum_{j=0}^{\infty} a_j p^j = (a_0, a_1, \ldots)$, where $0 \leq a_j \leq p-1$. Let $A_k = Z_{p^k} = A/p^k.A$. By induction on k one can construct projective systems (M_k) and (σ_k), $k \geq 1$, as follows: Assume M_l and σ_l have been defined for $1 \leq l \leq k$, and let ε, N be given. We shall construct $k' > k$, $M_{k'}$, $\sigma_{k'}$ such that $|\hat{\sigma}_{k'}(\chi)| < \varepsilon$ for every character on $A_{k'}$ which is not induced by a character of A_k, and such that the N-fold sum of $M_{k'} \cup -M_{k'} \cup \{0\}$ does not cover $A_{k'}$. As in Lemma 5.6 we apply (5.13) until we have reached $k_0 > k$ such that $\sigma_{k_0}(x) < \varepsilon/N$ for every $x \in A_{k_0}$, and define S_{k_0} by (5.17). Put $z = (p-1, p-1, \ldots)$ and denote by z_n the image of z in every A_n. Take $s_{k_0} \in S_{k_0}$, and proceed as in Lemma 5.6, but now one has to repeat the extension r times with the same element, where r is chosen with $p^r > N$, in order to be sure that $s_{k_0} \notin \pi_{k_0+r, k_0} S_{k_0+r}$. Otherwise the proofs of the Lemmas 5.6 and 5.7 are unchanged.

Proof of Theorem 5.1.

If A has a nontrivial connected subgroup C, there exists a continuous surjective homomorphism $\chi : A \to T$. Hence T is a quotient of A, and the theorem holds by Lemma 5.3 and 5.5.

If A has no nontrivial connected subgroup, i.e. if A is zero-dimensional, we can apply [2, Theorem 7.7] to find a compact open subgroup of A, which is infinite since A is not discrete. By lemma 5.4. we may assume that A itself is compact and zero-dimensional. By [2, Theorem 24.26], $\hat{A}$ is a torsion group. If $\hat{A}$ contains a countable direct sum of cyclic groups, A has a quotient which is a countable direct product of cyclic groups, and Theorem 5.1 follows from Lemma 5.3 and 5.7. If $\hat{A}$ contains no countable direct sum of cyclic groups, [2,A.3] shows that $\hat{A}$ is a direct sum of p-groups. By our assumptions only finitely many terms in this sum are different from zero, i.e. $\hat{A}$ contains an infinite p-group Γ_1. [2,A.24] shows that Γ_1 contains a subgroup Γ_2 such that Γ_2 is the direct sum of cyclic groups and Γ_1/Γ_2 is divisible. According to our assumptions, Γ_2 has to be finite, and hence $(\widehat{\Gamma_1/\Gamma_2})$ is an open subgroup of Γ_1, and Lemmas 5.3 and 5.4 show that we may assume $\hat{A} = \Gamma_1/\Gamma_2$. By [2,A.14], $\hat{A}$ contains $Z(p^\infty)$ as a subgroup, i.e. Δ_p is a quotient of A for some $p \geq 2$, and the theorem now follows from Lemma 5.3 and Lemma 5.8.

6. The existence of uncountable dense subgroups which do not carry full measures.

Theorem 6.1. Let A be an uncountable locally compact second countable abelian group. Then there exists an uncountable dense Borel subgroup B of A with $\sigma(B) < 1$ for every $\sigma \in F(A)$.

The proof of Theorem 6.1 is similar to that of Theorem 5.1. We start with a few lemmas.

Lemma 6.2. Suppose Theorem 6.1 holds for a quotient group A/C, where C is a closed subgroup of A. Then it holds for A.

Proof: Let B be an uncountable dense subgroup of A/C which is a Borel set and which satisfies $\sigma(B) < 1$ for every $\sigma \in F(A/C)$. We write $\pi: A \to A/C$ for the quotient map and put $B_1 = \pi^{-1}(B)$. If there exists a full measure σ on A with $\sigma(B_1) = 1$, then $\sigma\pi^{-1} \in F(A/C)$ and $\sigma\pi^{-1}(B) = 1$, contrary to our assumptions.

Lemma 6.3. Suppose Theorem 6.1 holds for an uncountable closed subgroup C of A. Then it holds for A.

Proof: Let $B \subset C$ be a subgroup satisfying the conditions of Theorem 6.1. for C. We choose a non-atomic measure $\sigma \in P(C)$ with support C and with $\sigma(B) = 1$. By assumption, $\sigma \notin F(C)$. Using Definition 3.2. and Propositions 3.2 and 3.5 in [3] we can find a sequence $(\chi_n : n \geq 1)$ in $\hat{A}$ and a Borel map $f: C \to T$ such that $\lim_n \chi_n = \infty$ in $\hat{A}$, but $\lim_n \chi_n(a) = f(a)$ for σ-a.e. $a \in C$. Let D be a countable dense subgroup in A. By choosing a subsequence of (χ_n) if necessary we may assume that $(\chi_n(d))$ converges for every $d \in D$. Let now

$$B_1 = \{a \in A : \lim_n \chi_n(a) \text{ exists}\}.$$

B_1 is an uncountable dense Borel subgroup of A, since $D \subset B_1$ and $\sigma(B_1) = 1$. Every measure $\sigma_1 \in P(A)$ with $\sigma_1(B_1) = 1$ has the property that (χ_n) converges σ_1-a.e. to some Borel map $f_1: A \to T$. Applying once again 3.1, 3.2 and 3.5 in [3] we conclude that $\sigma_1 \notin F(A)$.

Lemma 6.4. Theorem 6.1 holds for $A = \prod_{j=1}^{\infty} Z_{n_j}$, an infinite product of cyclic groups.

Proof: Let B be the set of $x = (x_1, x_2, \dots) \in A$, $x_j \in Z_{n_j}$, for which $\{j : x_j \neq 0\}$ has density zero in the set of natural numbers. B is clearly a dense Borel subgroup of A. We now define a sequence of characters $(\chi_k : k \geq 1)$ in $\hat{A}$ as follows: if $x = (x_1, x_2, \dots) \in A$, put $\chi_k(x) = \exp(2\pi i x_k/n_k)$. For every $x \in B$ we have $\lim_N N^{-1} . \sum_{k=1}^{N} \chi_k(x) = 1$, and hence we get $\lim_N N^{-1} . \sum_{k=1}^{N} \hat{\sigma}(\chi_k) = 1$ for every $\sigma \in P(A)$ with $\sigma(B) = 1$. Clearly this violates condition (2) of Proposition 1.1, so that $\sigma \notin F(A)$.

Lemma 6.5. Theorem 6.1 holds for $A = \Delta_p$, the group of p-adic integers.

Proof: We use the same notation as in Lemma 5.8. Let $B \subset A$ be the set of all $(a_0, a_1, \dots) = \sum_{j=0}^{\infty} a_j p^j \in A$, $0 \leq a_j \leq p-1$, for which there exists a number $\bar{a}$, $0 \leq \bar{a} \leq p-1$, such that $\{j : a_j \neq \bar{a}\}$ is a set of natural numbers of density zero. $\hat{A}$ is isomorphic to $Z(p^\infty) = \{\exp(2\pi i k/p^n) : k, n \geq 0\}$. If $\chi = \exp(2\pi i k/p^n) \in \hat{A}$ and $a = \sum_{j=0}^{\infty} a_j p^j \in A$, then $\chi(a) = \exp(2\pi i a k/p^n)$. Put

$$\chi_j = \exp(2\pi i . \sum_{k=j+1}^{j+p-1} p^{-k})$$

Then

$$\chi_j(a) = \exp(2\pi i . \sum_{l=1}^{j+p-1} \sum_{k=j+1}^{j+p-1} a_{k-l} p^{-l}),$$

where $a_m = 0$ for $m < 0$. If $\varepsilon > 0$ is given, one can find a positive integer k such that $a_m = \bar{a}$ for $j-k \le m \le j+p-1$ implies $|\chi_j(a)-1| < \varepsilon$. For every $a \in B$, we can find an $\bar{a} \in \{0,\ldots,p-1\}$ such that

$$\{j > 0 : a_m \neq \bar{a} \text{ for some } j-k \le m \le j+p-1\}$$

is a set of natural numbers of density zero. Consequently we get $\lim_N N^{-1} \cdot \sum_{j=1}^{N} \chi_j(a) = 1$ for every $a \in B$. The same argument as in the preceding lemma completes the proof.

Theorem 6.1 is now proved in exactly the same way as Theorem 5.1.

Corollary 6.6. If A is uncountable, there exist nonatomic probability measures on A whose support is equal to A, but which are not full.

7. Concluding remarks

Clearly this paper raises more questions than it is able to answer. One intriguing problem is the characterization of all Borel groups $B \subset A$, $B \neq A$, which are dense and which satisfy $E_\sigma \subset B$ for every $\sigma \in P(A)$ with $\sigma(B) = 1$. We shall call such groups *saturated*. In [3, Chapter 8] there are examples of saturated subgroups of T: These groups arise as groups of eigenvalues of certain ergodic transformations. The following proposition shows that we have constructed another example of a saturated group in the proof of Lemma 6.4.

Proposition 7.1. Let $A = \prod_{j=1}^{\infty} Z_2$, and let

$$B = \{(x_1, x_2, \ldots) \in A : \{j : x_j \neq 0\} \text{ is a set of density zero}\}.$$

Then B is saturated, i.e. B is a dense Borel subgroup of A, distinct from A, and $E_\sigma \subset B$ for every $\sigma \in P(A)$ with $\sigma(B) = 1$.

Proof: Define $\chi_j \in \hat{A}$ by $\chi_j(x_1, x_2, \ldots) = 1-2x_j$. If $\sigma \in P(A)$ satisfies $\sigma(B) = 1$, we have

$$\lim_N N^{-1}. \sum_{j=1}^{N} \chi_j(a) = 1 \quad \text{for } \sigma\text{-a.e. } a \in A,$$

and hence

$$\lim_N N^{-1}. \sum_{j=1}^{N} \int \chi_j \, d\sigma = 1.$$

s is well known, this implies the existence of a set D of natural numbers of density zero such that $\lim_{\substack{j\to\infty \\ j\notin D}} \int \chi_j \, d\sigma = 1$, or, equivalently, hat $\lim_{\substack{j\to\infty \\ j\notin D}} \int |\chi_j - 1| \, d\sigma = 0$. From (1.1) we see that every $a \in E_\sigma$ must satisfy $\lim_{\substack{j\to\infty \\ j\notin D}} x_j = 0$ $(a = (x_1, x_2, \ldots))$, which in turn shows hat $a \in B$.

Bearing in mind Theorem 4.1 it seems natural to ask under which con-itions a saturated subgroup $B \subset A$ arises as the group of eigenvalues f a suitable ergodic action of the dual group $\hat{A}$. As far as the uthors are aware, very little is known about this problem.

cknowledgements: The authors would like to thank Ch.Berg for some helpful references and H. Rindler for an explicit example of a dense subgroup of R which satisfies the conditions of Theorem 5.1. Thanks re also due to W. Parry for his contribution to the proof of Lemma .1.

1. Feldman, J., and Moore, C.C.: Ergodic Equivalence Relations, Cohomology, and von Neumann Algebras. I. Trans Amer. Math. Soc. 234 (1977), 289-324.

2. Hewitt, E., and Ross, K.A.: Abstract Harmonic Analysis. I. Springer, Berlin 1963.

3. Moore, C.C., and Schmidt, K.: Coboundaries and Homomorphisms for Non-singular Actions, and a Problem by H. Helson. Preprint.

4. Parry, W.: A Note on Cocycles in Ergodic Theory. Compositio Math. 28 (1974), 343-350.

5. Salem, R.: On Sets of Multiplicity for Trigonometrical Series. Amer. J. Math. 64 (1942), 531-538.

6. Schmidt, K.: Lectures on Cocycles of Ergodic Transformation Groups. MacMillan Lectures in Mathematics I, MacMillan India, 1977.

V. Losert
Mathematisches Institut
Universität Wien
A-1090 Wien
AUSTRIA.

K. Schmidt
Mathematics Institute
University of Warwick
Coventry CV4 7AL
GREAT BRITAIN.

SOME REMARKS ON LIMITS OF ITERATES OF PROBABILITY MEASURES ON GROUPS AND SEMIGROUPS

By

Arunava Mukherjea

The purpose of this paper is to mention certain asymptotic properties of infinite convolutions of probability measures on certain groups and semigroups and to discuss some related questions concerning them. Here we do not intend to go into the depth of any particular topic. We propose mainly to make the reader familiar with certain problems in the context of infinite convolutions, namely the purity properties of the limits and certain questions on weak*-convergence, and to discuss briefly what results can be obtained and what can be expected.

In what follows, S is always a locally compact Hausdorff topological semigroup and P(S) (respectively,B(S)) denotes all regular probability measures (respectively, subprobability measures) on the Borel sets (generated by the open sets) of S. We write Q_1Q_2 to denote the usual convolution of Q_1 and Q_2 in B(S), and Q^n to denote the nth iterate of Q in B(S). If (Q_n) is a sequence in B(S), then we will write $Q_{k,n}$ to denote the convolution product $Q_{k+1}Q_{k+2}\cdots Q_n$. By the weak*-convergence of a sequence Q_n in B(S) to some Q in B(S), we mean that for every function f in $C_c(S)$= the class of continuous func tions on S with compact support, $\int f dQ_n$ converges to $\int f dQ$ as n tends to infinity. A measure Q in P(S) is called an infinite convolution if there exists a sequence Q_n in P(S) such that the sequence $Q_{0,n}$ weak*-converges to Q as n tends to infinity. Uptil now, there has been no result giving necessary and sufficient conditions for the weak*-convergence of a sequence $Q_{0,n}$ as above even when S is a compact group. However, much is known in the discrete case. A very interesting result in this context is due to V.M. Maximov (see[3]) which can be stated as follows:

THEOREM 1. Suppose S is a finite group and (Q_n) is a sequence in P(S) such that for some positive number t and for all n, $Q_n(e) \geq t$. Then for each nonnegative integer k, the sequence $Q_{k,n}$ weak*-converges to some Q'_k in P(S). (Here e is the identity of S.) □

Another result in this context when S is an at most countable discrete group is due to B.Center and A.Mukherjea. This is given below.

THEOREM 2. Suppose S is an at most countable discrete group and (Q_n) is a sequence in P(S). Then for all nonnegative integers k, the sequence $Q_{k,n}$ weak*-converges to some Q'_k in P(S) if and only if there exists a finite subgroup G such that the series $\sum_{n=1}^{\infty} Q_n(S-G)$ is convergent and for any proper subgroup G' of G and any selection of elements g_n in S, n=0,1,2,...., the series $\sum_{n=1}^{\infty} Q_n(S- g_{n-1}G'g_n^{-1})$ is divergent.□

The interesting nature of Theorem 2, though not immediately apparent, is revealed by the following corollary.

COROLLARY 3. Suppose S is an at most countable discrete group and has no non-trivial proper finite subgroup. Let (Q_n) be a sequence in P(S). Let us write r_n to denote $1-Q_n(e)$, e being the identity of S. Then the following results are valid:

(a) If the series $\sum_{n=1}^{\infty} r_n$ is convergent, then for all nonnegative integers k, the sequence $Q_{k,n}$ weak*-converges to some Q'_k in P(S).

(b) Suppose S is infinite and the above series is divergent. Then the sequence $Q_{k,n}$ does not weak*-converge to a probability measure for any nonnegative integer k.

(c) Suppose S is finite and the above series is divergent. Then the series $\sum_{n=1}^{\infty}(1-\sup(Q_n(x): x \in S))$ is divergent if and only if for all nonnegative integers k, the sequence $Q_{k,n}$ weak*-converges to some Q'_k in P(S) (and in this case, each Q'_k is the uniform measure on S). □

The above problem of convergence in the non-discrete situation is far from being settled. Some nice references in this context are Heyer(3), Csiszar(4), Maximov(6) and Tortrat(8). Though the problem in the general situation is quite non-trivial, it is not difficult to make certain special remarks. In the case of a compact group S and a periodic sequence Q_n in P(S) (that means, for some positive integer p, $Q_n=Q_{n+p}$ for all n), a result of K.Urbanik states that the sequence $Q_{0,n}$ weak*-converges to some Q'_0 in P(S) if and only if the closed subgroup generated by $\bigcup_{n=1}^{\infty} S_{Q_n}$, S_{Q_n} = the support of Q_n, is the same as that generated by the infinite union $\bigcup_{n=1}^{\infty} S_{Q_{0,n}}(S_{Q_{0,n}})^{-1}$; moreover, in the case of convergence, Q'_0 is the normed Haar measure of the above closed subgroups. In the non-compact group case, one can easily obtain a convergence result in the same periodic situation as above by using a result of A.Mukherjea stating that in a non-compact group S generated as a closed subgroup by the support of some Q in P(S), the sequence Q^n weak*-converges to the zero measure.

<u>THEOREM 4</u>. Let S be a locally compact non-compact group generated by $\bigcup_{n=1}^{\infty} S_{Q_n}$, where (Q_n) is a periodic sequence with period p in P(S). Then the sequence $Q_{0,n}$ weak*-converges to the zero measure as n tends to infinity if and only if for all s with $1 \leq s \leq p$, the set $S_{Q_s} S_{Q_{s+1}} \dots S_{Q_{s+p-1}}$ is not contained in a compact subgroup. □

Now we recall an important theorem of I.Csiszar(4) (this theorem in a slightly different form also discovered independently by A.Tortrat(8)) stated as follows: In a second countable group S, either the sequence $\sup(Q_{0,n}(Kx):x \in S)$ tends to zero as n tends to infinity for every compact set K or there exist elements a_n in S such that for all non-negative integers k, the sequence $Q_{k,n}a_n$ weak*-converges to some Q'_k in P(S). It is easy to see from this theorem that the weak*-limit points of the sequence $Q_{k,n}$, where Q_n are in P(S) and S is a non-compact second countable group, are always either zero or elements of P(S). Also it is possible to state a simple, though quite restrictive, result in this context in the non-periodic situation as follows: Suppose S is a non-compact group and H is a compact normal subgroup of S such that S_{Q_n} is a subset of Ha_n for elements a_n in S and Q_n in P(S). Then the sequence $Q_{0,n}$ weak*-converges to the zero measure if and only if $a_1a_2...a_n$ tends to infinity as n tends to infinity.

The convergence questions for the sequence $Q_{k,n}$, though unsettled in the general non-discrete situation, are, as is well-known for a long time, completely settled when S is the real line or the circle and can be answered completely by the well-known three series theorem. One good reference on this is the classical van Kampen paper[9].

The next question that comes up naturally is what we can say about those measures which are infinite convolutions. The Brown-Moran[1] paper and the Hartman[2] paper are two recent references in this context. Both discuss purity properties of infinite convolutions in abelian groups. Let U be a class of Borel sets on S, which is closed under countable unions and with the property that whenever A is an element of U, the translate Ax, x in S, is also an element of U. Then Q in P(S) is called a pure probability measure if it has the following property with respect to every such class U: If Q(A) is positive for some A in U, then Q(B)=1 for some B in U. The measure Q is called continuous if Q(x)=0 for every singleton x in S, and purely discontinuous if Q(A)=1 for some countable set A. If m is a Haar measure on the group S, then Q is called absolutely continuous if Q(B)=0 whenever m(B)=0, and continuous singular if it is continuous and Q(B)=1 for some B with m(B)=0. Note that a pure probability measure on a group is always purely discontinuous or continuous singular or absolutely continuous. (This follows easily by considering U as the class of at most countable sets or the class of m-null sets.) Following essentially the same kind of proof as in van Kampen's paper or Hartman's paper, assertions similar to those made in these papers regarding the purity nature of the infinite convolutions can be made even in groups which are not necessarily abelian. To make this paper self-contained, we describe only a few of these results and sketch briefly their proofs(though the proofs run along the same lines as in the classical case).

THEOREM 5. Suppose that Q_n's are elements of P(S) and $Q_{0,n}$ weak*-converges to some Q in P(S). Then Q is discontinuous if $\prod_{n=1}^{\infty} \max_{x\in S} Q_n(x)$ is positive. □

Proof. Choose x_n in S so that $\prod_{n=1}^{\infty} Q_n(x_n) = d$ (positive). Write: $y_n = x_1x_2\cdots x_n$. Then $Q_{0,n}(y_n) \geq Q_1(x_1)Q_2(x_2)\cdots Q_n(x_n) \geq d$. This means that the sequence y_n must have a cluster point y and $Q(y) \geq d$. Q.E.D.

LEMMA 6. Suppose that S is a group and $Q_1=Q_2Q_3$ for Q_1, Q_2 and Q_3 in P(S). Suppose also that for some x in S, $Q_1(x)=d>0$; and V_n is an open neighborhood of e such that $V_n=V_n^{-1}$ and

$$d \leq Q_1(xV_n^2) < d+(1/n) \text{ and } Q_3(V_n) > 1-(1/n).$$

Then there exist elements y_n, z_n in S such that $x=y_nz_n$, $z_n \in V_n$ and

$$|Q_2(y_n)-d| < 2/n \text{ and } Q_3(z_n) > 1-(6/nd). \quad \square$$

Proof. Since $Q_1=Q_2Q_3$, we have

$$d = Q_1(x) \leq \int_{V_n} Q_2(xs^{-1})Q_3(ds) + (1/n).$$

Therefore, there is z_n in V_n such that $Q_2(xz_n^{-1}) \geq d-(1/n)$. Write: $y_n=xz_n^{-1}$. Notice that

$$d+(1/n) > Q_1(xV_n^2) \geq \int_{V_n} Q_2(xV_n^2s^{-1})Q_3(ds) \geq Q_2(y_n)-(1/n).$$

Finally,

$$d = \int Q_2(xs^{-1})Q_3(ds) \leq Q_2(y_n)Q_3(z_n) + \int_{V_n-z_n} Q_2(xs^{-1})Q_3(ds) +(1/n)$$

or

$$d \leq Q_2(y_n)Q_3(z_n) +(4/n) \leq (d+(2/n))Q_3(z_n)+(4/n).$$

The lemma now follows easily. Q.E.D.

THEOREM 7. Suppose that S is a group and $Q_n \in P(S)$. Suppose also that for every open neighborhood V of e, $\lim_k \inf_{n>k} Q_{k,n}(V) = 1$. Then $Q_{0,n}$ weak*-converges to some Q in P(S). If Q is discontinuous, then the infinite product $\prod_{n=1}^{\infty} \max_{x \in S} Q_n(x)$ is positive. □

Proof. It is easy to see that by the "lim inf" condition, the sequence $Q_{k,n}$ weak*-converges to some Q'_k in P(S) for each nonnegative integer k. Write: $Q=Q_{0,n}Q'_n$. Let V_n be a sequence of decreasing symmetric neighborhoods of e such that

$$d \leq Q(xV_n^2) < d+(1/n), \text{ where } Q(x)=d > 0.$$

Again, by the "lim inf" condition in the theorem there exist a sequence (increasing) of positive integers k_n such that for each positive integer $p \geq k_n$, we have:

$$Q'_p(V_n) > 1-(1/n).$$

By Lemma 6, we can find for each $p \geq k_n$ elements y_{np} and z_{np} with z_{np} in V_n such that

$$\left| Q_{0,p}(y_{np}) - d \right| < 2/n \text{ and } Q'_p(z_{np}) > 1-(6/nd).$$

Write for $k_n \leq p < k_{n+1}$, $z_p = z_{np}$. Then z_p converges to e and $Q'_p(z_p)$ converges to 1 as p tends to infinity. Writing P_k for $z_k^{-1}Q_k z_{k+1}$, we have: $P'_k(e)$ converges to 1 as k tends to infinity, where P'_k is the weak*-limit of the sequence $P_{k,n}$. Now for k sufficiently large,

$$\begin{aligned} P'_k(e) &= \int P'_{k+1}(es^{-1})P_k(ds) \\ &= P'_{k+1}(e)P_k(e) + \int_{S-e} P'_{k+1}(s^{-1})P_k(ds) \\ &\leq P'_{k+1}(e) + (1-2P'_{k+1}(e))(1-P_k(e)) \\ &\leq P'_{k+1}(e) - (1/3)(1-P_k(e)). \end{aligned}$$

Repeating this process n times, we have: the series $\sum_{n=1}^{\infty}(1-P_n(e))$ is convergent implying that the infinite product $\prod_{n=1}^{\infty} P_n(e)$ is positive. This means that the infinite product $\prod_{n=1}^{\infty} Q_n(z_n z_{n+1}^{-1})$ is positive. Q.E.D.

THEOREM 8. Suppose that S is a group and the Q_n's are as in Theorem 7 satisfying the "lim inf" condition there. Suppose also that each Q_n is purely discontinuous. Then the limit measure Q is discrete or continuous singular or absolutely continuous. □

Proof. Suppose Q is continuous. If Q is not absolutely continuous, then there exists a Borel set B with m(B)=0, but Q(B) > 0. Let x be an element of $S - \bigcup_{n=1}^{\infty} S_{Q_n}$. Let H be the countable group generated by x and the supports of the measures Q_n. Consider the infinite product measure space and P be the product of the Q_n's. Define the random variables $X_1, X_2, \ldots$ on this space such that for a typical point s= $(s_1, s_2, \ldots)$ in this space,

$$X_n(s)=s_n \text{ if } s_n \in S_{Q_n}, \ =x \text{ otherwise.}$$

Then the sequence X_n is independent and the sequence $Z_n = X_1 X_2 \ldots X_n$, by the "lim inf" condition, is stochastically convergent. It follows that Z_n converges almost surely to some Z with distribution Q. It is clear that the set $\{Z \in HB\}$ is shift-invariant and therefore, a remote event. Consequently, it has probability one by the zero-one law so that Q(HB)=1, whereas m(HB)=0 (m is assumed here a left-invariant Haar measure on S). Q.E.D.

Note that the same proof above shows that Q in Theorem 8 above is purely discontinuous when it is not continuous, and in fact, Q is pure.

One natural question now is whether Theorems 7 & 8 have analogs in semigroups. Theorem 5 is, of course, valid in semigroups. An easy example showing that Theorem 7 need not be valid even in compact semigroups is the following: Let S= [0,1] under multiplication and usual topology. Choose Q_n to be any probability measure (for example, any measure with an n-point support and having mass (1/n) at each point of its support) with its support contained in [0,1/n). Then it is clear that $S_{Q_{k,n}} \subset [0,1/n)$ and $Q_{k,n}$ weak*-converges to the unit mass at 0. So Q here is discontinuous whereas the infinite product $\prod_{n=1}^{\infty} \max_{x \in S} Q_n(x)$ need not be positive. This argument extends obviously to any compact semigroup with a zero. To understand what purity can mean in the context of semigroups, let us consider the semigroup of stochastic matrices of order two under multiplication and usual topology. Let us denote the matrix whose column elements are a and b by the point (a,b). Thus (1,0) represents the matrix $\binom{1\ 0}{0\ 1}$. Then the smallest two-sided ideal (called the kernel) of this compact semigroup consists of all elements of the form (a,a) where a is a number between 0 and 1 (0,1 inclusive). Let P be a probability measure on S whose support S_P contains a matrix where the entries are all nonzero. Then the closed subsemigroup generated by S_P intersects K (the kernel) and P^n weak*-converges to a probability measure Q with support contained in K. Noting that K is a right-zero semigroup, it is not difficult to verify that Q here is the unique solution of the convolution equation XP=X, where X is an element of P(K).

THEOREM 9. Suppose P is a purely discontinuous probability measure on the (above) compact semigroup of stochastic matrices such that the weak*-limit of P^n is the probability measure Q. Then $S_Q \subset K$. If all the points in S_P don't lie on the same line with (1,0), then Q is continuous. Moreover, the measure Q is either a point mass or continuous singular or absolutely continuous with respect to the Lebesgue measure on [0,1] (here identifying K with [0,1] and the semigroup S with the unit square). □

The proof of this theorem easily follows from the discussions on pages 114-115 of [7], even though P here has an at most countable (rather than two-point) support. Now let us examine what can happen in the above context if we consider the weak*-convergence of a sequence $Q_{0,n}$ where the individual probability measures are not all the same.

In the above semigroup of stochastic matrices, let us consider the sequence Q_n defined as follows:

(i) $Q_1(\frac{1}{4},\frac{1}{4})=\frac{1}{2}=Q_1(1,1)$;

(ii) for each positive integer n greater than 1, let

$$Q_n(1-3/(10)^n, 1/(10)^n)=\tfrac{1}{2}=Q_n(1,0).$$

Notice that $(p,p)(1-3/(10)^n,1/(10)^n)=(p+(1-4p)/(10)^n, p+(1-4p)/(10)^n)$.

Notice also that when $p=\frac{1}{4}$, $1-4p=0$. and when $p=1$, $(1-4p)/(10)^n = -3/(10)^n$. After doing some computations, it is then clear that the measure $Q_{0,n}$ has its support consisting of $1+2^{n-1}$ points contained in the kernel K; among these points, the point $(\frac{1}{4},\frac{1}{4})$ has mass $\frac{1}{2}$ and all other points have mass $1/2^n$. It is not difficult to see that $Q_{0,n}(\frac{1}{4},x))$ is an increasing function of x as n tends to infinity. (Here we are, of course, identifying K with $[0,1]$.) Defining Q on the open subintervals as the limit of the measures of $Q_{0,n}$ on these intervals, and then extending Q by the usual extension procedure to a Borel measure, we see that the sequence $Q_{0,n}$ weak*-converges to Q. The reason is as follows: Consider any cluster point of $Q_{0,n}$, say Q'; then it is clear that for any open interval I, $Q'(I) \leq \liminf Q_{0,n_1}(I)$ for some subsequence (n_1), which means that for any open interval (and therefore, for any open set) I, $Q'(I) \leq Q(I)$. Since Q and Q' are both regular probability measures and also since $Q'(C) \geq Q(C)$ for any closed set $C \subset [0,1]$, it follows that $Q=Q'$. Thus all weak*-cluster points of the sequence $Q_{0,n}$ are the same as Q, proving that $Q_{0,n}$ weak*-converges to Q. Now we claim that Q is not pure. To see this, first note that all the 2^{n-1} points in the support of $Q_{0,n}$ (excluding $(\frac{1}{4},\frac{1}{4})$) are contained in the interval $(3/4,1)$ because of the following **assertion:**

$$(1,1)(1-3/(10)^2,1/(10)^2).....(1-3/(10)^n,1/(10)^n) = (x,x)$$

where $x \geq 1-3/(10)^2-3/(10)^3-.... \geq 3/4$. It is clear that $Q(\frac{1}{4})= \frac{1}{2}$ and $Q([3/4,1])=\frac{1}{2}$. Our claim will be proven if we show that Q on $[3/4,1]$ is absolutely continuous with respect to the Lebesgue measure on this interval. To show this, let $t > 0$. Choose a positive integer N so that $1/2^{N-1} < t$. Let s be a positive number so small that an interval of length smaller than s **can contain** at the **most only** one point of $S_{Q_{0,N}}$. (It is enough to take s smaller than $3/(10)^N$.) Let I be such a subinterval of $[3/4,1]$. Suppose $a_j \in S_{Q_{0,N}} \cap I$. One can easily verify by writing the points of $S_{Q_{0,N}}$ in increasing order as

$a_1 < a_2 < \ldots < a_j < a_{j+1} < \ldots < a_{1+2^{N-1}}$ that for all positive integers k, $S_{Q_{0,N+k}} \cap I$ can contain only points from the set

$$A = \{a_j, a_{j+1}\} \cdot S_{Q_{N+1}} S_{Q_{N+2}} \ldots S_{Q_{N+k}}.$$

It follows that

$$Q_{0,N+k}(I) = \int Q_{N,N+k}(x^{-1}I) Q_{0,N}(dx)$$
$$\leq Q_{0,N}(a_j) + Q_{0,N}(a_{j+1}) = 1/2^N + 1/2^N < t.$$

This proves our claim. Hence our discussions above show the following: "In the compact semigroup of stochastic matrices where the kernel can be identified with $[0,1]$, there are sequences Q_n of purely discontinuous measures such that $S_{Q_{0,n}} \subset K$ and $Q_{0,n}$ weak*-converges to some probability measure Q, where Q is neither continuous nor purely discontinuous."

Finally, in this paper we again consider the problem of weak*-convergence of iterates of probability measures. Although not much is known about the weak*-convergence of the sequence $Q_{k,n}$ as we have mentioned in the beginning of this paper, relatively much more can be said on the weak*-convergence of the sequence Q^n even in the case of many general semigroups. The problem is completely solved in the case of discrete semigroups by Martin-Löf in his 1965 paper of Z.W. It is also solved for compact or completely simple semigroups- see Theorems 4.13, 4.14, 4.15, 4.17 & 4.18 in [7]. A result of the author states that in a completely simple semigroup S generated by the support of a probability measure Q, the sequence Q^n weak*-converges to the zero measure if and only if the group factor G in S is non-compact. It seems that the behavior of Q^n in the case of non-compact semigroups may very well depend upon the existence of certain completely simple subsemigroups. To be more precise, we conjecture the following: Let S be a locally compact non-compact semigroup generated by the support of a probability measure Q. Then either there exists a closed ideal I with $Q^n(I)=0$ for all n and $Q^n(K)$

converges to zero as n tends to infinity for all compact sets K such that $K \cap I = \emptyset$ or there is no such ideal in which case the sequence Q^n does not weak*-converge to zero if and only if there is a closed ideal of S which is completely simple with a compact group factor. This conjecture is supported by results of Martin-Löf in discrete semigroups and by recent results of Högnäs and the author in semigroups of matrices. We end this paper by describing briefly some of these results.

Let M_n be the multiplicative semigroup of $n \times n$ matrices with usual topology and let S be any locally compact subsemigroup of M_n. Suppose Q is a probability measure such that $S = \overline{\bigcup_{k=1} S_Q^k}$. Let us write: $Q' = \sum_{k=1}^{\infty} (Q)^k/2^k$, $T_r = \{x \in S: \text{rank}(x) \leq r\}$ and $a = \min\{r : Q'(T_r) > 0\}$. Let H be the set of all recurrent points of S, that is, points x in S such that $\sum_{k=1}^{\infty} Q^k(V_x) = \infty$ for every open set V_x containing x. Note that if Q^n does not weak*-converge to zero, then H is non-empty (in fact, H then contains positive recurrent points). Högnäs and Mukherjea have proven the following result: If $Q'(H) > 0$, then $H = T_a$, and T_a contains a dense completely 0-simple subsemigroup whose complement in T_a has Q' measure zero. In the case when S is abelian, one can say more. The following is a result of Högnäs and Mukherjea.

<u>THEOREM 10</u>. Let S be a locally compact abelian subsemigroup of M_n generated by S_Q for Q in P(S). Then either $H \subset T_{a-1}$ or $H = T_a$ in which case $T_a - T_{a-1}$ is a group. Moreover, if H' is the set of positive recurrent points x, that is, points x such that $\limsup Q^k(V_x) > 0$ for every open set V_x containing x, then either $H' \subset T_{a-1}$ or T_{a-1} is empty and $H' = T_a - T_{a-1} = T_a$ is a compact group. □

Proof. Let us sketch the arguments very briefly. T_a is a closed ideal of S and $\sum_{k=1}^{\infty} Q^k(T_a^c) < \infty$. Hence, $H \subset T_a$. Choose x in $T_a - T_{a-1}$ so that $Q'(x^{-1}T_{a-1}) = 0$. Write: $C_a = \{y \in T_a : xy \in T_a - T_{a-1}\}$. Then by going through some straightforward rank and null-space arguments, it can be shown that $T_a = C_a \cup T_{a-1}$ and C_a is a semigroup. Now if one assumes that $H \cap C_a$ is nonempty, then the group property of C_a follows by considering C_a embedded in the group $G = \{y \in M_n$: Range(y)=Range(x) and Null-space(y)=Null-space(x)$\}$ and then following some ideas as given on p.192 of [7]. To make the assertions on H', we assume that $H' \cap C_a$ is nonempty. Since $H' \cap C_a$ is an ideal of the group C_a, $H' \supset C_a$. Then we consider the measure $Q_e(=Q(.e^{-1}))$, where e is the identity of C_a. Then noting that $Q'(e^{-1}T_{a-1})=0$, we see that the restriction of of $(Q_e)^k$ to C_a is the kth convolution power of the restriction of Q_e to C_a. If C_a is a non-compact group, then by an earlier result of the author, $(Q_e)^k(K \cap C_a)$ converges to zero for every compact subset $K \subset C_a$. Since $(Q_e)^k = Q^k e$, this means that no point of C_a is positive recurrent, a contradiction. Since $T_{a-1} \subset \overline{C_a}$, T_{a-1} is empty. Q.E.D.

Acknowledgment: The author's research is supported by a National Science Foundation (USA) grant. Some of the discussions on the purity problem arise from some joint work with A.Nakassis and form a part of a future paper.

REFERENCES

1. Brown,G and Moran,W.:Z.Wahrscheinlichkeitstheorie verw. Gebiete 30,227-234(1974).

2. Hartman,Philip: Trans. Amer. Math. Soc.,Vol.214,215-231(1975).

3. Heyer,H.: Symposia Mathematica XVI,Academic Press,New York, 315-355(1975).

4. Csiszar,I.: Z.Wahrscheinlichkeitstheorie verw. Gebiete 5,279-295(1966).

5. Heyer,H.: Probability measures on locally compact groups (Ergebnisse series 94),Springer-Verlag,Berlin-Heidelberg-New York(1977).

6. Maximov,V.M.: Theory of Probability & its applications 16,No.1, 55-73(1971).

7. Mukherjea,Arunava and Tserpes,Nicolas: Lecture Notes in Math, Vol.547,Springer-Verlag,Berlin-Heidelberg-New York(1976).

8. Tortrat,A.: Ann.Inst.Henri Poincaré 1,217-237(1965).
9. van Kampen,E.R.: Amer.J.Math 62,417-448(1940).

Arunava Mukherjea
University of South Florida
Tampa
Florida 33620
USA

Infinitely divisible distributions in SL(k,ℂ) or SL(k,ℝ) may be imbedded in diadic convolution semigroups

by

K.R. Parthasarathy

Summary

It is shown that every infinitely divisible distribution λ in SL(k,ℂ) or SL(k,ℝ) can be imbedded in a diadic convolution semigroup if the smallest closed subgroup containing the support of λ is the whole group.

Let G = SL(k,ℂ) or SL(k,ℝ), $K \subset G$, the maximal compact subgroup of all unitary or orthogonal matrices, $A^+ \subset G$, the subgroup of all diagonal matrices with positive entries and $M(G)$ be the topological semigroup of probability measures (also called distributions) on G under convolution and weak topology. Elements of G and distributions degenerate at these elements will be denoted by the same symbol. Weak convergence will be denoted by $\Rightarrow$. By abuse of language conditionally compact sequences will be called compact sequences. Rest of the notations and terminology will be as in [1].

Lemma 1 Let $\{\lambda_n\}$ be a sequence in $M(G)$, $\lambda_n \Rightarrow \lambda$ and let the smallest closed subgroup containing the support of λ be G. Suppose $\lambda_n = \mu_n^{*2}$ for every n. Then $\{\mu_n\}$ has a weakly convergent subsequence.

Proof By the shift compactness theorem (Theorem 2.2, page 59, [1]) there exists a sequence $\{x_n\}$ in G such that the sequences

$\{\mu_n * x_n\}$ and $\{x_n^{-1} * \mu_n\}$ are compact. We can decompose x_n as $x_n = k_n a_n u_n$, where k_n and u_n belong to K and $a_n \in A^+$. Indeed, this follows from the polar decomposition of a nonsingular matrix into a positive definite and unitary matrix.

Let $y_n = k_n a_n$. Since K is compact it follows that $\{\mu_n * y_n\}$ and $\{y_n^{-1} * \mu_n\}$ are compact. The lemma will be proved if we show that $\{a_n\}$ is compact in G. To this end we put

$$\tilde{\lambda}_n = k_n^{-1} * \lambda_n * k_n$$

and assume that $k_n \to k$. Then $\tilde{\lambda}_n \Rightarrow \tilde{\lambda} = k^{-1} * \lambda * k$.

Further the smallest closed subgroup containing the support of $\tilde{\lambda}$ is G. Since

$$a_n^{-1} * \tilde{\lambda}_n * a_n = (y_n^{-1} * \mu_n) * (\mu_n * y_n)$$

it follows that the sequence $\{a_n^{-1} * \tilde{\lambda}_n * a_n\}$ is compact.

Suppose now that $\{a_n\}$ is not compact. We can write

$$a_n = \begin{pmatrix} \theta_{n1} & 0 \ldots\ldots\ldots\ldots 0 \\ 0 & \theta_{n2} \quad 0 \ldots\ldots 0 \\ \ldots\ldots\ldots\ldots & \ldots\ldots\ldots\ldots \\ 0 & 0 \ldots 0 \quad \theta_{nk} \end{pmatrix} \tag{1}$$

where $\theta_{n1}, \theta_{n2}, \ldots, \theta_{nk}$ are scalars satisfying the relation $\theta_{n1}\theta_{n2} \ldots \theta_{nk} = 1$. It follows from our assumption and (1) that at least one of the sequences $\{\theta_{nj}\}$ is bounded and at least one of them is unbounded. Without loss of generality we may assume the existence of an ℓ such that

$$\begin{aligned} \sup_n \theta_{nj} &< \infty \quad \text{for} \quad 1 \leq j \leq \ell, \\ \lim_{n\to\infty} \theta_{nj} &= \infty \quad \text{for} \quad \ell+1 \leq j \leq k. \end{aligned} \tag{2}$$

Let H denote the subgroup of all matrices $x = ((x_{ij}))$ in G for which $x_{ij} = 0$ for all $1 \leq i \leq \ell$, $\ell+1 \leq j \leq k$. It follows from the property of the measure $\tilde{\lambda}$ that

$$\tilde{\lambda}\,(H') = \delta > 0 \tag{3}$$

where H' is the complement of H in G. Since H' is open we can find a compact set $C \subset H'$ such that the interior C^o of C satisfies

$$\tilde{\lambda}\,(C) \geq \tilde{\lambda}\,(C^o) \geq \delta/2 \quad . \tag{4}$$

Since $\tilde{\lambda} \Rightarrow \tilde{\lambda}$ and therefore

$$\underline{\lim}\ \tilde{\lambda}_n(C^o) \geq \tilde{\lambda}(C^o)$$

there exists an n_o such that

$$\tilde{\lambda}_n(C) \geq \tilde{\lambda}_n(C^o) \geq \delta/4 \text{ for all } n \geq n_o. \tag{5}$$

The compactness of $\{a_n^{-1} * \tilde{\lambda}_n * a_n\}$ implies the existence of a compact set E such that

$$(a_n^{-1} * \tilde{\lambda}_n * a_n)\,(E) \geq 1 - \delta/8 \text{ for all } n. \tag{6}$$

Now (5) und (6) imply that

$$\tilde{\lambda}_n(C \cap a_n\, E\, a_n^{-1}) \geq \delta/8 \text{ for all } n \geq n_o \ .$$

Thus

$$C \cap a_n\, E\, a_n^{-1} \neq \phi \text{ for all } n \geq n_o \ .$$

In other words there exist $x_n \in E$ such that $a_n\, x_n\, a_n^{-1} \in C$ for every n. Since E and C are compact we may assume that

$$\lim_n \quad x_n = x \quad ,$$

$$\lim_n \quad a_n\, x_n\, a_n^{-1} = y \ , \quad y \in C$$

exist. But (1) implies that

$$(a_n x_n a_n^{-1})_{ij} = \epsilon_{ni} \epsilon_{nj}^{-1} (x_n)_{ij} .$$

Taking limits we conclude from (2) that

$$y_{ij} = 0 \quad \text{if } 1 \leq i \leq \ell , \ell+1 \leq j \leq k .$$

In other words $y \in H$. Since $C \subset H'$ this is a contradiction. This completes the proof of the lemma.

Corollary 2 Let λ be an infinitely divisible distribution on G and let the smallest closed subgroup containing the support of λ be G. Suppose $D = \{d : d \geq 0, d$ is a diadic rational number$\}$. Then there exists a convolution semigroup $\{\lambda_d, \ d \in D\}$ such that $\lambda_1 = \lambda$.

Proof Since λ is infinitely divisible we can write

$$\lambda = \lambda_n^{*2^n} \text{ for every } n .$$

Put $\mu_n = \lambda_n^{*2^{n-1}}$ and observe that $\lambda = \mu_n^{*2}$. By Lemma 1, μ_n has a convergent subsequence $\{\mu_m\}$ converging to a limit $\lambda_{2^{-1}}$ satisfying

$$\lambda = \lambda_{2^{-1}}^{*2} ,$$

and the smallest closed subgroup containing the support of $\lambda_{2^{-1}}$ is also G. Now

$$\lim_{m\to\infty} \lambda_m^{*2^{m-1}} = \lambda_{2^{-1}} .$$

Hence by Lemma 1 again we observe that $\{\lambda_m^{*2^{m-2}}\}$ is compact. We take a convergent subsequence of this with limit $\lambda_{2^{-2}}$.
Then

$$\lambda_{2^{-2}}^{*2} = \lambda_{2^{-1}} .$$

Continuing this procedure we construct $\{\lambda_{2^{-r}}\}$ such that

$$\lambda_{2^{-r}}^{*2} = \lambda_{2^{-(r-1)}} \text{ for } r = 1,2, \ldots .$$

For any $d = m2^{-r}$, where m and r are non-negative integers define

$$\lambda_d = \lambda_{2^{-r}}^{*m} \quad .$$

Then $\{\lambda_d\}$ is the required convolution semigroup.

Remark It is still an open problem whether λ satisfying the conditions of the Corollary can be imbedded in a one paramenter convolution semigroup.

Reference

1. K.R. Parthasarathy, Probability Measures on Metric Spaces, Academic Press, New York 1967.

K.R. Parthasarathy
Indian Statistical Institute
Delhi

THEOREME DE LA LIMITE CENTRALE POUR UN PRODUIT SEMI-DIRECT D'UN GROUPE DE LIE RESOLUBLE SIMPLEMENT CONNEXE DE TYPE RIGIDE PAR UN GROUPE COMPACT

par Albert RAUGI

Soit R un groupe de Lie résoluble connexe. Nous désignons par Ad la représentation adjointe de R. Nous disons que R est de type rigide si les valeurs propres des éléments de Ad R sont toutes de module égal à 1.

Soit G un produit semi-direct d'un groupe de Lie résoluble simplement connexe de type rigide par un groupe compact. Si λ est une mesure de probabilité sur les boréliens de G, nous savons que ([1]), lorsque G n'est pas compact, la $n^{\text{ième}}$ convolée, λ^n, de λ converge vaguement vers zéro. Nous nous proposons de trouver une suite d'homéomorphismes $\{U_n\}_{n\geq 1}$ de G, tels que la suite de mesures de probabilité $\{U_n(\lambda^n)\}_{n\geq 1}$ converge vaguement vers une mesure de probabilité sur G pas trop dégénérée.

Nous commençons (partie I) par étudier le cas d'un produit semi-direct d'un groupe de Lie nilpotent simplement connexe par un groupe compact. Nous montrerons par la suite (partie II) que le cas général se ramène à ce cas particulier.

I . Théorème de la limite centrale pour un produit semi-direct d'un groupe de Lie nilpotent simplement connexe par un groupe compact .

Soit $G = N \times_{\eta} K$ le produit semi-direct d'un groupe de Lie nilpotent simplement connexe N par un groupe compact K associé à un homomorphisme continu η de K dans le groupe des automorphismes de N. Le produit de deux éléments $g = (u,k)$ et $g' = (u',k')$ de G est défini par $gg' = (u\,[\eta(k)(u')],\, kk')$.

Soit λ une mesure de probabilité sur les boréliens de G. Nous savons ([1]) que la $n^{\text{ième}}$ convolée, λ^n, de λ converge vaguement vers zéro. Nous nous proposons de trouver une suite d'homéomorphismes(si possible d'automorphismes) $\{V_n\}_{n\geq 1}$ de G, tels que la suite de mesures de probabilité $\{V_n(\lambda^n)\}_{n\geq 1}$ converge vaguement vers une mesure de probabilité sur G, non dégénérée, en un sens à préciser.

1. Analyse du Problème

Nous désignons par π_1 et π_2 les projections de G respectivement sur N et K ; nous notons λ_1 et λ_2 les images de λ respectivement par π_1 et π_2 .

(1.1) La projection $\pi_2(\lambda^n)$ de λ^n sur K est la $n^{\text{ième}}$ convolée de λ_2 . Désignons par $K(\lambda_2)$ (resp. $\tilde{K}(\lambda_2)$) le sous-groupe compact de K engendré par le support $S(\lambda_2)$ de λ_2 (resp. par $S(\lambda_2)\ [S(\lambda_2)]^{-1}$). $\tilde{K}(\lambda_2)$ est un sous-groupe distingué de $K(\lambda_2)$; nous disons que λ_2 est apériodique si $\tilde{K}(\lambda_2) = K(\lambda_2)$. Nous savons que, ([7]), pour tout élément k de $S(\lambda_2)$, la suite de mesures de probabilité $\{\varepsilon_{k^{-n}} * \lambda_2^n\}_{n\geq 1}$ converge vaguement vers la mesure de Haar normalisée $\tilde{m}$ de $\tilde{K}(\lambda_2)$. En particulier quand λ_2 est apériodique, la suite de mesures de probabilité $\{\lambda_2^n\}_{n\geq 1}$ converge vaguement vers la mesure de Haar normalisée m de $K(\lambda_2)$.

D'autre part, nous avons, avec les notations précédentes :

(1.2) Proposition.

Soit $(X_i)_{i\geq 1}$ une suite de v.a. indépendantes définies sur un espace probabilisé (Ω, a, P) , à valeurs dans G , et de loi commune λ . Soit $\{U_n\}_{n\geq 1}$ une suite d'homéomorphismes de N tels que : la suite de mesures de probabilité sur N $\{U_n \circ \pi_1(\lambda^n)\}_{n\geq 1}$ converge en loi vers une mesure de probabilité ν ; et, pour tout entier naturel p , pour toute fonction f continue, à support compact sur N ,

$$\lim_n E[|f \circ U_n \circ \pi_1(X_1 \dots X_n) - f \circ U_n \circ \pi_1(X_1 \dots X_{n-p})|] = 0 .$$

Pour tout élément k de K, et tout entier naturel n , posons

$$V_n^k(g) = (U_n \circ \pi_1(g) , k^{-n}\pi_2(g)).$$

Alors, pour tout élément k de $S(\lambda_2)$, la suite de mesure de probabilité sur G $\{V_n^k(\lambda^n)\}_{n\geq 1}$ converge vaguement vers la mesure de probabilité produit $\nu \otimes \tilde{m}$. En particulier lorsque λ_2 est apériodique, $V_n^e(\lambda^n)$, (e désignant l'élément neutre de K), converge vaguement vers la mesure produit $\nu \otimes m$.

De (1.1) et (1.2), il résulte que le problème considéré se ramène à trouver une suite d'homéomorphismes $\{U_n\}_{n\geq 1}$ de N possèdant les deux propriétés énoncées dans la proposition (1.2) et pour laquelle la limite vague ν de $U_n \circ \pi_1(\lambda^n)$ soit non dégénérée (i.e. non portée par une sous variété propre de N).

(1.3) Pour toute mesure de probabilité μ sur G , nous désignerons par Q_μ l'opérateur de transition sur N , identifié à $G/_K$, associé à μ ; nous avons

$$Q_\mu f(v) = \int_G f(g.x)\ \mu(dg)$$
$$= \int_{N\times K} f(u[\eta(k)(v)])\ \mu(du,dk) \qquad (v \in N) ,$$

pour toute fonction borélienne f sur N pour laquelle, le second membre est défini pour tout $u \in N$. Pour une mesure de Dirac ε_g , $g \in G$, nous notons Q_g au lieu de Q_{ε_g} . Si μ et μ' sont deux mesures de probabilité sur G , nous avons $Q_\mu Q_{\mu'} = Q_{\mu' * \mu}$. En particulier si μ est une mesure de probabilité sur G idempotente, c'est-à-dire est la mesure de Haar normalisée d'un sous-groupe compact de G , alors Q_μ est un projecteur.

Avec ces notations, nous avons, pour toute fonction borélienne bornée (ou positive) sur N ,

$$\int_N f(u)\ \pi_1(\lambda^n * \varepsilon_v)(du) = Q_\lambda^n f(v) \quad , \qquad (v \in N) .$$

Pour étudier la suite de mesures de probabilité $\pi_1(\lambda^n)$, nous sommes ainsi amenés à étudier la suite d'opérateurs $\{Q_\lambda^n\}_{n\geq 1}$. Cette étude sera faite dans la section 2 dans une situation particulièrement favorable. Par la suite (section 5) nous montrerons que le cas général (moyennant des hypothèses de moment sur λ) se ramène toujours à cette situation particulière. Nous terminons cette section 1 en donnant une démonstration de la proposition (1.2)

Preuve de la proposition (1.2)

Nous commençons par établir un lemme.

(1.4) Lemme :

Soit $\{\mu_n\}_{n\geq 1}$ une suite de mesures de probabilité sur un groupe compact K, convergeant vaguement vers une mesure de probabilité μ. Alors pour toute fonction f continue sur K, nous avons

$$\lim_n \sup_{x\in K} \left|\int_K f(xk)\,\mu_n(dk) - \int_K f(xk)\,\mu(dk)\right| = 0 .$$

Preuve du lemme (1.4):

Si h est une fonction continue sur K, nous posons $\|h\|_\infty = \sup_{x\in K} |h(x)|$ et nous notons h^x, la translaté à gauche de h par l'élément x de K (i.e. $h^x(k) = h(xk)$, $\forall k \in K$).

Soit f une fonction continue sur K. Pour tout entier naturel non nul m, posons

$$f_m(x) = \sup_{n\geq m} \left|\int_K f(xk)\,\mu_n(dk) - \int_K f(xk)\,\mu(dk)\right| , \quad (x \in K).$$

On vérifie aisément que l'on a :

$$f_m(x) \leq f_m(y) + 2\, \| f^{xy^{-1}} - f\|_\infty , \quad \forall (x,y) \in K\times K ;$$

d'où

$$|f_m(x) - f_m(y)| \leq 2 \sup\{\|f^{xy^{-1}} - f\|_\infty , \|f^{yx^{-1}} - f\|_\infty\} , \quad \forall (x,y) \in K\times K ;$$

et par suite, pour tout entier $m \geq 1$, la fonction f_m est continue sur K.

Soit ε un réel > 0. $(\{f_m < \varepsilon\})_{m\geq 1}$ est une famille croissante d'ouverts de K. Comme la suite $\{\mu_n\}_{n\geq 1}$ converge vaguement vers μ, cette famille constitue un recouvrement ouvert de K. De la compacité de K, il résulte alors qu'il existe un entier $m(\varepsilon)$ pour lequel $K = \{f_{m(\varepsilon)} < \varepsilon\}$. Le lemme est prouvé.

La proposition (1.2) résulte alors du lemme suivant.

(1.5) Lemme :

Soit $(X_i)_{i\geq 1}$ une suite de v.a. indépendantes définies sur un espace probabilisé $(\Omega, a, \mathbb{P})$, à valeurs dans G, et de loi commune λ; pour tout entier naturel non nul n, nous posons $S_n = X_1 \dots X_n$.

Soit $\{U_n\}_{n\geq 1}$ une suite d'homéomorphismes de N tels que, pour tout entier naturel p, pour toute fonction f continue, à support compact sur N,

$$\lim_n E[|f \circ U_n \circ \pi_1(S_n) - f \circ U_n \circ \pi_1(S_{n-p})|] = 0 \quad .$$

Alors les v.a. $U_n \circ \pi_1(S_n)$ et $\pi_2(S_n)$ sont asymptotiquement indépendantes ; c'est-à-dire que pour toute fonction continue f, à support compact, sur N et pour toute fonction h continue sur K, nous avons

$$\lim_n \left|\mathbb{E}[f \circ U_n \circ \pi_1(S_n)\ h \circ \pi_2(S_n)] - \mathbb{E}[f \circ U_n \circ \pi_1(S_n)]E[h \circ \pi_2(S_n)]\right| = 0.$$

Preuve du lemme (1.5) :

Pour tout entier naturel p, nous avons

$$\left|\mathbb{E}[f \circ U_n \circ \pi_1(S_n)\ h \circ \pi_2(S_n)] - \mathbb{E}[f \circ U_n \circ \pi_1(S_n)]\ \mathbb{E}[h \circ \pi_2(S_n)]\right|$$

$$\leq \alpha(n,p) + \beta(n,p) \ .$$

avec,

$$\alpha(n,p) = \left| \mathbb{E}[(f \circ U_n \circ \pi_1(S_n) - f \circ U_n \circ \pi_1(S_{n-p}))\, h \circ \pi_2(S_n)] \right|$$

$$\leq \|h\|_\infty \, \mathbb{E}[|f \circ U_n \circ \pi_1(S_n) - f \circ U_n \circ \pi_1(S_{n-p})|]$$

$$\beta(n,p) = \left| \mathbb{E}[f \circ U_n \circ \pi_1(S_{n-p})(h \circ \pi_2(S_n) - \mathbb{E}[h \circ \pi_2(S_n)])] \right| .$$

Comme les v.a. $(X_i)_{i\geq 1}$ sont indépendantes et de loi λ, nous avons

$$\beta(n,p) \leq \|f\|_\infty \left| \mathbb{E}[\int_K h(\pi_2(S_{n-p})\, k)\lambda_2^p\,(dk)] - \int_K h(k)\, \lambda_2^n\,(dk) \right| .$$

Soit k_0 un élément de $S(\lambda_2)$. $k_0^{-n}\, S_{n-p}\, k_0^p$ s'écrit

$$[k_0^{-(n-1)}(k_0^{-1}\pi_2(X_1)k_0)k_0^{n-1}]\,[k_0^{-(n-2)}(k_0^{-1}\pi_2(X_2)k_0)k_0^{n-2}]\ldots[k_0^{-p}(k_0^{-1}\pi_2(X_{n-p})k_0)k_0^{p}],$$

et appartient donc, $\mathbb{P}$-p.s., à $\tilde{K}(\lambda_2)$. On en déduit que

$$\beta(n,p) \leq \|f\|_\infty \sup_{x\in\tilde{K}(\lambda_2)} \left| \int_K h(k_0^n\, xk)\, \mu_p(dk) - \int_K h(k_0^n k)\, \mu_n(dk) \right| ,$$

où $\{\mu_n\}_{n\geq 1}$ désigne la suite de mesure de probabilité $\{\varepsilon_{k_0^{-n}} * \lambda_2^n\}_{n\geq 1}$

Comme la suite $\{\mu_n\}_{n\geq 1}$ converge vaguement vers $\tilde{m}$, du lemme (1.4) il résulte que l'on a $\lim_p \sup_{n\geq p} \beta(n,p) = 0$.

D'autre part, l'hypothèse du lemme (1.5), nous assure que pour tout entier p fixé $\lim_n \alpha(n,p) = 0$.

Le lemme (1.5) est prouvé.

2. "Résolution" du Problème dans une situation particulière

(2.1) Nous pouvons toujours supposer, qu'en tant qu'espace topologique, N est un espace vectoriel réel de dimension finie et que pour tout élément k de K , $\eta(k)$ est un automorphisme d'espace vectoriel de N . [En effet, si ce n'est pas le cas, munissons l'algèbre de Lie $(\mathcal{L}(N), [\ ,\ ])$ du produit $\circ$ défini par la formule de Campbell-Hausdorff,

$$u \circ v = u + v + \frac{1}{2}[u,v] + \ldots \qquad (u,v \in N)\ ;$$

nous savons que l'application exponentielle de N est un isomorphisme de groupe analytique de $(\mathcal{L}(N),\circ)$ sur N ; le groupe G est alors isomorphe au produit semi-direct $(\mathcal{L}(N),\circ) \times_{\mathcal{L}(\eta)} K$, où pour tout $k \in K$, on désigne par $\mathcal{L}(\eta)(k)$ l'automorphisme de $(\mathcal{L}(N),[\ ,\ ])$ et par suite de $(\mathcal{L}(N),\circ)$, tangent à l'automorphisme $\eta(k)$ de N].

D'autre part, quitte à remplacer $G = N \times_\eta K$ par $N \times_\eta K(\lambda_2)$, nous pouvons supposer que $K(\lambda_2) = K$.

Nous faisons alors sur le couple (G,λ) l'hypothèse suivante

Il existe sur l'algèbre $\mathbb{C}[N]$ des fonctions polynomes sur N , à coefficients complexes, une notion de degré pour laquelle

i) $\forall T \in \mathbb{C}[N]$, $T(uv) = T(u) + T(v) + P_T(u,v)$, $(u,v \in N)$,

où P_T est une fonction polynome sur $N \times N$ dont le degré global, dg P_T , est inférieur ou égal à celui, dgT , de T ; et les valuations partielles de P , val P/u et val P/v , sont supérieures ou égales à 1

ii) $\forall T \in \mathbb{C}[N]$, $dgT \circ \eta(k) = dgT$, $\forall k \in K$.

iii) Pour toute forme linéaire A sur N de degré 1 ,

$$\int_N \sup_{k\in K} |A(\eta(k)(u))| \lambda_1(du) < +\infty$$

et

$$\int_N \int_K A(\eta(k)(u))\, m(dk)\, \lambda_1(du) = 0$$

(2.2) Remarque. Une notion de degré sur $\mathbb{C}[N]$, s'obtient obligatoirement en choisissant une base $\{f_i\}_{1\leq i\leq p}$ de l'espace vectoriel N, en attribuant un degré à chaque fonction coordonnée y_i , $1\leq i\leq p$, associée à cette base et en convenant que le degré (resp. la valuation) du polynome nul est $(-\infty)$ (resp. $(+\infty)$). Nous disons alors que $\{f_i\}_{1\leq i\leq p}$ est une base de référence pour cette notion de degré.

Pour qu'une notion de degré vérifie l'hypothèse i) resp ii)) de (2.1) il faut et il suffit que les relations de i) (resp. ii)) soient vérifiées par les fonctions coordonnées $(y_i)_{1\leq i\leq p}$ d'une de ses bases de référence. Dans ce cas, on remarquera que l'élément neutre et l'inverse d'un élément u du groupe N coïncident respectivement avec l'élément nul et l'opposé de u dans le groupe additif de l'espace vectoriel N . En outre, pour toute forme linéaire A de degré 1 sur N (i.e. pour toute combinaison linéaire des fonctions coordonnées y_i de degré 1), nous avons
$A(uv) = A(u) + A(v) = A(u+v)$.

(2.3) Pour tout élément u de N , désignons par L_u(resp. R_u) la multiplication à gauche (resp. à droite) par u sur N . Pour tous éléments $g = (u,k)$ de G et T de $\mathbb{C}[N]$, nous avons, (voir (1.4)),

$$Q_g T = T \circ L_u \circ \eta(k)) .$$

D'après les hypothèses ii) et iii) de (2.1), on voit qu'ains G opère à droite sur $\mathbb{C}[N]$ et les sous-espaces

$\mathbb{C}_\ell[N] = \{T \in \mathbb{C}[N] : dgT \leq \ell\}$, $\ell \in \mathbb{N}$; en outre, nous avons

$$(Q_g - Q_k)T(\cdot) = P_T(u, \eta(k)(\cdot)) + T(u) ,$$

qui montre que

$$dg(Q_g - Q_k)T \leq dgT - 1 \quad , \quad \forall g = (u,k) \in G \ , \ \forall T \in \mathbb{C}[N]$$

Soit μ une mesure de probabilité sur G . Posons

$$\ell(\mu) = \sup\{\ell \in \mathbb{N} : \int_N \sup_{k \in K} |T \circ \eta(k)(u)| \ \pi_1(\mu)(du) < +\infty \ , \ \forall T \in \mathbb{C}_\ell[N]\} .$$

De ce qui précède, il résulte que Q_μ opère sur $\mathbb{C}_{\ell(\mu)}[N]$, (on convient que $\mathbb{C}_{+\infty}[N] = \mathbb{C}[N]$!) , et nous avons :

(2.4) <u>Lemme</u>

Soit μ une mesure de probabilité sur G . Alors pour tout $T \in \mathbb{C}_{\ell(\mu)}[N]$,

$$dg(Q_\mu - Q_{\pi_2(\mu)}) T \leq dgT - 1 .$$

Si de plus $\mu = \pi_1(\mu) \otimes \pi_2(\mu)$ et $\int_N A(u) \ \pi_1(\mu)(du) = 0$, pour toute forme linéaire A de degré 1 sur N , alors

$$dg(Q_\mu - Q_{\pi_2(\mu)})T \leq dgT - 2$$

(2.5) L'opérateur $Q_{\lambda_2}(I - Q_m) = (I - Q_m)Q_{\lambda_2}$ de $\mathbb{C}[N]$ ne possède pas la valeur propre 1. En effet, si $T \in \mathbb{C}[N]$ vérifie $Q_{\lambda_2}(I - Q_m)T = T$, nous avons

$$T = \frac{1}{n} \left(\sum_{k=1}^{n} Q_{\lambda_2^{k}} \right) (I - Q_m) T \ ;$$

et par suite, puisque la suite de mesures de probabilité $\left\{ \frac{\lambda_2 + \ldots + \lambda_2^n}{n} \right\}_{n \geq 1}$ converge vaguement vers m , $T = Q_m(I - Q_m)T = 0$.

Il s'ensuit que $(I - Q_{\lambda_2}(I - Q_m))$ est un opérateur bijectif de $\mathbb{C}[N]$.

Désignons par $\overline{\lambda}$ la mesure de probabilité sur N définie par $\int_N n(k)(\lambda_1)\, m(dk)$. Appelons $\hat{\lambda}$ la mesure de probabilité sur N définie par

$$\hat{\lambda}(f) = \int_{N \times K} f(n(k^{-1})(u))\, \lambda(du, dk) \quad ,$$

pour toute fonction borélienne bornée f sur N .

Posons

$$\begin{aligned} L_\lambda &= Q_m(Q_\lambda - I)Q_m + Q_m(Q_\lambda - I)\left[I - Q_{\lambda_2}(I - Q_m)\right]^{-1}(Q_\lambda - I)Q_m \\ &= Q_m(Q_{\overline{\lambda}} - I)Q_m + Q_m(Q_{\lambda_1} - I)\left[I - Q_{\lambda_2}(I - Q_m)\right]^{-1}(Q_{\hat{\lambda}} - I) \ . \end{aligned}$$

D'après le lemme (2.4), l'opérateur L_λ baisse le degré de tout élément de $\mathbb{C}_{\ell(\lambda)}[N]$ d'au moins deux unités.

Nous avons alors :

(2.6) <u>Proposition</u> :

Soient $G = N \times_n K$ un produit semi-direct d'un groupe de Lie nilpotent simplement connexe par un groupe compact K et λ une mesure de probabilité sur G . Supposons que le couple (G, λ) vérifie les hypothèses de (2.1).

Alors, (avec les notations de (2.5)), pour toute fonction polynome T sur N de degré $\leq \ell(\lambda)$, la suite de fonctions polynomes $\{\frac{1}{n^{dgT/2}} Q_\lambda^n T\}_{n\geq 1}$ converge, uniformément sur tout compact de N, vers zéro si dgT est impair, vers la constante $\frac{1}{(dgT/2)!} L_\lambda^{dgT/2} T$ si dgT est pair.

Une démonstration de la proposition (2.6) sera donnée dans la section 3. On remarquera que cette proposition améliore un résultat de Guivarc'h ([2] corollaire de la proposition 2).

Nous allons à présent donner deux corollaires.

(2.7) Corollaire.

Soit (G,λ) comme dans la proposition (2.6). Supposons que λ vérifie, en outre, l'une des deux propriétés :

$\lambda = \lambda_1 \otimes \lambda_2$ ou $\int A(u)\, \lambda_1(du) = 0$, pour toute forme linéaire A de degré 1 sur N .

Alors, pour tout élément T de $\mathbb{C}_{\ell(\lambda)}[N]$ de degré pair, nous avons

$$L_\lambda^{dgT/2}\, T = (Q_{\overline{\lambda}} - I)^{dgT/2}\, T$$

Preuve.

Si $\lambda = \lambda_1 \otimes \lambda_2$, nous avons $(Q_\lambda - I)Q_m = Q_{\lambda_2}(Q_{\overline{\lambda}} - I)Q_m$. Par suite, dans les deux cas envisagés dans le corollaire, l'opérateur

$$Q_m(Q_{\lambda_1} - I)[I - Q_{\lambda_2}(I - Q_m)]^{-1}(Q_{\hat{\lambda}} - I)Q_m$$

baisse le degré de toute fonction polynome de degré $\leq \ell(\lambda)$ d'au moins trois unités. Nous avons alors

$$L_{\lambda}^{dgT/2}\ T = [Q_m(Q_{\overline{\lambda}} - I)Q_m]^{dgT/2}\ T = (Q_{\overline{\lambda}} - I)^{dgT/2}\ T \quad .$$

(2.8) Corollaire.

Soit (G,λ) comme dans la proposition (2.6). Soit a un élément de N vérifiant

(*) $$A(a) = ([I - Q_{\lambda_2}(I - Q_m)]^{-1}A)(\int_N u\ \lambda_1(du)) \quad ,$$

pour toute forme linéaire A sur N de degré 1 .

Alors, pour tout élément T de $\mathbb{C}_{\ell(\lambda)}[N]$ de degré pair, nous avons

$$L_{\lambda}^{dgT/2}\ T = (Q_{\overline{\lambda(a)}} - I)^{dgT/2}\ T \quad ,$$

où $\lambda(a) = \varepsilon_{(-a,e)} * \lambda * \varepsilon_{(a,e)}$ et $\overline{\lambda(a)} = \int_K \eta(k) \circ \pi_1(\lambda(a))\ m(dk)$.

Preuve.

Pour tout élément a de N vérifiant la relation (*), on voit facilement que l'on a

$$\int_N A(u)\ \pi_1(\varepsilon_{(-a,e)} * \lambda * \varepsilon_{(a,e)})(du) = 0 \quad , \text{ pour}$$

toute forme linéaire A de degré 1 sur N . Autrement dit la mesure de probabilité sur G , $\lambda(a)$, vérifie l'hypothèse du corollaire (2.7).

Or pour tout entier naturel $n \geq 1$, nous avons

$$Q_{\lambda}^{n} = Q_{(a,e)}^{-1}\ Q_{\lambda(a)}^{n}\ Q_{(a,e)}$$

On en déduit que

$$\lim_n \frac{1}{n^{dgT/2}} Q_\lambda^n T = \frac{1}{(\frac{dgT}{2})!} (Q_{\overline{\lambda(a)}} - I)^{dgT/2} Q_{(a,e)} T = \frac{1}{(\frac{dgT}{2})!} (Q_{\overline{\lambda(a)}} - I)^{dgT/2}$$

car $dg(Q_{(a,e)} - I)T \leq dgT - 1$; d'où le corollaire .

(2.9) Choisissons une base $b = \{e_i\}_{1 \leq i \leq p}$ de l'espace vectoriel N qui soit une base de référence pour la notion de degré sur $\mathbb{C}[N]$ considérée en (2.1), (voir remarque (2.2)). Notons $\{x_i\}_{1 \leq i \leq p}$ le système de fonctions coordonnées associé à cette base. Pour tout réel $t > 0$, nous posons

$$U_t^b(u) = \sum_{i=1}^{p} \frac{x_i(u)}{t^{dgx_i/2}} e_i \qquad (u \in N) \qquad ;$$

puis

$$U_t^b(u,k) = (U_t^b(u) \ , \ k) \qquad , \qquad ((u,k) \in N \times K) \quad .$$

(2.10) Soient u et v deux éléments de N . Pour tout $i \in \{1,\dots,p\}$, écrivons, (hypothèse ii) de (2.1)),

$$x_i(uv) = x_i(u) + x_i(v) + P_i(u,v) \qquad ,$$

où $P_i \in \mathbb{C}[N \times N]$ tel que $dg\ P_i \leq dg\ x_i$, $valP_i/u$ et $valP_i/v \geq 1$.

Alors, pour tout $i \in \{1,\dots,p\}$, nous avons

$$x_i(U_t(U_{1/t}(u)\ U_{1/t}(v))) = x_i(u) + x_i(v) + \frac{1}{t^{dgx_i/2}} P_i(U_{1/t}(u) \ , \ U_{1/t}(v))$$

et par suite,

$$\lim_{t\to+\infty} x_i(U_t(U_{1/t}(u)\ U_{1/t}(v))) = x_i(u) + x_i(v) + \overline{P_i}(u,v) \quad ,$$

où $\overline{P_i}$ est le polynome sur $N \times N$ qui se déduit de P_i en conservant uniquement les monomes de degré $dg\ x_i$.

En posant

$$u \cdot v = \lim_{t\to+\infty} U_t(U_{1/t}(u)\ U_{1/t}(v)) \quad ,$$

on vérifie facilement que l'on définit un produit nilpotent sur l'espace vectoriel N ; nous notons N^b le groupe de Lie nilpotent simplement connexe $(N,\cdot)$. Pour tout $t > 0$, U_t^b est un automorphisme de N^b .

(2.11) Pour tout élément $g = (u,k)$ de G , nous posons

$$\phi_b(g) = \phi_b(u) = \sup_{1\leq i\leq p}\ \sup_{k\in K} |x_i(n(k)(u))|^{1/dg\ x_i}$$

La condition (*) $\int_G \phi_b^2(g)\ \lambda(dg) < +\infty$,

est équivalente à

$$\int_N \sup_{k\in K} |T(n(k)(u))|^{2/dgT}\ \lambda_1(du) < +\infty \quad , \forall T \in \mathbb{C}[N]$$

et est par conséquent indépendante du choix de b .

Nous avons alors :

(2.12) Théorème :

Soit (G,λ) vérifiant les hypothèses de (2.1). Supposons qu'en outre λ vérifie la condition (*) de (2.11). Désignons par $b = \{e_i\}_{1\leq i\leq p}$ une base de référence pour la notion de degré sur $\mathbb{C}[N]$ considérée en (2.1) et par $\{x_i\}_{1\leq i\leq p}$ le système de fonctions coordonnées associé à cette base.

Alors, (avec les notations de (2.9) et (2.10)), la suite de mesures de probabilité $\{U_n^b \circ \pi_1(\lambda^n)\}_{n\geq 1}$ converge vaguement vers la loi au temps 1 du semi-groupe de convolution $\{\nu_t^b\}_{t>0}$ sur N^b de générateur infinitésimal

$$A = \sum_{\{i:dg\, x_i=2\}} (L_\lambda\, x_i) D_i + \frac{1}{2} \sum_{\{i,j:dg\, x_i=dg\, x_j=1\}} (L_\lambda\, x_i\, x_j)\, D_i\, D_j \ ,$$

où, pour toute fonction f de classe C^1 sur N,

$$D_i f(u) = \lim_{t\to 0} \frac{f(u\cdot te_i) - f(u)}{t} \ , \ u \in N \ , \quad i \in \{1,\ldots,p\} \ .$$

En outre la suite d'homéomorphismes $\{U_n^b\}_{n\geq 1}$ vérifie les hypothèses de la proposition (1.2) ; les conclusions de cette proposition s'appliquent donc.

Une démonstration du théorème (2.12) sera donnée dans la section 4 . Nous terminons cette section en faisant quelques commentaires.

(2.13) Remarque. Pour savoir dans quels cas nous avons effectivement résolu le problème posé, nous devons étudier le semi-groupe $\{\nu_t\}_{t>0}$ de g.i. A .

Chaque D_i , $1\leq i\leq p$, est un champ analytique de vecteurs tangents, invariant à gauche sur N^b ; c'est-à-dire un élément de l'algèbre de Lie, $\mathcal{L}(N^b)$, de N^b .

Si pour tout $i \in \{1,\dots,p\}$, $dg\, x_i \geq 2$, A est un élément de $\mathcal{L}(N^b)$. Nous avons alors $\nu_t = \varepsilon_{tf_0}$, où $f_0 = \sum_{\{i:dg\, x_i=2\}} (L_\lambda\, x_i)\, e_i \in N$; et par suite $U_n^b \circ \pi_1(\lambda^n)$ converge vaguement vers la mesure de Dirac ε_{f_0} .

S'il existe des coordonnées x_i de degré 1 , alors A s'écrit sous la forme

$$A = Z_0 + \frac{1}{2} \sum_{i=1}^{1} Z_i^2 \qquad \text{avec} \qquad s \in \mathbb{N}^*$$

où Z_i , $0 \leq i \leq s$, sont des éléments de $\mathcal{L}(N^b)$.

Désignons par $\mathcal{L}_1(A)$ la sous-algèbre de Lie de $\mathcal{L}(N^b)$ engendrée par les éléments $Z_0,\dots,Z_s$; et par $\mathcal{L}_2(A)$ l'idéal de $\mathcal{L}_1(A)$ formé par les éléments de $\mathcal{L}_1(A)$ de la forme

$$\sum_{i=1}^{s} \lambda_i\, Z_i + Y$$

avec $\lambda_i \in \mathbb{R}$, $1 \leq i \leq r$ et $Y \in [\mathcal{L}_1(A), \mathcal{L}_1(A)]$.

En omettant l'indice A , désignons par L_1 et L_2 les sous-groupes de Lie connexes de N^b possèdant respectivement $\mathcal{L}_1$ et $\mathcal{L}_2$ pour algèbre de Lie. L_2 est un sous-groupe distingué de L_1 .

Nous avons alors (voir [4]) :

1) si $\mathcal{L}_1 = \mathcal{L}_2$ (i.e. $L_1 = L_2$), le semi-groupe de convolution ν_t de g.i. A est absolument continu par rapport à la mesure de Haar de L_1 , avec une densité de classe C^∞ .

2) si $\mathcal{L}_1 \neq \mathcal{L}_2$, nous avons $\mathcal{L}_1 = \mathcal{L}_2 \oplus \mathbb{R}\, Z_0$ et par suite, puisque N^b est nilpotent simplement connexe, L_1 est un produit

semi-direct de L_2 par $(\mathbb{R},+)$. Le semi-groupe de convolution ν_t de g.i. A s'écrit alors $\nu_t = \mu_t \otimes \varepsilon_t$, où μ_t est absolument continue par rapport à la mesure de Haar de L_2 avec une densité de classe C^∞ .

Posons $I = \{i \in \{1,\dots,p\} : dg\, x_i = 1\}$. Dans le cas où la matrice carrée $((L_\lambda\, x_i\, x_j))_{i,j\in I}$ est définie positive, la sous-algèbre de Lie de $\mathcal{L}(N^b)$ engendrée par $Z_1,\dots,Z_s$ est égale à celle engendrée par les D_i , $i \in I$. $\mathcal{L}_1(A)$ est alors la sous-algèbre de $\mathcal{L}(N^b)$ engendrée par Z_0 et les D_i , $i \in I$; $\mathcal{L}_2(A)$ est l'idéal de $\mathcal{L}_1(A)$ formé par les éléments de $\mathcal{L}_1$ de la forme

$$\sum_{i\in I} \lambda_i\, D_i + Y \quad , \quad \lambda_i \in \mathbb{R} \quad \text{et} \quad Y \in [\mathcal{L}_1 , \mathcal{L}_1] \; .$$

(2.14) Lemme.

Soit (G,λ) vérifiant les hypothèses de (2.1). Supposons en outre que $\ell(\lambda) \geq 2$. Désignons par G_λ le sous-groupe fermé de G engendré par le support de λ et par a un élément de N vérifiant la relation (*) du corollaire (2.8).

Alors si la matrice $((L_\lambda\, x_i\, x_j))_{i,j\in I}$ n'est pas définie positive, nous avons, (en identifiant N et K à des sous-groupes de G)

$$G_\lambda \subset N_0(aka^{-1})$$

où N_0 est d'une part un sous-groupe fermé propre de N , distingué dans G , contenant le groupe dérivé de N , et d'autre part un sous-espace vectoriel propre de N.

Preuve : D'après le corollaire (2.8), la matrice considérée n'est pas définie positive si et seulement s'il existe une forme linéaire non nulle A sur N, de degré 1, telle que :

(1) pour $\lambda(a)$ -presque tout élément g de G, $A \circ \pi_1(kg) = 0$ $\forall k \in K$

Soit $N_0 = \{u \in N : A \circ \pi_1(ku) = 0 \quad \forall k \in K\}$. N_0 est un sous-groupe propre de N, distingué dans G et contenant le groupe dérivé de N ; en outre N_0 est un sous-espace vectoriel propre de N.

De (1), il résulte alors que $a^{-1}G_\lambda\, a \subset N_0K$; c'est-à-dire $G_\lambda \subset N_0\,(a\,K\,a^{-1})$.

(2.14) Pour $k \in K$ et $u \in N$, définissons

$$\eta^b(k)(u) = \lim_{t\to+\infty} U_t(\eta(k)(U_{1/t}(u))) ,$$

On voit facilement que cette limite existe et que l'on a ,

$$\eta^b(k)(e_i) = \sum_{\{j;\mathrm{dg}\,x_j=\mathrm{dg}\,x_i\}} x_j(\eta(k)(e_i))e_j \quad , \quad \forall i \in \{1,\dots,p\} .$$

D'où l'on déduit que η^b est un homomorphisme continu de K dans le groupe des automorphismes de N^b .

Nous notons $G^b = N^b \times_{\eta^b} K$ le produit semi-direct du groupe de Lie nilpotent simplement connexe N^b par le groupe compact K associé à η^b . Le produit $\cdot$ de deux éléments (u,k) et (v,h) de G^b est défini par

$$(u,k)\cdot(v,h) = \lim_{t\to+\infty} U_t(U_{1/t}\,(u,k)\;U_{1/t}(v,h)) .$$

Pour tout $t > 0$, U_t^b est un automorphisme de G^b .

Pour toute mesure de probabilité μ sur $N \times K$, désignons par Q_μ^b l'opérateur de transition sur N^b , identifié à G^b/K , associé à μ (voir (1.3)).

Alors, nous avons :

(2.15) Lemme :

Soit $(G = N \times_{\eta} K , \lambda)$ vérifiant les hypothèses de (2.1). Choisissons une base b de N et notons $G^b = N^b \times_{\eta^b} K$ le groupe associé à cette base par le procédé ci-dessus.

Alors, pour toute fonction polynome T sur N de degré $\leq \ell(\lambda)$, les limites des fonctions polynomes $\frac{1}{n^{dgT/2}} Q_\lambda^n T$ et $\frac{1}{n^{dgT/2}} (Q_\lambda^b)^n T$ sont les mêmes.

Preuve : Soit μ une mesure de probabilité sur N. D'après (2.10), on voit facilement que, pour tout élément T de $\mathbb{C}_{\ell(\mu)}[N]$,

$$dg(Q_\mu^b - Q_\mu)T \leq dgT - 2$$

et même

$$dg(Q_\mu^b - Q_\mu)T \leq dgT - 3 ,$$

si $\int_N A(u)\, \mu(du) = 0$, pour toute forme linéaire A sur N de degré 1.

D'autre part, pour tout $k \in K$, nous avons $dg(T \circ \eta(k) - T \circ \eta^b(k)) \leq dgT - 1 \quad \forall T \in C[N]$, d'où pour toute mesure de probabilité μ sur K,

$$dg(Q_\mu^b - Q_\mu)T \leq dgT - 1 \quad ;$$

et par suite

$$dg([I - Q_{\lambda_2}(I - Q_m)]^{-1} - [I - Q_{\lambda_2}^b(I - Q_m^b)]^{-1})T \leq dgT - 1 .$$

On en déduit alors que,

$$dg(L_\lambda^b - L_\lambda)T \leq dgT - 3 \quad , \qquad \forall T \in \mathbb{C}_{\ell(\lambda)}[N] .$$

Le lemme s'en déduit immédiatement.

(2.16) Remarque.

Désignons par $\lambda^{n\cdot}$ la $n^{\text{ième}}$ convolée de λ dans G^b. Du lemme (2.15), du corollaire (2.8) et du théorème (2.12), il résulte alors que les suites de mesures de probabilités,
$U_n^b \circ \pi_1(\lambda^n) = \pi_1 \circ U_n^b(\lambda^n)$, $U_n^b \circ \pi_1(\lambda^{n\cdot}) = \pi_1([U_n^b(\lambda)]^{n\cdot})$,
$U_n^b(\overline{\lambda(a)}^n)$ et $U_n^b(\overline{\lambda(a)}^{n\cdot}) = [U_n^b(\overline{\lambda(a)})]^{n\cdot}$, convergent vaguement vers la même mesure de probabilité, pour tout élément a de N vérifiant la condition (*) du corollaire (2.8).

En particulier, lorsque $\lambda = \lambda_1 \otimes \lambda_2$ ou $\int_N A(u)\lambda_1(du) = 0$, pour toute forme linéaire sur N de degré 1 ,

$U_n^b \circ \pi_1(\lambda^n)$, $\pi_1([U_n^b(\lambda)]^{n\cdot})$, $U_n^b(\overline{\lambda}^n)$ et $[U_n^b(\overline{\lambda})]^{n\cdot}$,

convergent vaguement vers la même mesure de probabilité.

3. Démonstration de la proposition (2.6)

Soit μ une mesure de probabilité sur G . Q_μ opère sur $\mathbb{C}_{\ell(\mu)}[N]$; nous notons $\Delta(\mu)$ l'ensemble des valeurs propres de Q_μ et pour $\alpha \in \Delta(\mu)$, nous posons

$$F_\alpha(\mu) = \{T \in \mathbb{C}_{\ell(\mu)}[N] : \exists\, k \in \mathbb{N} ,\ (Q_\mu - \alpha I)^k T = 0\} ,$$

$$E_\alpha(\mu) = \{T \in \mathbb{C}_{\ell(\mu)}[N] : Q_\mu T = \alpha T\} ;$$

nous avons

$$\mathbb{C}_{\ell(\mu)}[N] = \bigoplus_{\alpha\in\Delta(\mu)} F_\alpha(\mu)$$

(3.1) Lemme :

Soit μ une mesure de probabilité sur G , portée par K . Alors

a) Tout élément de $\Delta(\mu)$ est de module ≤ 1

b) Si $\alpha \in \Delta(\mu)$ avec $|\alpha| = 1$, nous avons

$$T \in F_\alpha(\mu) \iff T \in E_\alpha(\mu) \iff Q_k T = \alpha T , \quad \forall k \in S(\mu) .$$

c) $$\bigoplus_{\alpha\in\Delta(\mu)-\{1\}} F_\alpha(\mu) = E_0(m) , \quad F_1(\mu) = E_1(m)$$

$$\bigoplus_{\{\alpha\in\Delta(\mu):|\alpha|<1\}} F_\alpha(\mu) = E_0(\tilde{m}) , \quad \bigoplus_{\{\alpha\in\Delta(\mu):|\alpha|=1\}} F_\alpha(\mu) = E_1(\tilde{m})$$

où $S(\mu)$ désigne le support de μ , m(resp $\tilde{m}$) la mesure de Haar normalisée du sous-groupe fermé $K(\mu)$ (resp. $\tilde{K}(\mu)$) engendré par $S(\mu)$ (resp. par $S(\mu)[S(\mu)]^{-1}$).

Preuve du lemme (3.1) :

Soit $(\ ,\ )$ un produit scalaire quelconque sur l'espace vectoriel $\mathbb{C}[N]$; en posant

$<T,T'> = \int_K (Q_k T, Q_k T')dk \quad (T,T' \in \mathbb{C}[N])$, nous obtenons un produit scalaire sur $\mathbb{C}[N]$, pour lequel les opérateurs Q_k , $k \in K$, sont unitaires (i.e. $||Q_kT||^2 = <Q_kT,Q_kT> = <T,T> = ||T||^2$,

$\forall T \in \mathbb{C}[N]$).

Soit T un élément de $\mathbb{C}[N]$, nous avons

$Q_\mu T = \int_K Q_k T\ \mu(dk)$;

comme dans l'espace de Hilbert de dimension finie $(\mathbb{C}_{dgT}[N], <\ ,\ >)$ toute sphère de centre 0 est strictement convexe, il s'ensuit que

$||Q_\nu T|| < ||T||$,

à moins que $Q_k T = T_0$, $\forall k \in S(\mu)$, avec $||T_0|| = ||T||$.

L'affirmation a) du lemme s'en déduit immédiatement.

Soit $T \in F_\alpha(\mu)$. Des relations

$Q_{\tilde{m}} Q_\mu = Q_\mu Q_{\tilde{m}} = Q_{\tilde{m}} Q_k = Q_k Q_{\tilde{m}}$,

vérifiées pour tout élément k de $S(\mu)$, il résulte que les fonctions polynômes $Q_{\tilde{m}}T$ et $(I - Q_{\tilde{m}})T$ appartiennent respectivement à $F_\alpha(\varepsilon_k) = E_\alpha(\varepsilon_k)$ et $F_\alpha(\mu)$. Si $|\alpha| < 1$, nous avons $E_\alpha(\varepsilon_k) = 0$ car Q_k est unitaire ; par suite $Q_{\tilde{m}}T = 0$. Supposons que $|\alpha| = 1$ et désignons par ℓ le plus petit entier tel que $(Q_\mu-\alpha I)^\ell(I-Q_{\tilde{m}})T=0$. Si $\ell \geq 1$, le polynôme $T' = (Q_\mu-\alpha I)^{\ell-1}(I-Q_{\tilde{m}})T$ appartient à $E_\alpha(\mu)$, d'après ce qui précède nous avons alors $Q_k T' = \alpha T'$ $\forall k \in S(\mu)$; d'où $(I - Q_{\tilde{m}})T' = T' = 0$. Il s'ensuit donc que $\ell = 0$; c'est-à-dire $Q_k T = T$ $\forall k \in \tilde{K}(\mu)$. Comme nous savons en outre que $Q_{\tilde{m}} T$ (qui est égal à T) appartient à $E_\alpha(\varepsilon_k)$, l'affirmation b) ainsi que les deux dernières égalités de c) sont prouvées .

Enfin si $T \in F_\alpha(\mu)$, nous avons $(Q_\mu - \alpha I)^\ell T = 0$, pour un certain entier ℓ . D'où $(1 - \alpha)^\ell Q_m T = 0$ et $Q_m T = 0$ si $\alpha \neq 1$. Si $T \in F_1(\mu)$, d'après b) nous avons $Q_k T = T$, $\forall k \in S(\mu)$, et par suite $Q_m T = T$. Le lemme est ainsi démontré .

Pour simplifier l'écriture , nous notons Q (resp. P) l'opérateur Q_λ (resp. Q_{λ_2}) . Si $r \in \mathbb{N}^*$ et $\alpha \in \mathbb{N}^r$, nous notons $\alpha_1, \ldots, \alpha_r$ les composantes de α et nous posons $||\alpha|| = \alpha_1 + \ldots + \alpha_r$. D'autre part , si $t \in \mathbb{R}_+$, nous désignons par $[t]$ la partie entière de t .

D'après le lemme (2.4) , nous avons $dg(Q - P)T \leq dgT - 1$ $\forall\, T \in \mathbb{C}_{\ell(\lambda)}[N]$; par suite en écrivant $Q = (Q - P) + P$, nous avons

$$\frac{1}{n^{dgT/2}} Q^n T = \frac{1}{n^{dgT/2}} \sum_{j=0}^{dgT} \sum_{\{\alpha \in \{0,\ldots,n-j\}^{j+1} : ||\alpha|| = n-j\}} P^{\alpha_1}(Q-P)P^{\alpha_2}\ldots(Q-P)P^{\alpha_{j+1}}T$$

Posons

$$R_0^0 = I - Q_m \quad , \quad R_0 = (I - Q_m)\; P\; (I - Q_m) \; ,$$
$$R_1^0 = R_1 = Q_m \quad \text{et} \quad P^0 = Q^0 = I \quad ;$$

pour tout entier naturel p , nous avons

$$Q^p = R_0^p + R_1^p \quad , \quad R_0^p = (I - Q_m)\; P^p\; (I - Q_m) \quad \text{et} \quad R_1^p = Q_m \; .$$

$\frac{1}{n^{dgT/2}} Q^n T$ s'écrit alors $\sum_{j=0}^{dgT} \sum_{i \in \{0,1\}^{j+1}} \gamma_n(j,i)$,

où $\gamma_n(j,i) = \frac{1}{n^{dgT/2}} \sum_{\{\alpha \in \{0,\ldots,n-j\}^{j+1} : ||\alpha|| = n-j\}} \phi(j,i,\alpha)$

avec $\phi(j,i,\alpha) = R_{i_1}^{\alpha_1}(Q - P)R_{i_2}^{\alpha_2}\ldots(Q - P)R_{i_{j+1}}^{\alpha_{j+1}}T$.

3.2 Lemme.

Soient $j \in \{0,\dots,dgT\}$ et $i \in \{0,1\}^{j+1}$ avec $||i|| \leq [(dgT+1)/2]$. Alors la suite de fonctions polynomes $\gamma_n(j,i)$ converge vers zéro , uniformément sur tout compact .

Preuve. Notons $\sigma(1),\dots,\sigma(s)$, (resp. $\sigma(s+1),\dots,\sigma(j+1)$), avec $s \leq [(dgT+1)/2]$, les entiers ℓ de $\{1,\dots,j+1\}$ tels que $i_\ell = 1$ (resp. $i_\ell = 0$) . Nous avons

$$\gamma_n(j,i) = \frac{1}{n^{dgT/2}} \sum_{0\leq\alpha_{\sigma(s+1)},\dots,\alpha_{\sigma(j+1)}\leq n-j} \xi_{n-j-(\alpha_{\sigma(s+1)}+\dots+\alpha_{\sigma(j+1)})}(s) \, \phi(j,i,\hat{\alpha})$$

où , pour tout $\ell \in \mathbb{N}$ et tout $s \in \mathbb{N}^*$, $\xi_\ell(s) = \text{card}\,\{\beta\in\mathbb{N}^s ; ||\beta|| = \ell\}$ et $\hat{\alpha}$ est l'élément de $\mathbb{N}^{j+1}$ qui se déduit de α en remplaçant les composantes $\alpha_{\sigma(i)}$, $i \in \{1,\dots,s\}$, par zéro .

D'après le lemme (3.1) , les valeurs propres de l'opérateur R_0 sont de module ≤ 1 et différentes de 1 ; il s'ensuit que la suite d'opérateurs de $\mathbb{C}[N]$, $\{\sum_{p=0}^{n} R_0^p\}_{n\geq 1}$, est bornée . D'autre part pour tout $\ell \in \mathbb{N}$ et tout $s \in \mathbb{N}^*$, $\xi_\ell(s)$ est un polynome de degré $(s-1)$ en ℓ dont le terme de plus haut degré est $\ell^{s-1}/(s-1)!$. Le lemme (3.2) est alors clair .

3.3 Lemme

Soient $j \in \{0,\dots,dgT\}$ et $i \in \{0,1\}^{j+1}$ avec $||i|| \geq [(dgT+1)/2] + 1$. Alors $\gamma_n(j,i)$ est une constante qui est non nulle seulement si dgT est pair , $||i|| = dgT/2 + 1$, $i_1 = i_{j+1} = 1$ et i ne possède pas deux composantes nulles consécutives .

Preuve. Désignons par $N(i)$ le nombre de couples (i_k, i_{k+1}) , $0\leq k\leq j$, égaux à $(1,1)$; nous avons

$$N(i) = \sum_{1\leq k\leq j} [(i_k+i_{k+1})/2] \geq 2\,||i|| - j - 2 \ .$$

Comme l'opérateur $R_1(Q-P)R_1 = Q_m(Q_{\overline{\lambda}} - I)Q_m$ baisse le degré de tout polynome de $\mathbb{C}_{\ell(\lambda)}[N]$ de deux unités , nous avons , pour tout

$\alpha \in \{0,\dots,n-j\}^{j+1}$,

$$dg\ \phi(j,i,\alpha) \leq dgT - (j+N(i))$$
$$\leq dgT - 2||i|| + 2$$

qui montre que $\phi(j,i,\alpha)$ est une constante qui est non nulle seulement si dgT est pair , $||i|| = dgT/2 + 1$ et $N(i) = 2||i|| - j - 2$ Le lemme (3.3) s'en déduit immédiatement .

Des lemmes (3.2) et (3.3) , il résulte que si dgT est impair la suite de fonctions polynomes $\{\frac{1}{n^{dgT/2}} Q^n T\}_{n\geq 1}$ converge , uniformément sur tout compact , vers zéro.

Supposons que dgT soit pair . Pour tout entier k , notons S^k (resp. V , V^0) l'opérateur $Q_m(Q - I)R_0^k(Q - I)Q_m$ (resp. $Q_m(Q - I)Q_m$, I) . Des lemmes (3.2) et (3.3) , il résulte que $\frac{1}{n^{dgT/2}} Q^n T$ s'écrit $\delta_1(n) + \delta_2(n)$, où $\delta_1(n)$ est une suite de fonctions polynomes convergeant , uniformément sur tout compact , vers zéro et

$$\delta_2(n) = \sum_{s=1}^{dgT/2} \sum_{\{\beta\in\mathbb{N}^{s+1} : ||\beta|| = dgT/2 - s\}} \psi_n(s,\beta)$$

avec

$$\psi_n(s,\beta) = \frac{1}{n^{dgT/2}} \sum_{\{\alpha\in\mathbb{N}^s : ||\alpha|| \leq n-(s+dgT/2)\}} \xi_{n-||\ell||}(dgT/2)$$
$$V^{\beta_1} S^{\alpha_1} V^{\beta_2} \dots V^{\beta_s} S^{\alpha_s} V^{\beta_{s+1}} T .$$

Désignons par $a_1,\dots,a_q$ les différentes valeurs propres de P de module 1 et distinctes de 1 . Nous avons (lemme (3.1))

$$R_0 = (I - Q_m)P(I - Q_m) = (I - Q_{\tilde{m}})P(I - Q_{\tilde{m}}) + \sum_{i=1}^{q} a_i \Pi_i$$

où Π_i désigne la projection de $\mathbb{C}[N]$ sur $F_{a_i}(\lambda_2) = E_{a_i}(\lambda_2)$; et par suite , pour tout élément k de $\mathbb{N}^*$,

$$R_0^k = (I - Q_{\tilde{m}})P^k(I - Q_{\tilde{m}}) + \sum_{i=1}^{q} a_i^k \Pi_i .$$

Pour tout entier naturel k , nous posons

$$S_0^k = Q_m(Q - I)(I - Q_{\tilde{m}})P^k(I - Q_{\tilde{m}})(Q - I)Q_m$$

$$S_i^k = a_i^k\, Q_m(Q - I)\Pi_i(Q - I)Q_m \ , \quad 1\le i\le q \ .$$

Soient $s \in \{0,\dots,dgT/2\}$ et $\beta \in \mathbb{N}^{s+1}$ avec $||\beta|| = dgT/2 - s$. Nous avons

$$\psi_n(s,\beta) = \sum_{j\in\{0,\dots,q\}^s} \zeta_n(s,\beta,j) \ ,$$

avec

$$\zeta_n(s,\beta,j) = \frac{1}{n^{dgT/2}} \sum_{\{\alpha\in\mathbb{N}^s : ||\alpha|| \le n-(s+dgT/2)\}} \xi_{n-||\alpha||}(dgT/2)\, V^{\beta_1} S_{j_1}^{\alpha_1} \dots V^{\beta_s} S_{j_s}^{\alpha_s} V^{\beta_{s+1}} T$$

$$= \frac{1}{n^{dgT/2}} \sum_{k=0}^{n-(s+dgT/2)} \zeta_{n-k}(dgT/2)\ \eta_k(s,\beta,j)$$

où $\eta_k(s,\beta,j) = \sum_{\{\alpha\in\mathbb{N}^s : ||\alpha|| = k\}} V^{\beta_1} S_{j_1}^{\alpha_1} \dots V^{\beta_s} S_{j_s}^{\alpha_s} V^{\beta_{s+1}} T$.

En notant que

$$(I - P(I - Q_m))^{-1} = Q_m + \sum_{l\ge 0} (I - Q_{\tilde m})P^l(I - Q_{\tilde m}) + \sum_{i=1}^{q} \Pi_i \ / \ (1-a_i) \ ,$$

la proposition (2.6) résulte alors du lemme suivant :

3.4 <u>Lemme</u>

Soient p et q deux entiers ≥ 1 ; $x_1,\dots,x_q$ des nombres complexes différents de 1 et de modules 1 ; u une fonction réelle de N^p telle que $\sum_{l\in\mathbb{N}^p} u(l) < +\infty$. Alors , pour tout entier $r \ge 1$,

$$\lim_n \frac{1}{n^r} \sum_{k=0} \xi_{n-k}(r) \sum_{\{l\in\mathbb{N}^{p+q} : ||l|| = k\}} x_1^{l_1} \dots x_q^{l_q}\, u(l_{q+1},\dots,l_{p+q}) = \frac{1}{1-x_1} \dots \frac{1}{1-x_q} \frac{1}{r!} \sum_{l\in\mathbb{N}^p} u(l) \ ;$$

$$\lim_n \frac{1}{n^r} \sum_{k=0} \xi_{n-k}(r) \sum_{\{l\in\mathbb{N}^p : ||l|| = k\}} x_1^{l_1} \dots x_q^{l_q} = \frac{1}{r!} \frac{1}{1-x_1} \dots \frac{1}{1-x_q}$$

$$\lim_n \frac{1}{n^r} \sum_{k=0} \xi_{n-k}(r) \sum_{\{l\in\mathbb{N}^p : ||l|| = k\}} u(l_1,\dots,l_p) = \frac{1}{r!} \sum_{l\in\mathbb{N}^p} u(l) \ .$$

<u>Preuve</u>. Posons

$$\beta(k) = \sum_{\{l\in\mathbb{N}^{p+q} : ||l|| = k\}} x_1^{l_1} \dots x_q^{l_q}\, u(l_{q+1},\dots,l_{p+q}) \ ;$$

nous avons $\beta(k) = \sum_{s=0}^{k} \sigma(s)\, \tau(k-s)$, en posant , pour tout entier s ,

$$\sigma(s) = \sum_{\{l \in \mathbb{N}^p : \|l\| = s\}} u(l) \quad \text{et} \quad \tau(s) = \sum_{\{l \in \mathbb{N}^q : \|l\| = s\}} x_1^{l_1} \cdots x_q^{l_q} .$$

Considérons le polynome $Q(x) = \prod_{i=1}^{q} (1 - x x_i)$; il est facile de voir que $\tau(s) = \sum_{i=1}^{q} x_i^{s+1}/Q'(1/x_i)$ et par suite

$$\beta(k) = \sum_{i=1}^{q} (x_i/Q'(1/x_i)) \sum_{s=0}^{k} \sigma(s)\, x_i^{k-s} .$$

D'où

$$\sum_{k>0} \beta(k) = \sum_{i=1}^{q} (x_i/(1-x_i)Q'(1/x_i)) \sum_{s\geq 0} \sigma(s)$$

$$= (1/Q(1)) \sum_{s\geq 0} \sigma(s) .$$

Le lemme (3.4) résulte alors du lemme élémentaire suivant.

3.5 Lemme

Soient x un nombre complexe différent de 1 , de module 1 et $\{\beta_k\}_{k\geq 0}$ une série convergente . Alors , pour tout entier $r > 0$,

$$\lim_n \frac{1}{n^r} \sum_{k=0}^{n} (n-k)^r x^k = \frac{1}{r!} \frac{1}{1-x}$$

$$\lim_n \frac{1}{n^r} \sum_{k=0}^{n} (n-k)^r \beta_k = \frac{1}{r!} \sum_{k\geq 0} \beta_k .$$

4. Démonstration du théorème (2.12)

Nous posons $r = \sup\{dg\ x_i\ ,\ 1 \leq i \leq p\}$; r est indépendant du choix du système de fonctions coordonnées $\{x_i\}_{1 \leq i \leq p}$ de l'e.v. N Pour faciliter la lecture, nous distinguons deux cas.

1er cas

Nous supposons que λ vérifie, outre l'hypothèse iii) de (2.1), la condition $m(\lambda) \geq 2 \sup(r,2)$. [On notera que d'après l'inégalité de Hölder, cette dernière hypothèse entraîne la condition (*) de (2.11)] .

La démonstration du théorème se fait en trois étapes.

1ère étape . Elle est suggérée par la remarque (2.14).

Pour prouver le théorème (2.12), nous pouvons supposer que pour tout $t > 0$, U_t^b est un automorphisme de G (i.e. $G = G^b$, voir (2.14)). En effet, nous avons :

(4.1) Proposition :

Soit $(X_i)_{i \geq 1}$ une suite de v.a. indépendantes, à valeurs dans $N \times K$, et de loi commune λ . Alors la suite de v.a.

$$\Delta_n = U_n^b \circ \pi_1(X_1 \ldots X_n) - U_n^b \circ \pi_1(X_1 \cdot \ldots \cdot X_n)$$

converge vers zéro dans L^2 .

Preuve de la proposition (4.1) :

Désignons par M le produit direct des groupes de Lie nilpotent simplement connexe N et N^b ; par $n \otimes n^b$ l'homomorphisme de K dans le groupe des automorphismes de M défini par

$[\eta \otimes \eta^b(k)](u,v) = (\eta(k)(u)\ ,\ \eta^b(k)(v))\quad ,\quad (u,v) \in N \times N^b\ ;$

et par $H = M \times_{\eta\otimes\eta^b} K$ le produit semi-direct de M par K associé à $\eta \otimes \eta^b$.

Si μ est une mesure de probabilité sur $N \times N \times K$, nous notons Q_μ l'opérateur de transition sur M, identifié à $H/_K$, associé à μ (voir (1.3)). Soit $X = (Y,U)$ une v.a., à valeurs dans dans $N \times K$, de loi λ; nous notons λ' la loi de la v.a. (Y,Y,U), à valeurs dans $N \times N \times K$.

Avec ces notations, nous avons, pour tout $i \in \{1,\dots,p\}$,

$$\mathbb{E}[x_i^2(\Delta_n)] = \frac{1}{n^{dg\ xi}} Q^n_{\lambda'}\ T_i(0)\quad ,$$

où T_i désigne la fonction polynome sur $N \times N$ définie par
$T_i(u,v) = (x_i(u) - x_i(v))^2\quad ,\quad u$ et $v \in N$.

Nous définissons une notion de degré sur $\mathbb{C}[M] = \mathbb{C}[N \times N]$, en attribuant un degré à toute fonction polynome T de $\mathbb{C}[N \times N]$ de la forme $T_1 \otimes T_2$, où T_1 et T_2 sont des éléments de $\mathbb{C}[N]$, (i.e. $T_1 \otimes T_2(u,v) = T_1(u)\ T_2(v)$, $\forall (u,v) \in N \times N$) ; le degré de T, noté dgT, est par définition égal à $dgT_1 + dgT_2$. Le couple (H,λ') vérifie alors les hypothèses de (2.1).

Considérons la base $b \otimes b = \{(e_i,0),(0,e_j),\ 1 \le i,j \le p\}$ de $N \times N$. Avec les notations de la remarque (2.14), $M^{b\otimes b}$ est le produit direct de N^b par lui-même et $H^{b\otimes b}$ est le produit semi-direct de $M^{b\otimes b}$ par K associé à $\eta^b \otimes \eta^b$.

Le lemme (2.15), appliqué au couple (H,λ'), nous dit que

$$\lim_n \frac{1}{n^{dgx_i}} Q^n_{\lambda'}\ T_i = \lim_n \frac{1}{n^{dgx_i}} (Q^{b\otimes b}_{\lambda'})^n\ T_i\quad ,\quad \forall i \in \{1,\dots,p\}\ ,$$

où $Q^{b\otimes b}_{\lambda'}$ désigne l'opérateur de transition sur $M^{b\otimes b}$, identifié à

$^{b\otimes b}/K$, associé à λ' . Or nous avons, pour $u \in N$ et $i \in \{1,\dots,p\}$

$$(Q^{b\otimes b}_{\lambda'})^n\ T_i(u) = \mathbf{E}\Big[[x_i \circ \pi_1(X_1\cdot\dots\cdot X_n\cdot(u,e)) - x_i \circ \pi_1(X_1\cdot\dots\cdot X_n\cdot(u,e))]^2\Big] = 0$$

l s'ensuit que $\lim_n \frac{1}{n^{dgx_i}} Q^n_{\lambda'}\, T_i = 0$ et la proposition (4.1) est rouvée.

<u>2è étape</u>. Nous supposons donc que $\forall t > 0$, U^b_t est un automorphisme de G .

Soit $(X_i)_{i\geq 1}$ une suite de v.a. indépendantes, à valeurs dans $N \times K$, de loi commune λ . Nous posons, pour tout entier naturel ≥ 1 ,

$$S_n(0) = 0$$

$$S_n(\tfrac{k}{n}) = \pi_1 \circ U_n(X_1\dots X_k) = \pi_1(U_n(X_1)\dots U_n(X_n)) \ , \ \forall k \in \{1,\dots,n\};$$

uis

$$S_n(t) = (1-\{nt\})\, S_n(\tfrac{[nt]}{n}) + \{nt\}\ S_n(\tfrac{[nt]+1}{n}) \ , \ \forall t \in [0,1] \ ,$$

où $[nt]$ désigne la partie entière de nt et $\{nt\} = nt - [nt]$.

Pour tout $u \in N$, nous posons $||u||^2 = \sum_{i=1}^{p} x_i^2(u)$.

(4.2) <u>Proposition</u>.

Pour tous réels s et t de $[0,1]$,

$$\mathbf{E}\Big[||S_n(t) - S_n(s)||^2\Big] \leq C\ |t - s| \ ,$$

où C est une constante > 0 indépendante de n , t et s .

<u>Preuve de la proposition</u> (4.2) -

Nous commençons par prouver qu'il suffit de montrer la proposition pour des réels s et t de la forme $\frac{k}{n}$ avec $k \in \{0,\dots,n\}$.

Admettons donc un instant que la proposition soit vraie pour les réels de cette forme et considérons deux réels s et t de $[0,1]$ vérifiant $s < t$.

Nous avons

$$S_n(t) - S_n(s) = \{nt\}u + v + (1 - \{ns\})\,w ,$$

avec

$$u = S_n\left(\frac{[nt]+1}{n}\right) - S_n\left(\frac{[nt]}{n}\right) ,$$

$$v = S_n\left(\frac{[nt]}{n}\right) - S_n\left(\frac{[ns]+1}{n}\right) ,$$

et

$$w = S_n\left(\frac{[ns]+1}{n}\right) - S_n\left(\frac{[ns]}{n}\right) .$$

De la relation

$$||ax + by + cz||^2 \leq (a + b + c)(a||x||^2 + b||y||^2 + c||z||^2) , \quad a,b,c>0 , \quad x,y,z\in N ,$$

il résulte que l'on a

$$\mathbf{E}\left[||S_n(t) - S_n(s)||^2\right] \leq \frac{3C}{n}\left(\{nt\} + \left|[nt] - [ns] - 1\right| + 1 - \{ns\}\right) .$$

D'où

$$\mathbf{E}\left[||S_n(t) - S_n(s)||^2\right] \leq 3C(t - s) \quad , \quad \text{si} \quad [nt] - [ns] \geq 1 .$$

Si $[nt] = [ns]$, nous avons

$$S_n(t) - S_n(s) = n(t-s)\left(S_n\left(\frac{[nt]+1}{n}\right) - S_n\left(\frac{[nt]}{n}\right)\right)$$

et

$$\mathbf{E}\left[||S_n(t) - S_n(s)||^2\right] \leq C\, n(t-s)^2$$
$$\leq C\,(t-s)$$

car $[nt] = [ns]$ implique que $n(t-s) \leq 1$.

Ceci dit soient k et ℓ deux entiers de $\{0,\dots,n\}$ tels que $\ell < k$. Pour tout $i \in \{1,\dots,p\}$, posons

$$T_i(u,v) = x_i(uv) - x_i(u) \ , \quad (u,v \in N) \quad ;$$

nous avons

$$T_i(u,v) = x_i(v) + P_i(u,v) \quad ,$$

avec $dg\, P_i \leq dg\, x_i$, $val\, P_i/u$ et $val\, P_i/v \geq 1$.

Par suite

$$\mathbb{E}\left[[x_i(S_n(\tfrac{k}{n}) - S_n(\tfrac{\ell}{n}))]^2\right] = \mathbb{E}\Big[[x_i([S_n(\tfrac{\ell}{n})]^{-1} S_n(\tfrac{k}{n})) + P_i(S_n(\tfrac{\ell}{n}),[S_n(\tfrac{\ell}{n})]^{-1} S_n(\tfrac{k}{n}))]^2\Big] .$$

Or $[S_n(\frac{\ell}{n})]^{-1} S_n(\frac{k}{n})$ s'écrit $\eta \circ \pi_2(X_1 \dots X_\ell)(\Sigma_{\ell,k}(n))$,

où $\Sigma_{\ell,k}(n)$ est une v.a. indépendante de $S_n(\frac{\ell}{n})$ et ayant la même loi que $S_n(\frac{k-\ell}{n})$. La proposition (4.2) résulte alors du lemme suivant

(4.3) Lemme.

Pour tout élément T de $\mathbb{C}_{\ell(\lambda)}[N]$ de la forme $x_1^{\alpha_1} \dots x_p^{\alpha_p}$, où $\alpha_1,\dots,\alpha_p \in \mathbb{N}$, et pour tout élément k de $\{0,\dots,n\}$, nous avons

$$\left|\mathbb{E}[T(S_n(\tfrac{k}{n}))]\right| \leq C(T) \ (\tfrac{k}{n})^{dgT/2}$$

où $C(T)$ est une constante > 0 indépendante de k et n .

Preuve du lemme (4.3).

Nous avons

$$E[T(S_n(\frac{k}{n}))] = \frac{1}{n^{dgT/2}} Q_\lambda^k \; T\;(0)$$

$$= (\frac{k}{n})^{dgT/2} \; (\frac{1}{k^{dgT/2}} Q_\lambda^k \; T\;(0)) \; .$$

D'après la proposition (2.6), la suite $\frac{1}{k^{dgT/2}} Q_\lambda^k \; T\;(0)$ est convergente. Le lemme est ainsi prouvé.

(4.4) De la proposition (4.2), il résulte que le processus $S_n(t)$ vérifie les hypothèses du corollaire 1 du chapitre I, sec. 6 de [8] Désignons par $\mathcal{B}([0,1])$ la tribu des boréliens de $[0,1]$ et par ρ la mesure de Lebesgue de $[0,1]$. Nous savons alors que pour toute suite d'entiers, on peut extraire une sous-suite $\{n_k\}_{k\geq 1}$ pour laquelle il est possible de construire des processus $x_{n_k}(t,\omega')$ et $x_\infty(t,\omega')$ sur l'espace probabilisé $(\Omega' = [0,1], \mathcal{B}([0,1]) , \rho)$ tels que

i) pour tout entier ℓ et tous réels $t_1,\dots,t_\ell$, $(x_{n_k}(t_1),\dots,x_{n_k}(t_\ell))$ et $(S_{n_k}(t_1),\dots,S_{n_k}(t_\ell))$ ont même loi, pour tout entier k .

ii) pour tout $t \in [0,1]$, $x_{n_k}(t)$ converge en probabilité vers $x_\infty(t)$, quand k tend vers l'infini .

<u>3ème étape</u>. Désignons par $\{n_k\}_{k\geq 1}$ une suite d'entiers pour laquelle il est possible de construire des processus $x_{n_t}(t)$ et $x_\infty(t)$ ayant les propriétés i) et ii) de (4.4). Alors le processus $x_\infty(t)$ vérifie la proposition suivante :

(4.5) <u>Proposition</u>.

Pour toute fonction sur N , de classe C^2 , à support

compact,

$\mathbb{E}_\rho[f(x_\infty(t))] = f(0) + \int_0^t \mathbb{E}_\rho[Af(x_\infty(s))]ds$, $\forall t \in [0,1]$, où $\mathbb{E}_\rho$ désigne l'espérance associée à la mesure ρ.

Preuve de la proposition (4.5).

En posant $S'_k(t) = S_{n_k}(t)$ et $x'_k(t) = x_{n_k}(t)$, on se ramène au cas où $\{n_k\}_{k\geq 1}$ est la suite des entiers naturels. Pour alléger l'écriture, nous nous plaçons donc dans cette situation.

Pour toute fonction f sur N , de classe C^1, à support compact, et pour tout élément v de N , nous posons

$$D_v f(u) = \lim_{t\to 0} \frac{f(u(tv)) - f(u)}{t} , \quad (u \in N)$$

Nous avons

$$D_v f = \sum_{i=1}^{p} x_i(v)\, D_i f , \quad \text{où} \quad D_i = D_{e_i} , \quad i \in \{1,\dots,p\} ;$$

et (formule de Taylor), pour tous $u,v \in N$, il existe $\theta \in \,]0,1[$ tel que

$$f(u\,v) = f(u) + D_v f(u) + \frac{1}{2} D_v^2 f(u(\theta v)) .$$

Choisissons une partition $t_0 = 0 < t_1 < \dots < t_q = t$ de $[0,t]$ et écrivons, pour tout entier naturel n ,

$$f(S_n(t)) - f(0) = f(S_n(t)) - f(S_n(\tfrac{[nt]}{n})) + \sum_{j=1}^{q} \{f(S_n(\tfrac{[nt_j]}{n})) - f(S_n(\tfrac{[nt_{j-1}]}{n}))\}.$$

D'après la proposition (4.2), il est clair que

$$\lim_n \left|\mathbb{E}[f(S_n(t)) - f(S_n(\tfrac{[nt]}{n}))]\right| = 0 .$$

Posons $\Sigma_j(n) = [S_n(\frac{[nt_{j-1}]}{n})]^{-1} S_n(\frac{[nt_j]}{n})$, $j \in \{1,\dots q\}$;

$\Sigma_j(n)$ s'écrit $\eta \circ \pi_2(X_1 \dots X_{[nt_{j-1}]})(\Sigma'_j(n))$ où $\Sigma'_j(n)$ est une v.a.

indépendante de $S_n(\frac{[nt_{j-1}]}{n})$ et de même loi que $S_n(\frac{[nt_j]-[nt_{j-1}]}{n})$

D'après la formule de Taylor, nous avons

$$f(S_n(\frac{[nt_j]}{n})) - f(S_n(\frac{[nt_{j-1}]}{n})) = \alpha_j(n) + \frac{1}{2}\beta_j(n) \quad ,$$

avec

$$\alpha_{j(n)} = \sum_{i=1}^{p} x_i(\Sigma_j(n))\ D_i f(S_n(\frac{[nt_{j-1}]}{n})) + \frac{1}{2} \sum_{1\le i,\ell\le p} x_i(\Sigma_j(n)),\ x_\ell(\Sigma_j(n))\ D_i D_\ell f(S_n(\frac{[nt_{j-1}]}{n}))$$

et

$$\beta_j(n) = \sum_{1\le i,\ell\le p} x_i(\Sigma_j(n))x_\ell(\Sigma_j(n))\left[D_i D_\ell f[S_n(\frac{[nt_{j-1}]}{n})(\theta\Sigma_j(n))] - D_i D_\ell f[S_n(\frac{[nt_{j-1}]}{n})]\right]$$

D'après le lemme (1.5), via la proposition (4.2), les v.a. $\sum_j(n)$ et $S_n(\frac{[nt_{j-1}]}{n})$ sont asymptotiquement indépendantes. Compte tenu de la proposition (2.6), nous avons alors

$$(1)\qquad \lim_n \mathbb{E}[\alpha_j(n)] = \sum_{\{i\in\{1,\dots,p\}:dgx_i\ \text{pair}\}} \gamma_i(t_i - t_{j-1})^{dgx_i/2}\ \mathbb{E}_\rho[D_i f(x_\infty(t_{j-1}))]$$

$$+ \frac{1}{2} \sum_{\{1\le i,\ell\le p:dgx_i x_\ell\ \text{pair}\}} \sigma_{i,\ell}\ (t_j - t_{j-1})^{dgx_i x_\ell/2}\ \mathbb{E}_\rho[D_i D_\ell f(x_\infty(t_{j-1}))] \quad ,$$

où $\gamma_i = \frac{1}{(dgx_i/2)!} L_\lambda^{dgx_i/2}\ x_i$ et $\sigma_{i,\ell} = \frac{1}{(dgx_i x_\ell/2)!} L_\lambda^{dgx_i x_\ell/2}\ x_i x_\ell$

D'autre part, pour $i,\ell \in \{1,\dots,p\}$, $j \in \{1,\dots,q\}$, posons,

$$\delta^j_{i,\ell}(n) = \mathbb{E}\Big[x_i(\Sigma_j(n))x_\ell(\Sigma_j(n)) \ [D_iD_\ell f[S_n(\frac{[nt_{j-1}]}{n})(\theta\Sigma_j(n))] - - D_i\ D_\ell f(S_n(\frac{[nt_{j-1}]}{n}))]\Big]$$

Pour $dgx_i + dgx_\ell \geq 3$, nous avons, en appliquant l'inégalité de Schwartz et le lemme (4.3),

$$|\delta^j_{i,\ell}(n)| \leq \Big[2\ C_{i,\ell}||D_iD_\ell f||_\infty \sup_{1\leq j\leq q} (\frac{[nt_j]-[nt_{j-1}]}{n})^{1/2}\Big](\frac{[nt_j]-[nt_{j-1}]}{n}),$$

où $C_{i,\ell}$ est une constante > 0 indépendante de n et de la partition $(t_j)_{1\leq j\leq q}$ de $[0,t]$.

Soient $i,\ell \in \{1,\ldots,p\}$ avec $dgx_i = dgx_\ell = 1$. Pour tout $\varepsilon > 0$, il existe $\alpha > 0$ tel que pour tout $u \in N$ vérifiant $||u|| < \alpha$, on ait

$$\sup_{u\in N} |D_i\ D_\ell\ f(u\ v) - D_i\ D_\ell\ f(u)| < \varepsilon$$

Nous avons alors

$$|\delta^j_{i,\ell}(n)| \leq \varepsilon\mathbb{E}\Big[|x_i(\Sigma_j(n))x_\ell(\Sigma_j(n))|\Big] + 2||D_i\ D_\ell\ f||_\infty \ \mathbb{E}\Big[|x_i(\Sigma_j(n))\ x_\ell(\Sigma_j(n))|1_{\{||\Sigma_j(n)||>\alpha\}}\Big]$$

$$\leq \varepsilon C_{i,\ell}\ (\frac{[nt_j]-[nt_{j-1}]}{n}) + 2||D_iD_\ell f||_\infty \sqrt{\mathbb{E}[x_i^2(\Sigma_j(n))x_\ell^2(\Sigma_j(n))]} \ (\frac{\sqrt{\mathbb{E}[||\Sigma_j(n)||^2]}}{\alpha})$$

$$\leq \Big[\varepsilon C_{i,\ell} + \frac{2||D_iD_\ell f||_\infty}{\alpha} C'_{i,\ell} \sup_{1\leq j\leq q} (\frac{[nt_j]-[nt_{j-1}]}{n})^{1/2}\Big](\frac{[nt_j]-[nt_{j-1}]}{n}) ,$$

où $C_{i,\ell}$ et $C'_{i,\ell}$ sont des constantes > 0 indépendantes de n et de la partition $(t_j)_{1 \leq j \leq q}$ de $[0,t]$.

On en déduit que pour tout $\varepsilon > 0$, il existe $\alpha > 0$ tel que

$$\left|\sum_{j=1}^{q} \beta_j(n)\right| \leq \sum_{j=1}^{q} \sum_{1 \leq i,\ell \leq p} |\delta^j_{i,\ell}(n)| \leq (C_1 + \frac{C_2}{\alpha}) \sup_{1 \leq j \leq q} \left(\frac{[nt_j]-[nt_{j-1}]}{n}\right)^{1/2} + \varepsilon C_3$$

où C_1 , C_2 et C_3 sont des constantes > 0 indépendantes de n et de la partition $(t_j)_{1 \leq j \leq q}$ de $[0,t]$.

De l'inégalité

$$\left(\frac{[nt_j]-[nt_{j-1}]}{n}\right)^{1/2} \leq \left[(t_j - t_{j-1}) + \frac{1}{n}\right]^{1/2} \leq (t_j - t_{j-1})^{1/2} + \frac{1}{\sqrt{n}} ,$$

il résulte que,

$$\forall \varepsilon > 0 , \exists \alpha > 0 , \left|\sum_{j=1}^{q} \beta_j(n)\right| \leq \varepsilon C_3 + (C_1 + \frac{C_2}{\alpha})\left(\sup_{1 \leq j \leq q} (t_j - t_{j-1})^{1/2} + \frac{1}{\sqrt{n}}\right).$$

On en déduit que, pour tout $\varepsilon > 0$, il existe $\alpha > 0$ tel que, pour tout entier naturel n ,

$$\left|\mathbb{E}[f(S_n(t)) - f(0) - \sum_{i=1}^{q} \alpha_i(n)]\right| \leq \left|\mathbb{E}[f(S_n(t)) - f(S_n(\frac{[nt]}{n}))]\right| + \frac{1}{2}\left|\sum_{i=1}^{p} \beta_i(n)\right|$$

$$\leq \varepsilon C_3 + (C_1 + \frac{C_2}{\alpha}) \sup_{1 \leq j \leq q} (t_j - t_{j-1})^{1/2} + \delta(n) ,$$

où $\delta(n)$ est une suite de réels positifs, indépendante de la partition $(t_j)_{1 \leq j \leq q}$, convergeant vers zéro et C_1 , C_2 , C_3 sont des réels > 0 indépendants de n et de la partition $(t_j)_{1 \leq j \leq q}$ de $[0,t]$.

En faisant tendre n vers l'infini, nous voyons que :

(2) pour tout $\varepsilon > 0$, il existe $\alpha > 0$ tel que

$$\left|\mathbf{E}[f(x_\infty(t))] - f(0) - \sum_{i=1}^{q} \lim_n \mathbf{E}[\alpha_j(n)]\right| \leq \varepsilon C_3 + (C_1 + \frac{C_2}{\alpha}) \sup_{1\leq j\leq q} (t_j - t_{j-1})^{1/2}$$

D'après la proposition (4.2), pour toute fonction continue h sur N, à support compact, la fonction, $t \to \mathbf{E}[h(x_\infty(t))]$, est continue. De (1) il s'ensuit que, lorsque $\sup_{1\leq j\leq q} (t_j - t_{j-1})$ tend vers zéro, $\lim \sum_{j=1}^{q} E|\alpha_j(n)|$ tend vers $\int_0^t E_\rho[Af(x_\infty(s))]ds$. La proposition (4.5) résulte alors de (2).

<u>Fin de la preuve du théorème</u>.

D'après la proposition(4.2), pour tout $t \in [0,1]$, la suite $\{\mathbf{E}\left[||S_n(t)||^2\right]\}_{n\geq 1}$ est bornée ; l'ensemble des lois des v.a. $\{S_n(t)\}_{n\geq 1}$ est donc relativement compact pour la topologie de la convergence étroite. De (4.4) et de la proposition (4.5), il résulte que la seule valeur d'adhérence de cette famille de probabilité est la loi au temps t du semi-groupe $(\nu_t)_{t>0}$ sur N^b de g.i. A. Il s'ensuit que, pour tout $t \in [0,1]$, $S_n(t)$ converge en loi vers ν_t et la première assertion du théorème est démontrée.

D'autre part, des propositions (4.1) et (4.2) il résulte que pour tout entier naturel p, la v.a.

$$U_n^b \circ \pi_1(X_1 \dots X_n) - U_n^b \circ \pi_1(X_1 \dots X_{n-p})$$

converge vers zéro dans L^2.

On en déduit que la suite d'homéomorphismes $\{U_n^b\}_{n\geq 1}$ de N vérifie les hypothèses de la proposition (1.2). La démonstration du théorème, dans ce premier cas, est donc achevée.

Cas Général

Le passage du cas précédent au cas général se fait par un procédé de troncature analogue à celui utilisé dans [5]. Nous indiquons la marche à suivre, sans entrer dans les détails.

Pour tout entier naturel $n \geq 1$, désignons par α_n l'élément de N défini par :

$$\forall i \in \{1,\ldots,p\} , \quad x_i(\alpha_n) = \begin{cases} 0 & \text{si } dg\ x_i \geq 2 \\ \dfrac{\int_N x_i(u)\, 1_{\{\phi_b(u) \leq \sqrt{n}\}}\, \bar{\lambda}(du)}{\lambda_1(\{\phi_b > \sqrt{n}\})} & \text{si } dg\ x_i = 1 \end{cases}$$

Nous posons alors, pour $g = (u,k) \in N \times K$ et pour tout entier naturel $n \geq 1$

$$T_n^b(g) = \begin{cases} (u,k) & \text{si } \phi_b(u) \leq \sqrt{n} \\ (\alpha_n,k) & \text{si } \phi_b(u) > \sqrt{n} \end{cases}$$

Pour tout entier naturel $n \geq 1$, la mesure de probabilité $T_n^b(\lambda)$ sur G vérifie l'hypothèse iii) de (2.1) et $\ell(T_n^b(\lambda)) = +\infty$.

Nous avons :

(4.6) Lemme

Avec les notations précédentes,

$$\lim_n \| \lambda^n - [T_n^b(\lambda)]^n \| = 0$$

Preuve : On voit facilement que

$$\| \lambda^n - [T_n^b(\lambda)]^n \| \leq n\lambda(\{\phi_b > \sqrt{n}\}) \leq \int_G \phi_b^2(g) 1_{\{\phi_b > \sqrt{n}\}}(g) \lambda(dg) ,$$

qui tend vers zéro, quand n tend vers l'infini.

(4.7) Lemme

Pour toute fonction polynome T sur N , nous avons

$$\lim_n \frac{1}{n^{dgT/2-1}} Q_{T_n^b(\lambda)} T = 0 \quad \text{si} \quad dgT \geq 3$$

$$\lim_n Q_{T_n^b(\lambda)} T = Q_\lambda T \quad \text{si} \quad dgT \leq 2$$

Preuve : voir lemme (4.11) de [5] .

Soit T une fonction polynome sur N . Soit λ une mesure de probabilité sur G vérifiant l'hypothèse iii) de (2.1) et $\ell(\lambda) \geq dgT$. L'expresssion $L_\lambda^{dgT/2}T$ (voir (2.5)), ne dépend de λ que par λ_2 et les moments :

$$\int_N P(u)\, \lambda_1(du) \;,\; \int_N P(u)\, \hat{\lambda}(du) \quad , \quad P \in \mathbb{C}_1[N] \;;$$

$$\int_N P(u)\ \overline{\lambda}(du) \quad , \quad P \in \mathbb{C}_2[N] \ .$$

Pour toute mesure de probabilité λ sur G vérifiant l'hypothèse iii) de (2.1) et $\ell(\lambda) \geq 2$, nous pouvons donc donner un sens à l'expression $L_\lambda^{dgT/2}T$, pour toute fonction polynome T sur N .

En tenant compte du lemme (4.6) et en procèdant comme pour la preuve de la proposition (2.6), on montre alors la proposition :

(4.8) <u>Proposition</u>.

Soit (G,λ) vérifiant les hypothèses du théorème (2.12) . Alors pour toute fonction polynome T sur N , la suite de fonctions polynomes $\{\frac{1}{n^{dgT/2}}\ [Q_{T_n^b(\lambda)}]^n\ T\}_{n\geq 1}$ converge, uniformément sur tout compact de N , vers zéro si dgT est impair, vers la constante $\frac{1}{(dgT/2)!}\ L_\lambda^{dgT/2}T$ si dgT est pair.

D'après le lemme (4.6), nous sommes amenés, pour prouver le théorème (2.12), à étudier la suite de mesures de probabilité. $\{U_n \circ \pi_1([T_n^b(\lambda^n)]^n)\}_{n\geq 1}$. A l'aide du lemme (4.6) et de la proposition (4.8) on procède alors comme dans le cas étudié précédemment en remplaçant les v.a. $(X_i)_{i\geq 1}$ par les v.a. $(T_n^b(X_i))_{i\geq 1}$.

5. Théorème de la limite centrale pour un produit semi-direct d'un groupe de Lie nilpotent simplement connexe par un groupe compact

Dans cette section, nous allons utiliser les résultats de la section 2 pour établir un théorème de la limite centrale pour un produit semi-direct d'un groupe de Lie nilpotent simplement connexe par un groupe compact.

Définitions et Notations

(5.0) Soit $G = N \times_\eta K$ un produit semi-direct d'un groupe de Lie nilpotent simplement connexe N par un groupe compact K.

Pour simplifier l'écriture, nous identifions N et K à des sous-groupes de G ; tout élément g de G s'écrit donc de façon unique $g = uk$ avec $u \in N$ et $k \in K$; N est un sous-groupe distingué de G et, pour tout couple $(u,k) \in N \times K$, $\eta(k)(u)$ s'écrit kuk^{-1}.

D'autre part, nous identifions N à son algèbre de Lie $(N,[\ ,\ ])$ munie du produit $\circ$ défini par la formule de Campbell-Hausdorff (voir (2.1)) ; le produit de deux éléments $g = uk$ et $g' = u'k'$ de G s'écrit donc $gg' = [u(ku'k^{-1})]kk' = [u\circ Adk(u')]kk'$, où pour tout $k \in K$, Adk désigne l'automorphisme d'algèbre de Lie de $(N,[\ ,\ ])$ (et par suite l'automorphisme de groupe de $(N,\circ)$) tangent à l'automorphisme $u \to kuk^{-1}$ de N.

(5.1) Moments d'une mesure de probabilité sur un groupe L.C.D. compactement engendré ([2])

Soit H un groupe localement compact à base dénombrable, compactement engendré. Une application borélienne δ de H dans $\mathbb{R}_+$ est appelé jauge (resp. fonction sous-additive) si elle vérifie.

$$\forall g_1, g_2 \in H \quad \delta(g_1 g_2) \leq \delta(g_1) + \delta(g_2) + C \quad (\text{resp } \delta(g_1 g_2) \leq \delta(g_1) + \delta(g_2)),$$

où C est une constante > 0 indépendante de g_1 et g_2.

Une jauge δ de H est dite principale s'il existe un voisinage compact V de l'unité engendrant H tel que
$\{x; \delta(x) \leq n\} \subset V^n$, $\forall n \in \mathbb{N}^*$.

Sur un groupe L.C.D. compactement engendré, il existe des jauges principales et si δ_0 est l'une d'elles, pour toute autre jauge δ nous avons : $\forall g \in H \quad \delta(g) \leq C_1 \, \delta_0(g) + C_2$
où C_1 et C_2 sont des constantes > 0 indépendantes de $g \in H$.
Il s'ensuit que si μ est une mesure de probabilité sur H, l'expression $\int [\delta_0(g)]^\alpha \mu(dg) < +\infty$, pour $\alpha > 0$, est indépendante du choix de δ_0 ; dans ce cas nous disons que μ possède un moment d'ordre α.

(5.2) Moments d'une mesure de probabilité sur G.

Appelons r la longueur de l'algèbre de Lie nilpotente $(N, [\ ,\ ])$ et désignons par
$N^1 = N \supset N^2 = [N,N] \supset \ldots \supset N^r = [N, N^{r-1}] \supset N^{r+1} = (0)$, la suite centrale descendante de N. Pour tout $i \in \{1,\ldots,r\}$, désignons par m^i un supplémentaire de N^{i+1} dans N^i. Nous avons

$N = \bigoplus_{i=1}^{r} m^i$; si $u \in N$, nous notons $u^{(i)}$ sa composante sur m^i .
Supposons les sous-espaces m^i , $i \in \{1,...r\}$, de N normés par $\|\ \|$, on définit une fonction ϕ sur N par

$$\phi(u) = \sup_{1 \leq i \leq r} \|u^{(i)}\|^{1/i} \quad (u \in N) .$$

On sait ([3], lemme II.1) que, quitte à remplacer les normes données par des normes homothétiques, ϕ est une jauge principale sur N .

En posant, pour $g = (u,x) \in N \times K$,

$$\psi(g) = \sup_{k \in K} \phi(\mathrm{Ad}\,\mathbf{k}(u))$$

nous obtenons une jauge principale sur G . Nous disons donc qu'une mesure de probabilité μ sur G possède un moment d'ordre $\alpha > 0$ si

$$\int_G \psi^\alpha(g)\ \mu(dg) < +\infty$$

(5.3) <u>Définition</u>.

Soit $(\mathfrak{J}^i)_{i \geq 1}$ une suite décroissante d'idéaux de N telle que $\mathfrak{J}^1 = N$ et $\mathfrak{J}^i = (0)$ pour $i \geq s+1$, $s \in \mathbb{N}^*$. Pour $i \in \{0,...,s\}$, notons q_i la dimension de l'espace vectoriel $N/_{\mathfrak{J}^{i+1}}$. Nous disons qu'une base ordonnée $\{e_k\}_{1 \leq k \leq p}$, $(p=q_s)$, de N est adaptée à la suite d'idéaux $(\mathfrak{J}^i)_{i \geq 1}$ si pour tout $i \in \{1,...,s\}$, vérifiant $q_i \neq q_{i-1}$, $\{e_{q_{i-1}+1} ,..., e_{q_i}\}$ est une base d'un supplémentaire de $\mathfrak{J}^{i+1}$ dans $\mathfrak{J}^i$.

(5.4) Soit λ une mesure de probabilité sur G. Nous désignons par π_1(resp. π_2) la projection de G sur N (resp. K) et nous écrivons $\lambda_i = \pi_i(\lambda)$, $i = 1,2$. Nous notons m(resp. $\tilde{m}$) la mesure de Haar normalisée du sous-groupe compact $K(\lambda_2)$(resp. $\tilde{K}(\lambda_2)$) de K engendré par le support, $S(\lambda_2)$, de λ_2 (resp. par $S(\lambda_2)[S(\lambda_2)]^{-1}$). Quitte à remplacer $G = NK$ par $G = NK(\lambda_2)$, nous pouvons supposer que $K(\lambda_2) = K$. Nous disons que λ_2 est apériodique si $\tilde{K}(\lambda_2) = K$. Nous posons $\overline{\lambda} = \int_K \mathrm{Ad}k(\lambda_1)\, m(dk)$; $\overline{\lambda}$ est une mesure de probabilité sur N .

(5.5) Soit μ une mesure de probabilité sur N possèdant un moment d'ordre 1 (i.e. $\int_N \phi(u)\, \mu(du) < +\infty$, voir (5.2)). Appelons ζ l'application naturelle de N sur $N/[N,N]$. D'après (5.2) l'intégrale $\int_N \zeta(u)\, \mu(du)$ a un sens et défini un élément $e(\mu)$ de $N/[N,N]$. Nous disons qu'une mesure de probabilité μ sur N est <u>centrée</u> si μ possède un moment d'ordre 1 et si $e(\mu) = 0$.

<u>Théorème de la limite centrale dans le cas centré</u>

Nous supposons que la mesure de probabilité sur N , $\overline{\lambda}$, est centrée (voir (5.4) et (5.5)).

(5.6) <u>Notion de degré sur l'algèbre des fonctions polynomes sur N</u> ([2]).

Désignons par $\{e_k\}_{1\leq k\leq p}$ une base de N adaptée à la suite centrale descendante, $(N^i)_{i\geq 1}$ de N (voir (5.2) et définition (5.3)); et par $\{x_k\}_{1\leq k\leq p}$ le système de fonctions coordonnées associé à cette base. Nous définissons une notion de degré sur $\mathbb{C}[N]$, en attribuant un degré à chaque générateur x_k , $1 \leq k \leq p$; le degré de x_k , noté $dg\, x_k$, est par définition égal au plus grand entier i tel que e_k appartienne à N^i . On convient que le degré (resp. la valuation) du polynome nul est $(-\infty)$ (resp. $(+\infty)$). Il est facile de voir que cette notion est indépendante du choix de la base adaptée.

(5.7) Crochet de Lie associé à une base de N adaptée à sa suite centrale descendante

Soit $b = \{e_i\}_{1\leq i\leq p}$ une base de N adaptée à sa suite centrale descendante, $(N^i)_{1\leq i\leq r}$. Pour tout $i \in \{1,\dots,r\}$, $\{e_{p_{i-1}+1},\dots,e_{p_i}\}$, où $p_i = \dim(N/N^{i+1})$ et $P_0 = 0$, est une base d'un supplémentaire m^i de N^{i+1} dans N^i.

Pour tous i et j appartenant à $\{1,\dots,r\}$, nous avons

$$[m^i,m^j] \subset N^{i+j} = \bigoplus_{\{\ell:\ell\geq i+j\}} m^\ell$$

Pour $u \in m^i$ et $v \in m^j$, nous notons $[u,v]_b$ la composante de $[u,v]$ sur m^{i+j} ; puis pour u et v appartenant à N , nous posons

$$[u,v]_b = \sum_{1\leq i,j\leq r} [u_i,v_j]_b \quad ,$$

où $u = \sum_{i=1}^{r} u_i$ avec $u_i \in m^i$ et $v = \sum_{j=1}^{r} v_j$ avec $v_j \in m^j$.

Il est clair que l'application de N x N dans N qui au couple (u,v) associe $[u,v]_b$ est bilinéaire alternée. D'autre part, pour $u \in m^i$, $v \in m^j$, $w \in m^\ell$, avec $i,j,\ell \in \{1,\dots,r\}$, $[u,[v,w]_b]_b$ n'est autre autre que la composante de $[u,[v,w]]$ sur $m^{i+j+\ell}$; il s'ensuit que $[\ ,\ ]_b$ vérifie l'identité de Jacobi. On en déduit donc que $[\ ,\ ]_b$ est un crochet de Lie sur N , vérifiant

$[m^{i_1},[m^{i_2},\ldots[m^{i_{s-1}},m^{i_s}]_b\ldots]_b]_b = m^{\sum_{j=1}^{s} i_j}$, pour $i_j \in \{1,\ldots,r\}$, $j \in \{1,\ldots,s\}$; et $(N,[\ ,\ ]_b)$ est une algèbre de Lie nilpotente de longueur inférieure ou égale à r . Nous notons $\circ_b$ le produit sur N associé au crochet de Lie $[\ ,\ ]_b$ par la formule de Campbell-Hausdorff.

(5.8) Soit $b = \{e_i\}_{1\leq i\leq p}$ une base de N adaptée à sa suite centrale descendante. Si f est une fonction de classe C^1 sur N et v un élément de N , nous définissons

$$D_v f(u) = \lim_{t\to 0} \frac{f(u \circ_b tv) - f(u)}{t} .$$

Si λ possède un moment d'ordre 2, nous désignons par A^b le générateur infinitésimal,

$$A^b = \sum_{i=p_1+1}^{p_2} (L_\lambda\, x_i) D_i + \frac{1}{2} \sum_{1\leq i,j\leq p_1} (L_\lambda\, x_i\, x_j)\, D_i\, D_j ,$$

où L_λ est défini en (2.5) et $D_i = D_{e_i}$, $i \in \{1,\ldots,p\}$.

Pour tout $v \in N$, le champ analytique de vecteurs tangents invariant à gauche, D_v , sur $(N,\circ_b)$, s'identifie à l'élément v de l'algèbre de Lie $(N,[\ ,\ ]_b)$ de $(N,\circ_b)$.

La sous-algèbre de Lie de $(N,[\ ,\ ])$ engendrée par les éléments e_i , $1\leq i\leq p_1$, est égale à $(N,[\ ,\ ])$. D'après la définition du crochet de Lie $[\ ,\ ]_b$ (voir (5.7)), il est alors clair, que la sous-algèbre de Lie de $(N,[\ ,\ ]_b)$ engendrée par ces éléments est égale à $(N,[\ ,\ ]_b)$.

Lorsque la matrice $((L_\lambda\, x_i\, x_j))_{i,j\in\{1,\ldots,p_1\}}$, nous savons donc (remarque (2.13)) que le semi-groupe de convolution $(\nu_t^b)_{t>0}$ sur

$(N,\circ_b)$ de g.i. A^b est absolument continu par rapport à la mesure de Haar de $(N,\circ_b)$, avec une densité de classe C^∞ .

L'adaptation du lemme (2.4) à la situation précédente, nous dit que la matrice $((L_\lambda x_i x_j))_{i,j\in\{1,\dots,p_1\}}$ est définie positive si et seulement si la mesure λ n'est pas portée par un sous-groupe fermé de G de la forme $N_0(a K^{-1}a)$ où : N_0 est un idéal propre de $(N,[\ ,\ ])$, K-invariant (i.e. $Adk(N_0) \subset N_0$, $\forall k \in K$), contenant $[N,N]$; et a est un élément de N vérifiant

(1) $x_i(a) = ([I - Q_{\lambda_2}(I-Q_m)]^{-1}x_i)(\int_N u\ \lambda_1(du))$, $\forall i \in \{1,\dots,p_1\}$.

[On notera que des propriétés de N_0 , il résulte que le groupe $N_0(a\ k^{-1}a)$ est le même pour tous les éléments a de N vérifiant (1)].

Si $\lambda = \lambda_1 \otimes \lambda_2$ ou λ_1 est centrée, on peut prendre pour a l'élément neutre de G .

Nous avons alors :

(5.9) <u>Théorème</u>.

Soit λ une mesure de probabilité sur G possèdant un moment d'ordre 2 et telle que $\bar{\lambda}$ soit centrée. Soit $b = \{e_i\}_{1\leq i\leq p}$ une base de N , adaptée à sa suite centrale descendante. Pour tout $(u,k) \in N \times K$ et pour tout $k_0 \in K$, posons

$$V_n^{b,k_0}(u,k) = \left(\sum_{i=1}^{p} \frac{x_i(u)}{n^{dgx_i/2}} e_i \quad , \quad k_0^{-n} k\right) .$$

Alors, pour tout $k_0 \in S(\lambda_2)$, la suite de mesure de probabilité sur $N \times K$, $\{V^{b,k_0}(\lambda^n)\}_{n\geq 1}$ converge vaguement vers la mesure de probabilité produit $\nu_1^b \otimes \tilde{m}$. En particulier lorsque λ_2

est apériodique , $\{V_n^{b,e}(\lambda^n)\}_{n\geq 1}$, converge vaguement vers $\nu_1^b \otimes m$.

Preuve du théorème (5.9)

Pour la notion de degré sur $\mathbb{C}[N]$ définie en (5.6), le couple (G,λ) vérifie les hypothèses de (2.1) : compte tenu de la formule de Campbell-Hausdorff, l'hypothèse i) (resp. ii)) est une conséquence des relations $[N^i,N^j] \subset N^{i+j}$, $\forall i,j \geq 1$, (resp. du fait que les N^i , $i \geq 1$, sont stables par les éléments Adk , $k \in K$) ; enfin l'hypothèse iii) signifie que $\overline{\lambda}$ est centrée.

En remarquant d'une part que la condition (*) de (2.11) équivaut à l'existence d'un moment d'ordre 2 pour λ et d'autre part que les groupes $(N,\circ_b)$ et N^b associés à b respectivement en (5.7) et (2.9) coïncident, nous voyons que le théorème (5.9) résulte du théorème (2.12).

Théorème de la limite centrale dans le cas général

λ est une mesure de probabilité sur G possèdant un moment d'ordre 1 .

(5.10) Suite graduée d'idéaux de N associée à λ .

Considérons l'ensemble

$E = \{(\ell,k) \in \mathbb{N} \times \mathbb{N} : 0 \leq k < \ell\} \cup \{(1,1)\}$ muni de la relation d'ordre total $\geq$ définie par

$$(\ell,k) \geq (\ell',k') \iff \begin{cases} \ell > \ell' \\ \text{ou} \\ \ell = \ell' \text{ et } k \geq k' \end{cases} .$$

On vérifie aisément que l'ordre ainsi défini sur E est celui qui est induit par la bijection :

$$\sigma : E \to \mathbb{N}^*$$
$$(\ell,k) \to \sigma(\ell,k) = \begin{cases} \ell(\ell-1)/2 + k + 2 & \text{si } \ell \geq 2 \\ k + 1 & \text{si } \ell = 1 \end{cases}$$

Nous appelons suite graduée d'idéaux de N associée à λ, la suite décroissante d'idéaux de N définie de la façon suivante : $\mathfrak{J}^{1,0}(\lambda) = N$, $\mathfrak{J}^{1,1}(\lambda) = \zeta^{-1}(e(\overline{\lambda}))$ et $\mathfrak{J}^{\ell,k}(\lambda)$, $(\ell,k) \in E$ avec $\ell \geq 2$, est l'idéal de N^{ℓ} formé des éléments de N^{ℓ} qui s'écrivent $[u_1,[u_2,\ldots,[u_{\ell-1},u_{\ell}]\ldots]$, $u_1,\ldots,u_{\ell} \in N$, où <u>au moins</u> k éléments parmi les u_i, $i \in \{1,\ldots,\ell\}$, appartiennent à $\zeta^{-1}(e(\overline{\lambda}))$. Nous avons $\mathfrak{J}^{\ell,0}(\lambda) = N^{\ell}$, $\forall \ell \geq 1$; $\mathfrak{J}^{\ell,k}(\lambda) = 0$ si $\ell > r$ et $[\mathfrak{J}^{\ell,k}(\lambda),\mathfrak{J}^{\ell',k'}(\lambda)] \subset \mathfrak{J}^{\ell+\ell',k+k'}(\lambda)$ pour (ℓ,k) et $(\ell',k') \in E$.

(5.11) <u>Notion de "degré suivant λ" sur $\mathbb{C}[N]$</u>.

Désignons par $\{e_k\}_{1\leq k\leq p}$ une base de N adaptée à la suite graduée d'idéaux de N associée à λ (voir définitions (5.3) et (5.10)) ; et par $\{x_k\}_{1\leq k\leq p}$ le système de fonctions coordonnées associée à cette base. Nous définissons une notion de dégré sur $\mathbb{C}[N]$ en attribuant un dégré à chaque générateur x_k, $1\leq k\leq p$; le <u>degré suivant λ</u> de x_k, noté $dg^{\lambda} x_k$, est par définition égal à 1 si e_k n'appartient pas à N^2, égal à $\sup\{s+\ell : (s,\ell) \in E , e_k \in \mathfrak{J}^{s,\ell}(\lambda)\}$ si e_k appartient à N^2. On convient que le degré (resp. la valuation) suivant λ du polynome nul est $(-\infty)$ (resp. $(+\infty)$). Il est facile de voir que cette notion de degré sur $\mathbb{C}[N]$ est indépendante du choix de la base de N adaptée à la suite graduée d'idéaux associée à λ.

Dans le cas où la mesure de probabilité $\overline{\lambda}$ est centrée, la suite graduée d'idéaux associée à λ n'est autre que la suite centrale descendante de N ; nous retrouvons alors la notion de degré définie en (5.6).

(5.12) Crochets de Lie associés à une base de N adaptée à $\{\mathfrak{J}^{\ell,k}(\lambda),(\ell,k)\in E\}$.

Soit $b = \{e_i\}_{1\leq i\leq p}$ une base de N adaptée à la suite d'idéaux $\{\mathfrak{J}^{\ell,k}(\lambda),(\ell,k) \in E\}$. Pour tout élément (ℓ,k) de E, posons

$$q_{\sigma(\ell,k)-1} = \dim N/\mathfrak{J}^{\ell,k}(\lambda) \quad \text{(voir la définition de } \sigma \text{ en (5.11))} ;$$

si $\bar{\lambda}$ est centrée $q_2 = q_1$, si $\bar{\lambda}$ n'est pas centrée $q_2 = q_1 + 1$.

Pour tout élément (ℓ,k) de E vérifiant $q_{\sigma(\ell,k)-1} \neq q_{\sigma(\ell,k)}$ $\{e_{q_{\sigma(\ell,k)-1}+1}, \ldots, e_{q_{\sigma(\ell,k)}}\}$ est une base d'un supplémentaire $m^{\ell,k}$ de $\mathfrak{J}^{\ell_0,k_0}(\lambda)$ dans $\mathfrak{J}^{\ell,k}(\lambda)$, où $(\ell_0,k_0) = \sigma^{-1}(\sigma(\ell,k)+1)$.

Pour tous couples (ℓ,k) et (ℓ',k') de E, nous avons

$$[m^{\ell,k},m^{\ell',k'}] \subset \mathfrak{J}^{\ell+\ell',k+k'}(\lambda) = \bigoplus_{\{(s,j)\in E:(s,j)\geq(\ell+\ell',k+k')\}} m^{s,j}$$

Pour $u \in m^{\ell,k}$ et $v \in m^{\ell',k'}$, nous notons $[u,v]_b$ la composante de $[u,v]$ sur $m^{\ell+\ell',k+k'}$. Si $u \in N$, pour tout couple (ℓ,k) de E, nous notons $u^{(\ell,k)}$ la composante de u sur $m^{\ell,k}$ nous avons

$$u = \sum_{(\ell,k)\in E} u^{(\ell,k)} = \sum_{\{(\ell,k)\in E:\ell\leq r\}} u^{(\ell,k)}$$

Pour u et v appartenant à N, posons alors

$$[u,v]_b = \sum_{\{(\ell,k),(\ell',k')\in E:\ell,\ell'\leq r\}} [u^{(\ell,k)},v^{(\ell',k')}]_b$$

Il est clair que l'application de $N \times N$ dans N qui au couple (u,v) associe $[u,v]_b$ est bilinéaire alternée. D'autre part, pour $u \in m^{\ell,k}$, $v \in m^{\ell'k'}$, $w \in m^{\ell'',k''}$, avec $(\ell,k),(\ell',k')$, $(\ell'',k'') \in E$, $[u,[v,w]_b]_b$ n'est autre que la composante de $[u,[v,w]]$ sur $m^{\ell+\ell'+\ell'',k+k'+k''}$; il s'ensuit que $[\ ,\ ]_b$ vérifie l'identité de Jacobi. On en déduit que $[\ ,\ ]_b$ est un crochet de Lie de N, vérifiant

$$[m^{\ell_1,k_1},[m^{\ell_2,k_2},\ldots,[m^{\ell_{s-1},k_{s-1}},m^{\ell_s,k_s}]_b\ldots]_b]_b \subset m^{\sum_{i=1}^{s}\ell_i,\sum_{i=1}^{s}k_i} ,$$

pour $(\ell_i,k_i) \in E$, $i \in \{1,\ldots,s\}$; et $(N,[\ ,\ ]_b)$ est une algèbre de Lie nilpotente de longueur inférieure ou égale à r. Nous notons $\circ_b$ le produit sur N associé au crochet de Lie $[\ ,\ ]_b$ par la formule de Campbell-Hausdorff.

D'autre part, nous définissons un nouveau crochet de Lie $[\ ,\ ]'_b$ sur N en posant, pour u et $v \in N$,

$$[u,v]'_b = [u-u^{(1,1)},v-v^{(1,1)}]_b .$$

Nous notons $\circ'_b$ le produit sur N associé à ce nouveau crochet de Lie par la formule de Campbell-Hausdorff. Nous avons

$$u \circ'_b v = (u-u^{(1,1)}) \circ_b (v-v^{(1,1)}) + u^{(1,1)} + v^{(1,1)} ,$$

pour $u,v \in N$.

Nous notons ad(resp. ad^b et ad'^b) la représentation adjointe de l'algèbre de Lie $(N,[\ ,\ ])$(resp.$(N,[\ ,\ ]_b)$ et $(N,[\ ,\ ]'_b)$).

(5.13) Soit $b = \{e_i\}_{1\leq i\leq p}$ la base de N considérée en (5.12).
Dans le cas où $\overline{\lambda}$ n'est pas centrée, posons

$$f = \int_K \mathrm{Ad}k(e_{q_2})\, m(dk) \;;$$

f est un élément K-invariant de N et $f - e_{q_2} \in [N,N]$. Nous désignons par $\overline{b}$ la base de N qui se déduit de b en remplaçant e_{q_2} par f.

Désignons par τ_b l'lément de N défini par

$$\tau_b = \begin{cases} [\int_N x_{q_2}(u)\ \overline{\lambda}(du)]f & \text{si } \overline{\lambda} \text{ n'est pas centrée} \\ 0 & \text{sinon} \end{cases}$$

τ_b est un élément K-invariant de N.

Nous munissons l'espace $N \times \mathbb{R}$, (resp. $N \times K \times \mathbb{R}$), du produit, noté $\circ_b'$, (resp. noté multiplicativement), défini par

$(u,t) \circ_b' (v,s) = (u \circ_b' \mathrm{Exp}\, \mathrm{ad}^{\overline{b}}\, t\tau_b(v)\ ,\ t + s)$ pour $u,v \in N$ et $s,t \in \mathbb{R}$,

$$[\text{resp. } (u,k,t)(v,k',s) = (u\circ_b' \mathrm{Expad}^{\overline{b}}\ t\tau_b(\mathrm{Ad}k(v)), kk',\ t+s)$$
$$= (u\circ_b' \mathrm{Ad}\ k\ (\mathrm{Exp}\ \mathrm{ad}^{\overline{b}}\ t\tau_b(v))\ ,\ kk'\ ,\ t+s)\ ,$$

(car τ_b est K-invariant), pour $u,v \in N$, $k,k' \in K$ et $t,s \in \mathbb{R}$].

Nous notons $M(b)$ (resp. $H(b)$) les groupes ainsi obtenus. $M(b)$ est un groupe de Lie nilpotent simplement connexe qui est un produit semi-direct de $(N,\circ_b')$ par $(\mathbb{R},+)$. $H(b)$ est un produit semi-direct de $(N,\circ_b')$ par le produit direct des groupes K et $(\mathbb{R},+)$; c'est donc aussi le produit semi-direct du groupe de Lie nilpotent simplement connexe $M(b)$ par le groupe compact K.

(5.14) Soit $b = \{e_i\}_{1 \leq i \leq p}$ la base de N considérée en (5.12) et (5.13). Si f est une fonction de classe C^1 sur $N \times \mathbb{R}$ et z est un élément de $N \times \mathbb{R}$ nous posons

$$D_z f(y) = \lim_{t \to 0} \frac{f(y \circ_b' tz) - f(y)}{t}$$

pour $y \in N \times \mathbb{R}$.

Désignons par λ^b la mesure de probabilité produit $[\lambda * \varepsilon_{\tau_b^{-1}}] \otimes \varepsilon_1$ sur $G \times \mathbb{R}$. Lorsque λ possède un moment d'ordre 2 sur G, λ^b possède un moment d'ordre 2 sur $H(b)$. Nous considérons alors le générateur infinitésimal,

$$A^b = \sum_{i=q_2+1}^{q_3} (L_{\lambda^b}\, x_i)\, D_{(e_i,0)} + \frac{1}{2} \sum_{1 \leq i,j \leq q_2} (L_{\lambda^b}\, x_i x_j)\, D_{(e_i,0)} D_{(e_j,0)} + D_{(0,1)},$$

où L_{λ^b} est défini en (2.5).

Définissons sur $N \times \mathbb{R}$ le crochet de Lie suivant :

$[(u,t),(v,s)]_b = ([u,v]_b' + t[\tau_b,v]_{\overline{b}} - s[\tau_b,u]_{\overline{b}}, 0)$, pour $u,v \in N$ et $t,s \in \mathbb{R}$, où $\overline{b}$ désigne la base de N qui se déduit de b en remplaçant e_{q_2} par $f = \int_K \mathrm{Ad}k(e_{q_2})\, m(dk)$, dans le cas où $\overline{\lambda}$ n'est pas centrée et $\overline{b} = b$ si $\overline{\lambda}$ est centrée. L'algèbre de Lie du groupe $M(b)$ est alors $(N \times \mathbb{R}, [\ ,\ ]_b)$. Notons $\exp_{M(b)}$ l'application exponentielle du groupe $M(b)$; pour tout élément z de $N \times \mathbb{R}$, le champ analytique de vecteurs tangents invariant à gauche D_z sur $M(b)$ s'identifie à l'élément $\exp_{M(b)}^{-1} z$ de l'algèbre de Lie $(N \times \mathbb{R}, [\ ,\ ]_b)$ de $M(b)$; en particulier si z est de la forme $(u,0)$ avec $u \in N$ ou $(0,t)$ avec $t \in \mathbb{R}$, D_z s'identifie à z.

Dans le cas où la matrice $((L_{\lambda^b} x_i\ x_j))_{1\leq i,j\leq q_2}$ est définie positive, on voit alors que l'on a (avec les notations de la remarque (2.13)) ,

$$\mathcal{L}_1(A^b) = N \times \mathbb{R} \qquad \text{et} \qquad \mathcal{L}_2(A^b) = N \ .$$

Si bien que le semi-groupe de convolution $(\nu_t^b)_{t>0}$ sur $M(b)$ de g.i. A^b , s'écrit $(\mu_t^b \otimes \varepsilon_t)_{t>0}$ où μ_t^b est absolument continue par rapport à la mesure de Haar de N avec une densité de classe C^∞ .

L'adaptation du lemme (2.14) à la situation présente, nous dit que la matrice précédente est définie positive si et seulement si la mesure λ n'est pas portée par la classe de τ_b modulo un sous-espace fermé de G de la forme $N_0(aKa^{-1})$, où : N_0 est un idéal propre de $(N,[\ ,\ ])$, K-invariant, contenant $[N,N]$; et a est un élément de N vérifiant

$$(1)\quad x_i(a) = (\ I - Q_2(I-Q_m)^{-1}x_i)(\textstyle\int_N u\ \lambda_1 * \varepsilon_{\tau_b^{-1}}(du)) \quad \forall i \in \{1,\dots,q_2$$

[On notera que des propriétés de N_0 , il résulte que, pour tous éléments b et c de N , $bN_0Kc = b'N_0Kc' = N_0b'Kc'$, où

$$x_i(b') = x_i(b) \quad \text{et} \quad x_i(c') = x_i(c) \qquad \forall i \in \{1,\dots,q_2\}$$

et

$$x_i(b') = x_i(c') = 0 \qquad\qquad \forall i \in \{q_2+1,\dots,p\}\ .$$

Si bien que d'une part le groupe $N_0(aKa^{-1})$ est le même pour tous les éléments a de N vérifiant (1) et d'autre part $\tau_b\ N_0(aKa^{-1}) = N_0(aKa^{-1})\ \tau_b$].

Si $\lambda_1 * \varepsilon_{\tau_b^{-1}}$ est centrée ou si $\lambda = \lambda_1 \otimes \lambda_2$, on peut prendre pour a l'élément neutre e de G .

(5.15) Théorème

Soit λ une mesure de probabilité sur G possèdant un moment d'ordre 2. Soit $b = \{e_i\}_{1\leq i\leq p}$ une base de N adaptée à la suite graduée d'idéaux de N associée à λ (voir (5.10)). Pour tout $(u,k) \in N \times K$ et pour tout $k_0 \in K$, posons

$$V_n^{b,k_0}(u,k) = \left(\sum_{i=1}^{p} \frac{x_i(u o(-n\tau_b))}{{}_n dg^\lambda x_i/2} e_i \ , \ k_0^{-n} k\right) .$$

Alors, pour tout $k_0 \in S(\lambda_2)$, la suite de mesures de probabilité sur $N \times K$, $\{V_n^{b,k_0}(\lambda^n)\}_{n\geq 1}$ converge vaguement vers la mesure de probabilité produit $\mu_1^b \otimes \tilde{m}$. En particulier, lorsque λ_2 est apériodique, $\{V_n^{b,e}(\lambda^n)\}_{n\geq 1}$ converge vaguement vers $\mu_1^b \otimes m$.

Preuve :

Munissons l'espace $N \times \mathbb{R}$ (resp. $G \times \mathbb{R} = N \times K \times \mathbb{R}$) du produit, noté $\circ$ (resp. multiplicativement), défini par

$$(u,t) \circ (v,s) = (u \circ \operatorname{Exp} \operatorname{ad} t\tau_b (v) \ , \ t + s) \ ,$$

pour $u,v \in N$ et $s,t \in \mathbb{R}$.

[resp. $(u,k,t)(v,k',s) = (u\circ\operatorname{Exp} \operatorname{ad} t\tau_b (\operatorname{Ad}k(v)),\ kk',\ t + s)$

$= (u\circ\operatorname{Ad}k(\operatorname{Exp} \operatorname{ad} t\tau_b (v)),\ kk',\ t + s)$,

pour $u,v \in N$, $k,k' \in K$ et $s,t \in \mathbb{R}$).

Nous notons M et H les groupes ainsi obtenus. M est un groupe de Lie nilpotent simplement connexe qui est un produit semi-direct de $(N,\circ)$ par $(\mathbb{R},+)$. H est un produit semi-direct de $(N,\circ)$ par le produit direct des groupes K et $(\mathbb{R},+)$; c'est donc aussi le produit semi-direct du groupe de Lie nilpotent simplement connexe M par le groupe compact K.

Soit $(X_i)_{i\geq 1}$ une suite de v.a. à valeurs dans G, indépendantes et de loi commune λ. Ecrivons $X_i = Y_i k_i$, $i \geq 1$, où Y_i et k_i sont des v.a. à valeurs respectivement dans N et K. En tenant compte que l'élément de N, τ_b, est K-invariant, le produit $X_1 \ldots X_n \tau_b^{-n} = X_1 \ldots X_n(-n\tau_b)$ s'écrit alors $N_n K_n$ avec

$$N_n = Z_1 \circ [\tau_b \circ Adk_1(Z_1) \circ (-\tau_b)] \circ \ldots \circ [(n-1)\tau_b \circ Adk_1 \ldots k_{n-1}(Z_n) \circ (-(n-1)\tau_b)]$$

$$= Z_1 \circ \mathrm{Exp}\, ad\tau_b\, (Ak_1(Z_2)) \circ \ldots \circ \mathrm{Exp}\, ad(n-1)\tau_b\, (Adk_1 \ldots k_{n-1}(Z_n)) ,$$

en posant $Z_i = Y_i \circ (-\tau_b)$, $i \geq 1$,

et $K_n = k_1 \ldots k_n$.

Pour tout entier naturel $n \geq 1$, nous voyons donc que (N_n, K_n, n) est égal au produit dans H des v.a. $(Z_i, k_i, 1)$, $1 \leq i \leq n$, à valeurs dans $N \times K \times \mathbb{R}$, indépendantes et de loi commune λ^b .

Nous définissons une notion de degré sur $\mathbb{C}[M] = \mathbb{C}[N\times\mathbb{R}]$, en attribuant un degré à tout élément T de $\mathbb{C}[N\times\mathbb{R}]$ de la forme $T_1 \otimes T_2$ où $T_1 \in \mathbb{C}[N]$ et $T_2 \in \mathbb{C}[\mathbb{R}]$; le degré de T est par définition égal à $dg^\lambda T_1 + 2\, dg\, T_2$, où $dg^\lambda T_1$ est défini en (5.11) et $dg\, T_2$ est le degré habituel d'un élément T_2 de $\mathbb{C}[\mathbb{R}]$.

Pour cette notion de degré sur $\mathbb{C}[M]$, le couple (H, λ^b) vérifie les hypothèses de (2.1) : l'hypothèse i) est une conséquence des relations $[\mathfrak{J}^{\ell,k}(\lambda) , \mathfrak{J}^{\ell',k'}(\lambda)] \subset \mathfrak{J}^{\ell+\ell,k+k'}(\lambda)$, $\forall(\ell,k)$ et $(\ell',k') \in E$; l'hypothèse ii) résulte du fait que les idéaux $\mathfrak{J}^{\ell,k}(\lambda)$ de N , $(\ell,k) \in E$, sont stables par les éléments Adk , $k \in K$; enfin l'hypothèse iii) signifie que $\overline{\lambda} * \varepsilon_{\tau_b^{-1}}$ est centrée sur N

Puisque la mesure de probabilité λ possède un moment d'or-
re 2 sur G , il en est de même de la mesure λ^b sur H ; λ^b véri-
ie alors la condition (*) de (2.11). En remarquant alors que le grou-
e $M^{b'}$ associée à la base $b' = \{(e_i,0) : 1 \leq i \leq p\} \cup \{(0,1)\}$ de
'espace vectoriel $N \times \mathbb{R}$ (voir (2.9)) coïncide avec le groupe
(b) construit en (5.13), le théorème (5.15) résulte du théorème
2.12).

II . Théorème de la limite centrale pour un produit semi-direct d'un groupe de Lie résoluble simplement connexe de type rigide par un groupe compact

Soit R un groupe de Lie résoluble connexe ayant $(\mathcal{R}, [\ ,\ ])$ pour algèbre de Lie. Désignons par Ad la représentation adjointe de R ; d'après le théorème de Lie nous savons qu'il existe une base de $\mathcal{R}$ qui triangularise simultanément tous les éléments de AdR. Nous disons que R est de type rigide si les valeurs propres des éléments de AdR sont toutes de module égal à 1 .

Nous avons :

(6.1) Théorème

Soit R un groupe de Lie résoluble connexe. Désignons par D(R) l'ensemble des parties semi-simples des éléments de AdR ; D(R) est un groupe de Lie connexe abélien qui est en outre compact si R est de type rigide. On munit alors l'espace produit R x D(R) d'un produit $\cdot$ explicitement décrit tel que :

i) $(R,\cdot)$ est un groupe de Lie nilpotent connexe et $(R \times D(R),\cdot)$ est le produit semi-direct du groupe $(R,\cdot)$ et du groupe de Lie abélien connexe D(R) .

ii) l'application $\begin{array}{l} R \to (R \times D(R),\cdot) \\ r \mapsto (r,D(r)) \end{array}$, où D(r) désigne la partie semi-simple de Adr , est un isomorphisme de groupes de Lie de R sur son image.

Preuve du théorème.

Choisissons une sous-algèbre de Cartan $\mathcal{P}$ de $\mathcal{R}$ (i.e. une sous-algèbre nilpotente de $\mathcal{R}$ qui est son propre normalisateur).

Désignons par ad la représentation adjointe de $\mathcal{R}$. Pour toute forme linéaire complexe α sur $\mathcal{P}$, posons

$$\mathcal{R}_\alpha = \{X \in \mathcal{R}^{\mathbb{C}} : \exists\, \ell \in \mathbb{N} \,/\, (\mathrm{ad}P - \alpha(P)I)^{\ell}(X) = 0 \quad \forall P \in \mathcal{P}\}$$

Notons Δ l'ensemble des formes linéaires α telles que $\mathcal{R}_\alpha \neq (0)$.

Nous avons

$$\mathcal{P} = \mathcal{R}_0 \quad , \quad \mathcal{R}^{\mathbb{C}} = \bigoplus_{\alpha \in \Delta} \mathcal{R}_\alpha$$

et

$$[\mathcal{R}_\alpha, \mathcal{R}_\beta] \subset \mathcal{R}_{\alpha+\beta} \quad , \qquad \forall \alpha, \beta \in \Delta$$

Si $u \in \mathcal{R}$, nous notons u_α sa composante sur $\mathcal{R}_\alpha$, $\alpha \in \Delta$; nous avons $u = \sum_{\alpha \in \Delta} u_\alpha$. Posons

$$\alpha(u) = \alpha(u_0) \ , \ u \in \mathcal{R} \ , \ \alpha \in \Delta \quad ;$$

nous savons ([6]) que :

1) α est une forme linéaire sur $\mathcal{R}$ nulle sur le nilradical $\mathcal{N}$ de $\mathcal{R}$ (i.e. le plus grand idéal nilpotent de $\mathcal{R}$) ; et tout élément u de $\mathcal{R}$ tel que $\alpha(u) = 0 \quad \forall \alpha \in \Delta$, appartient à $\mathcal{N}$.

2) Tout élément α de Δ est la différentielle en e d'un homomorphisme ϕ_α de $\mathbb{R}$ dans $(\mathbb{C}^*, \times)$. Pour tout $\alpha \in \Delta$, ϕ_α prend la valeur 1 sur le nilradical N de R (i.e. le plus grand sous-groupe distingué nilpotent de R) ; et tout élément r de R pour lequel $\phi_\alpha(r) = 1 \quad \forall \alpha \in \Delta$, appartient à N.

[D'après le théorème de Lie, nous pouvons trouver une base de $\mathfrak{R}$ qui triangularise simultanément les éléments de AdR ; alors la famille $\{\phi_\alpha, \alpha \in \Delta\}$ n'est autre que la famille d'homomorphismes définie par les éléments diagonaux].

Pour tout $\alpha \in \Delta - \{0\}$, $\mathfrak{R}_\alpha \subset [\mathfrak{R},\mathfrak{R}]^{\mathbb{C}} \subset \mathcal{N}^{\mathbb{C}}$,

Par suite,

$$\mathfrak{R} = \mathfrak{P} + \mathcal{N} \qquad \text{(somme non nécessairement directe).}$$

Désignons par P le groupe de Lie connexe de R ayant $\mathfrak{P}$ pour algèbre de Lie ; nous avons R = NP (décomposition non nécessairement unique).

(6.2) Lemme

Pour tous éléments r de R et A de D(R), posons

$$\eta(A)(r) = (\exp A u)\, p \quad ,$$

où r s'écrit $(\exp u)p$ avec $u \in \mathcal{N}$ et $p \in P$.

On définit ainsi, sans ambiguité, un élément $\eta(A)(r)$ de R ; η est un homomorphisme continu de D(R) dans le groupe des automorphismes de R .

Preuve :

Pour tout $r \in R$, désignons par D(r) la partie semi-simple de Adr ; nous avons

$$D(r)\, u = \sum_{\alpha\in\Delta} \phi_\alpha(r)\, u_\alpha \quad , \qquad (u \in \mathfrak{R})$$

D(r) possède les propriétés suivantes :

i) D(r) = D(p) , pour tout $p \in P$ tel que $rp^{-1} \in N$ [car $\phi_\alpha(N) = \{1\}$ $\forall \alpha \in \Delta$].

ii) $D(r)\,u = u$, $\forall u \in \mathcal{N} \cap \mathcal{G}$ [car $\phi_0 \equiv 1$]

iii) $D(r)$ est un automorphisme d'algèbre de Lie de $\mathcal{R}$ [car $\phi_{\alpha+\beta} = \phi_\alpha \phi_\beta$ et $[\mathcal{R}_\alpha, \mathcal{R}_\beta] \subset \mathcal{R}_{\alpha+\beta}$, pour $\alpha, \beta \in \Delta$ avec $\alpha + \beta \in \Delta$]

iv) $D(r)$ commute avec tous les éléments de AdP [car les sous-espaces $\mathcal{R}_\alpha$, $\alpha \in \Delta$, sont stables pour les éléments de AdP].

A l'aide de ces quatre propriétés, on vérifie aisément : out d'abord que :

$(\exp A\,u)\,p = (\exp A\,u')\,p'$

our $A \in D(R)$ et pour $u, u' \in \mathcal{N}$, $p, p' \in P$ avec exp $u)p = (\exp u')p'$; et ensuite que η est un homomorphisme conti-u de $D(R)$ dans le groupe des automorphismes de R .

Le lemme est prouvé.

Posons alors

$_1 \cdot r_2 = r_1[\eta(D(r_1^{-1}))(r_2)]$, pour tout r_1 , $r_2 \in R$.

On voit facilement que l'on définit ainsi un produit nilpo-ent sur R ; η est un homorphisme continu de $D(R)$ dans le groupe es automorphismes de $(R,\cdot)$; et l'application

$\to (R,\cdot) \times_\eta D(R)$ est un isomorphisme de groupes analytiques

$\mapsto (r, D(r))$

e R sur son image .

Désignons par $D(\mathcal{R})$ l'ensemble des parties semi-simples des éléments de $\mathrm{ad}\mathcal{R}$; nous avons

$$D(u)\ v = \sum_{\alpha \in \Delta} \alpha(u)\ v_\alpha \quad , \quad (u,v \in \mathcal{R}) .$$

Posons alors

$$[u,v]' = [u,v] - (D(u)\ v - D(v)\ u) \ , \ (u,v \in \mathcal{R}) \ ;$$

puis

$$[(u,A),(v,B)]' = ([u,v]' + Av - Bu \ , \ 0) \ , \ (u,v \in \mathcal{R} \ ; \ A,B \in D(\mathcal{R})) .$$

$[\ ,\ ]'$ est un crochet de Lie sur $\mathcal{R} \oplus D(\mathcal{R})$ et le groupe de Lie $(R,\cdot) \times_\eta D(R)$ possède $(\mathcal{R} \oplus D(\mathcal{R}),[\ ,\ ]')$ pour algèbre de Lie.

Le théorème (6.1) est ainsi démontré.

(6.2) Soit $G = R \times_\xi K$ un produit semi-direct d'un groupe de Lie résoluble simplement connexe de type rigide R par un groupe compact K.

Munissons l'espace produit $R \times D(R) \times K$ du produit $\cdot$ défini par (voir lemme (6.2))

$$(r,D(s),k) \cdot (r',D(s'),k') = (r[\eta(D(r^{-1}s))(\xi(k)(r'))] \ , \ D(s)\ D(\xi(k)(s')),kk')$$

pour $r,r',s,s' \in R$ et $k,k' \in K$; nous notons H le groupe ainsi obtenu. H est un produit semi-direct du groupe nilpotent simplement connexe $(R,\cdot)$ par un groupe compact. L'application ψ de G dans H qui au couple (r,k) associe le triplet $(r,D(r),k)$ est un isomorphisme de G sur son image.

Nous sommes ainsi ramené à la situation de la section 5 ;

les résultats de cette section nous donnent donc un théorème de la limite centrale pour G .

(6.3) Exemple

Soit G le groupe constitué par l'espace produit $\mathbb{C} \times \mathbb{R}$ muni du produit

$$(z,t)(z',t') = (z + e^{it}z',\ t + t').$$

G est le revêtement simplement connexe du groupe des déplacements du plan.

Le plongement qui résulte du théorème (6.1) est le suivant :

$$\psi : G \to H$$
$$(z,t) \mapsto ((z,t),e^{it})$$

où H est le groupe constitué par l'espace produit $\mathbb{C} \times \mathbb{R} \times T$ muni du produit

$$(z,t,e^{is})(z',t',e^{is'}) = (z + e^{is}z'\ ,\ t + t'\ ,\ e^{i(s+s')})\quad ;$$

H est donc un produit semi-direct de $(\mathbb{R}^3,+)$ par le tore T .

Désignons par λ une mesure de probabilité possèdant un moment d'ordre 2 sur G ; et par $\{(z_i,t_i)\}_{i\geq 1}$ une suite de v.a. à valeurs dans $\mathbb{C} \times \mathbb{R}$, indépendantes et de loi commune λ. Si g est un élément de G , nous notons $z(g) = x_1(g) + ix_2(g)$ et $t(g)$ ses composantes respectivement sur $\mathbb{C}$ et $\mathbb{R}$.

D'après la section 5 , nous avons :

1er cas : Nous supposons que $\int_G t(g)\, \lambda(dg) = 0$ (i.e. que la mesure $\overline{\psi(\lambda)}$ est centrée dans R^3 (notations de la section 5)).

Supposons que λ ne soit pas portée par le sous-groupe $\mathbb{C} \times \{0\}$ de G . Désignons par a l'élément de G défini par

$$z(a) = \frac{\int_G z(g)\ \lambda(dg)}{1-\int_G e^{it(g)}\ \lambda(dg)} \quad \text{et} \quad t(a) = 0 \ .$$

La mesure de probabilité $\varepsilon_{a^{-1}} * \lambda * \varepsilon_a$ est centrée dans $\mathbb{R}^3$.

Alors, d'après le théorème (5.9), la suite de v.a.

$$S_n = \left(\frac{z_1 + e^{it_1} z_2 + \ldots + e^{i(t_1+\ldots+t_{n-1})} z_n}{\sqrt{n}} \ , \ \frac{t_1 + \ldots + t_n}{\sqrt{n}}\right)$$

converge en loi vers la loi gaussienne centrée de matrice de covarian covariance $\begin{bmatrix} \sigma_1^2 & 0 & 0 \\ 0 & \sigma_1^2 & 0 \\ 0 & 0 & \sigma_2^2 \end{bmatrix}$, où

$$\sigma_1^2 = \int_G |z(g)|^2\ \varepsilon_{a^{-1}} * \lambda * \varepsilon_a(dg)$$

$$= \int_G \left| z(g) - \frac{1 - e^{it(g)}}{1-\int_G e^{it(g)}\ \lambda(dg)} \int_G z(g)\ \lambda(dg)\right|^2 \lambda(dg)$$

et

$$\sigma_2^2 = \text{var } t_1 = \int_G t^2(g)\ \lambda(dg)$$

Cette loi limite est non dégénérée si et seulement si λ n'est portée ni par $\mathbb{C} \times \{0\}$, ni par $a(\{0\} \times \mathbb{R})a^{-1}$.

2ème cas : (cas général)

Désignons par τ l'élément de G défini par $z(\tau) = 0$ et $t(\tau) = \int_G t(g)\ \lambda(dg)$.

Supposons que la mesure $\lambda * \varepsilon_{\tau^{-1}}$ ne soit pas portée par le sous-groupe $\mathbb{C} \times \{0\}$ de G et notons a l'élément de G défini par

$$z(a) = \frac{\int_G z(g)\, \lambda * \varepsilon_{\tau^{-1}}(dg)}{1-\int_G e^{it(g)}\, \lambda * \varepsilon_{\tau^{-1}}(dg)} \quad , \quad t(a) = 0 .$$

La mesure de probabilité $\varepsilon_{a^{-1}} * \lambda * \varepsilon_{\tau^{-1}a}$ est centrée dans $\mathbb{R}^3$.

Alors, d'après le théorème (5.15), la suite de v.a.

$$S_n = \left(\frac{z_1 + e^{it_1} z_2 + \ldots + e^{i(t_1+\ldots+t_{n-1})} z_n}{\sqrt{n}} , \frac{t_1 + \ldots + t_n - n \int_G t(g)\lambda(dg)}{\sqrt{n}}\right)$$

converge en loi vers la loi gaussienne centrée de matrice de covariance $\begin{bmatrix} \sigma_1^2 & 0 & 0 \\ 0 & \sigma_1^2 & 0 \\ 0 & 0 & \sigma_2^2 \end{bmatrix}$, où

$$\sigma_1^2 = \int_G |z(g)|^2\, \varepsilon_{a^{-1}} * \lambda * \varepsilon_{\tau^{-1}a}(dg)$$

$$= \int_G \left| z(g) - \frac{1-e^{i[t(g)-\int_G t(g)\lambda(dg)]}}{1-\int_G e^{i[t(g)-\int_G t(g)\lambda(dg)]}\lambda(dg)} \int_G z(g)\lambda(dg)\right|^2 \lambda(dg)$$

et

$$\sigma_2^2 = \operatorname{var} t_1 = \int_G t^2(g)\, \lambda(dg) - \left(\int_G t(g)\, \lambda(dg)\right)^2 .$$

Cette loi limite est non dégénérée si et seulement si λ n'est portée ni par $\mathbb{C} \times \{\tau\}$, ni par $a(\{0\} \times \mathbb{R})\, a^{-1}\, \tau$.

BIBLIOGRAPHIE

[1] DERRIENNIC Y. : "Lois "zéro ou deux" pour les processus de Markov. Applications aux marches aléatoires". Ann. Inst. Henri Poincarré, vol XII, n°2, 1976, p.111-129.

[2] GUIVARC'H Y. : "Loi des grands nombres et rayon spectral pour une marche aléatoire sur un groupe de Lie". Bull. Soc. math. France, (à paraître).

[3] GUIVARC'H Y. : "Croissance polynomiale et périodes des fonctions harmoniques". Bull. Soc. Math. France, 101, 1973, 333-379.

[4] ICHIHARA K., KUNITA H. : "A classification of the second order degenerate elliptic operators and its probabilistic characterization". Zeit. für Wahr. 30, 1974, p.235.

[5] RAUGI A. : "Théorème de la limite centrale pour les groupes de Lie nilpotents simplement connexes". Zeit. für Wahr. (à paraître).

[6] RAUGI A. : "Fonctions harmoniques et théorèmes limites pour les marches aléatoires sur les groupes". Bull. Soc. Math. France, Mémoire 54, 1977, 127 p.

[7] SAZONOV V.V. and TUTUBALIN V.N. "Probability distributions on topological groups". Theory Prob. Applications, 11(1966), pp. 3-55.

[8] SKOROKHOD A.V. : "Studies in the theory of Random Process", Addison-Wesley publishing company, Inc. Massachussets, 1965.

Université de Rennes I
U.E.R. Mathématiques et Informatique
Avenue du Général Leclerc
35042 RENNES CEDEX
(FRANCE)

A Continuity Theorem for Weakly Stationary Stochastic Processes on an Arbitrary l.c.a. Group

Paul Ressel

Let G be an arbitrary l.c.a. group, then $\hat{G}$ denotes the associated dual group. Let further H be an arbitrary complex Hilbert space where we have in mind $H = L^2(P)$ for some probability space (Ω, A, P). A weakly stationary stochastic process indexed by $\hat{G}$ is a continuous mapping

$$X : \hat{G} \longrightarrow H$$

such that $\langle X(y_1), X(y_2)\rangle$ depends only on $y_1 - y_2$. It is well known that these stationary processes are exactly the Fourier transforms of a special class of H - valued measures, denoted and defined by $M \in$ caos (G,H) iff M is an H - valued Borel measure on G such that $\langle M(A), M(B)\rangle = 0$ whenever $A, B \in \mathcal{B}(G)$ [the Borel sets in G] are disjoint and the associated finite measure $m := ||M||^2$ is a Radon measure on G.

Our main purpose will be to establish a Lévy-type continuity theorem for weakly stationary processes. In the special case where $G = \mathbb{R}^n$ some of the results have already been proved in [3].

First of all we have to equip the space caos (G,H) with a suitable topology. For this purpose G needn't carry an algebraic structure. We only assume that G is a completely regular Hausdorff space. Let caos (G,H) be defined as above with the only alteration that the finite measure $||M||^2$ associated with M should only be τ-smooth (a finite nonnegative Borel measure μ on an topological space is called τ-smooth iff $\mu(U) = \sup_{\lambda \in \Lambda} \mu(U_\lambda)$ for every family of open sets $(U_\lambda)_{\lambda \in \Lambda}$ filtering upwards to U; if G is locally compact then every τ-smooth measure is a Radon measure, cf. [5],P 16). The weak topology on caos (G,H) is by definition the topology generated by the H-valued functions

$$M \longrightarrow \int f dM$$

where $f \in C_{\mathbb{R}}(G) := \{f : G \longrightarrow \mathbb{R},\ f \text{ is continuous and bounded}\}$.

Plainly we may replace $C_{\mathbb{R}}(G)$ here by $C_{\mathbb{C}}(G)$. It is not difficult to see that caos (G,H) in this topology is again a completely regular Hausdorff space. The following result is a kind of Portmanteau theorem for caos-measures.

<u>Theorem 1.</u> A net (M_α) in caos(G,H) converges weakly to $M \in$ caos(G,H) iff $M_\alpha(B) \longrightarrow M(B)$ for all $B \in \mathcal{B}(G)$ with $M(\partial B)=0$.

<u>Proof.</u> Suppose $M_\alpha \longrightarrow M$ weakly. Then for all $f \in C_{\mathbb{R}}(G)$

$$||\int f dM_\alpha||^2 = \int f^2 dm_\alpha \longrightarrow \int f^2 dm = ||\int f dM||^2$$

hence $m_\alpha \longrightarrow m$ weakly. Let $B \in \mathcal{B}(G)$, $M(\partial B)=0$, and $\varepsilon > 0$. From

$$1_{\overline{B}} = \inf \{f : f \text{ uniformly cont.}, 1_{\overline{B}} \leq f \leq 1\}$$

$$1_{\overset{o}{B}} = \sup \{g : g \text{ uniformly cont.}, 0 \leq g \leq 1_{\overset{o}{B}}\}$$

and τ-smoothness of m we get uniformly continuous functions f and g such that

$$0 \leq g \leq 1_{\overset{o}{B}} \leq 1_{\overline{B}} \leq f \leq 1$$

and

$$\int f dm - m(\overline{B}) < \frac{\varepsilon^2}{32}$$
$$m(\overset{o}{B}) - \int g dm < \frac{\varepsilon^2}{32}$$

Hence $\int (f-g)\, dm < \frac{\varepsilon^2}{16}$ implying

$$\int (f-g)\, dm_\alpha < \frac{\varepsilon^2}{16} \quad \forall \alpha \geq \alpha_o.$$

Now choose $\alpha_1 \geq \alpha_o$ such that

$$m_\alpha(\partial B) < \frac{\varepsilon^2}{16} \text{ and } ||\int g dM_\alpha - \int g dM|| < \frac{\varepsilon}{4}$$

for all $\alpha \geq \alpha_1$. Then, if $\alpha \geq \alpha_1$, we have

$$||M_\alpha(B)-M(B)|| \leq ||M_\alpha(B) - M_\alpha(\overline{B})|| + ||M_\alpha(\overline{B}) - \int g dM_\alpha|| +$$
$$+ ||\int g dM_\alpha - \int g dM|| + ||\int g dM - M(B)||$$

$\leq \sqrt{m_\alpha(\partial B)} + \sqrt{\int (f-g)\,dm_\alpha} + \frac{\varepsilon}{4} + \sqrt{\int (f-g)\,dm} < \varepsilon$

and we are done.

For the other direction let $f \in C_{\mathbb{R}}(G)$ and $\varepsilon > 0$ be given. There is α_0 such that

$$a := \sup_{\alpha \geq \alpha_0} ||M_\alpha(G)|| < \infty$$

and of course we may assume $a > 0$. We find $\{t_0, t_1, \ldots, t_k\} \subseteq \mathbb{R}$ with the following properties

$$m^f(\{t_j\}) = 0 \quad \forall j = 0,1,\ldots,k$$

$$\inf f(G) = t_0 < t_1 < \ldots < t_k = \sup f(G)$$

$$t_j - t_{j-1} < \varepsilon' \quad \forall j = 1,\ldots,k$$

where $\varepsilon' := \varepsilon/4a$. Put $B_j := f^{-1}(]t_{j-1}, t_j])$, $j = 1,\ldots,k$, then $M(\partial B_j) = 0$ and the function

$$g := \sum_{j=1}^{k} t_j \, 1_{B_j}$$

has a uniform distance of at most ε' from f. Then

$$|| \int f dM_\alpha - \int f dM || \leq || \int (f-g) dM_\alpha || + || \int g dM_\alpha - \int g dM || +$$

$$+ || \int (f-g) dM ||$$

$$\leq 2\,\varepsilon' a + \sum_{j=1}^{k} |t_j| \cdot ||M_\alpha(B_j) - M(B_j)|| < \varepsilon$$

for α large enough. $\rule{1em}{0.4pt}\!|$

Let again G be a locally compact abelian group. The Fourier transform of a measure $M \in$ caos (G,H) is given by

$$X_M : \hat{G} \longrightarrow H$$

$$y \longmapsto \int_G y(x)\, dM(x)$$

and as $M \longmapsto X_M$ is an injective transformation, the pointwise topology of $H^{\hat{G}}$ induces a Hausdorff topology - let's call it Fourier topology - on caos (G,H) which of course is weaker than the weak topology introduced above. However the following holds:

Theorem 2. Let $\emptyset \neq K \subseteq$ caos (G,H) be uniformly bounded, i.e. $\sup_{M\in K} ||M(G)|| < \infty$. Then K is compact in the weak topology iff it is compact in the Fourier topology and in this case the two topologies coincide on K.

Proof. We may assume that $||M(G)|| \leq 1$ for all $M\in K$. Let K be compact in the Fourier topology and suppose $\{M_\alpha\} \subseteq K$ is a net converging to $M_0 \in K$ in that topology. We have to show that

$$\int f dM_\alpha \longrightarrow \int f dM_0$$

for any $f\in C_{\mathbb{C}}(G)$ and in a first step we take f of the special form

$$f(x) = \int_{\hat{G}} y(x)\, d\mu(y)$$

μ being a complex Radon measure on $\hat{G}$.

From

$$< \int y(x) dM_\alpha(x), \int 1\, dM_\alpha > = \int y(x) dm_\alpha(x)$$

$$\longrightarrow \int y(x) dm_0(x) = < \int y(x) dM_0(x), \int 1\, dM_0 >$$

we get $\lim m_\alpha = m_0$ in the usual weak topology for finite non negative measures. Therefore

$$|| \int f dM_\alpha ||^2 = \int |f|^2 dm_\alpha \longrightarrow \int |f|^2 dm_0 = || \int f dM_0 ||^2$$

for every $f\in C_{\mathbb{C}}(G)$. We fix $h\in H$, $||h|| \leq 1$, and have to prove that

$$< \int f dM_\alpha, h > \longrightarrow < \int f dM_0, h >$$

and now we use the special form of f. We define $\mathbb{C}$-valued Radon measures by

$$\nu_\alpha(B) := < M_\alpha(B), h >$$

$$\nu_0(B) := < M_0(B), h >$$

and the joint continuity, hence Borel measurability of $(x,y)\mapsto y(x)$ gives us

$$< \int f dM_\alpha, h > = \int f d\nu_\alpha = \int_G \int_{\hat{G}} y(x) d\mu(y) d\nu_\alpha(x)$$

$$= \int_G \int_{\hat{G}} y(x) d\nu_\alpha(x) d\mu(y) .$$

We choose a compact subset $\hat{K} \subseteq \hat{G}$ such that $|\mu|(\hat{G} \setminus \hat{K}) < \varepsilon/3$, given $\varepsilon > 0$.

Consider the mapping

$$K \longrightarrow C_{\mathbb{C}}(\hat{K})$$

$$M \longmapsto (y \longmapsto \int_G y(x)d\nu(x))$$

($\nu := \langle M,h\rangle$); it is continuous from the Fourier topology on K to the pointwise topology on $C_{\mathbb{C}}(\hat{K})$. The well known property that a uniformly bounded subset of $C_{\mathbb{C}}(\hat{K})$ is compact in the pointwise topology iff it is compact in the weak topology of $C_{\mathbb{C}}(\hat{K})$ regarded as a Banach space, cf. ([2], Théorème 5), allows the conclusion

$$\int_{\hat{K}} \int_G y(x)\ d\nu_\alpha(x)\ d\mu(y) \longrightarrow \int_{\hat{K}} \int_G y(x)\ d\nu(x)\ d\mu(y)$$

and putting $g_\alpha(y) := \int y(x)\ d\nu_\alpha(x)$, g_o similarly, we get

$$\left|\int_{\hat{G}} g_\alpha d\mu - \int_{\hat{G}} gd\mu\right| \leq \left|\int_{\hat{G}\setminus\hat{K}} g_\alpha d\mu\right| + \left|\int_{\hat{K}} g_\alpha d\mu - \int_{\hat{K}} gd\mu\right| + \left|\int_{\hat{G}\setminus\hat{K}} gd\mu\right|$$

$$\leq \frac{\varepsilon}{3} + \left|\int_{\hat{K}} g_\alpha d\mu - \int_{\hat{K}} gd\mu\right| + \frac{\varepsilon}{3}$$

$$< \varepsilon$$

for α sufficiently large.

In the second step let $f \in C_{\mathbb{C}}(G)$ be vanishing at infinity, i.e. $f \in C_o(G)$. Then given $\varepsilon > o$ there exists g of the type considered above (even with μ absolutely continuous, see [4], Theorem 1.2.4.) such that $\| f - g \|_\infty < \varepsilon/3$. We get

$$\left\| \int fdM_o - \int fdM_\alpha \right\| \leq \left\| \int (f-g)dM_o \right\| + \left\| \int gdM_o - \int gdM_\alpha \right\| +$$

$$+ \left\| \int (f-g)dM_\alpha \right\|$$

$$< \varepsilon$$

for α large enough.

Finally let $0 \neq f \in C_{\mathbb{C}}(G)$ be arbitrary and $\varepsilon > 0$. The continuity of

$$\begin{aligned} \text{caos } (G,H) &\longrightarrow ca_+(G) \\ M &\longmapsto m \end{aligned}$$

$ca_+(G)$ denoting the non negative finite Radon measures on G, and both spaces being equipped with the Fourier topology, shows that

$$\{m : M \in K\}$$

is compact also in the weak topology of $ca_+(G)$ and this implies - any locally compact space being a Prohorov space - the existence of a compact set $K \subseteq G$ such that

$$\sup_{M \in K} m(G \setminus K) < \frac{\varepsilon}{3 \cdot \|f\|_\infty} .$$

We find a continuous function u with compact support fulfilling

$$1_K \leq u \leq 1 ;$$

then fu belongs to $C_0(G)$ and finally

$$\begin{aligned} \|\int f dM_0 - \int f dM_\alpha\| &\leq \|\int f(1-u) dM_0\| + \|\int fu dM_0 - \int fu dM_\alpha\| + \\ &+ \|\int f(1-u) dM_\alpha\| \\ &< \varepsilon \end{aligned}$$

for α large enough. This proves the theorem. ┘

Corollary 1. Let $X_0, X_1, X_2, \ldots$ be a sequence of weakly stationary stochastic processes indexed by the l.c.a. group $\hat{G}$. Let further X_n be the Fourier transform of M_n. Then $X_n(y) \longrightarrow X_0(y)$ for all $y \in \hat{G}$ iff $M_n \longrightarrow M_0$ weakly.

Corollary 2. Let $\{M_n : n \geq 1\} \subseteq$ caos(G,H) have the Fourier transforms $\{X_n : n \geq 1\}$. Suppose that $X_o(y) := \lim X_n(y)$ exists for all $y \in \hat{G}$ and that X_o is continuous at the origin. Then there is $M_o \in$ caos(G,H) whose Fourier transform is X_o and $M_n \longrightarrow M_o$ weakly.

Proof. It follows easily that $< X_o(y_1), X_o(y_2) >$ depends only on $y_1 - y_2$ and is therefore continuous on all of G because of

$$\| X_o(y_1) - X_o(y_2) \|^2 = 2 \| X_o(o) \|^2 - 2 \operatorname{Re} < X_o(y_1 - y_2), X_o(0) >. \quad \lrcorner$$

Let $X : \hat{G} \longrightarrow H$ be a weakly stationary process. Then there is a uniquely determined unitary representation U of $\hat{G}$ on $H_X := \overline{\text{lin}}\,(X(\hat{G})) \subseteq H$ such that

$$U_y(X(o)) = X(y) \quad \forall\ y \in \hat{G}\ .$$

Assume for a moment that $H_X = H$, then (cf. [1], Theorem 1) U is the Fourier transform of a Radon spectral measure M on G, which by definition is a mapping

$$M : \mathcal{B}(G) \longrightarrow \text{Orthogonal projections on } H$$

σ-additive in the strong topology, such that

$$M(A) \circ M(B) = M(A \cap B) \quad \forall\ A, B \in \mathcal{B}(G)$$

and

$$M(G) = id_H\ .$$

Furthermore we require that the non negative measures

$$A \longmapsto < M(A)(h), h > \quad , h \in H$$

are all τ-smooth measures. Here again we first let G be an arbitrary completely regular Hausdorff space and should perhaps call M a τ-smooth spectral measure. The set of all τ-smooth spectral measures on G is denoted spec (G,H). It is easy to see that for a spectral measure M the condition of τ-smoothness is equivalent with

$$M(D_\lambda) \uparrow M(D)$$

for every family of open subsets $D_\lambda \subseteq G$ filtering up to D. The weakest topology on spec (G,H) such that the mappings

$$\begin{aligned} \text{spec } (G,H) &\longrightarrow L(H) \\ M &\longmapsto \int f\, dM \end{aligned}$$

are all continuous for $f \in C_{\mathbb{R}}(G)$ (or equivalently for all $f \in C_{\mathbb{C}}(G)$), makes spec (G,H) a completely regular Hausdorff space. Here $L(H)$ denotes all linear continuous operators from H to H. The integral appearing above may be defined as

$$\left(\int f dM\right)(h) := \int f dM^h$$

where $M^h(B) := M(B)(h) \quad \forall\, h \in H \quad \forall\, B \in \mathcal{B}(G)$. The connection with the previously considered class of vector measures is given by the fact that $M^h \in$ caos(G,H) for all $h \in H$, if $M \in$ spec (G,H). Using the natural order structure for the set of orthogonal projections, the following theorem may be proved, which is a complete analogue of the classical Portmanteau theorem for probability measures:

<u>Theorem 3.</u> Let G be a completely regular Hausdorff space and let H be a Hilbert space. Then for a net (M_α) and M_0 in spec (G,H) the following conditions are equivalent:

(i) $\lim \int f dM_\alpha = \int f dM_0 \qquad \forall\, f \in C_{\mathbb{R}}(G)$

(ii) $\lim\sup \int f dM_\alpha \leq \int f dM_0 \qquad \forall\, f$ bounded, u.s.c.

(iii) $\lim\sup M_\alpha(F) \leq M_0(F) \qquad \forall\, F \subseteq G$, F closed

(iv) $\lim\inf M_\alpha(D) \geq M_0(D) \qquad \forall\, D \subseteq G$, D open

(v) $\lim M_\alpha(B) = M_0(B) \qquad \forall\, B \in \mathcal{B}(G),\ M_0(\partial B) = 0.$

If G is a l.c.a. group, then for $M \in$ spec (G,H) its ourier transform is defined by

$$U_M : \hat{G} \longrightarrow L(H)$$

$$y \longmapsto \int_G y(x)\, dM(x)$$

nd turns out to be a unitary representation of $\hat{G}$. Again we obain in this way - $M \longmapsto U_M$ being an injective transformation - socalled Fourier topology on spec (G,H).

heorem 4. A subset $K \subseteq$ spec (G,H) is compact in the weak topoogy iff it is compact in the Fourier topology and in this case he two topologies coincide.

roof. Let K be "Fourier compact" and suppose (M_α) is a net in converging to $M_o \in K$ in the Fourier topology. Then for any $h \in H$

$$U_{M_\alpha}(h) = \int_G y(x)\, dM_\alpha^h(x) \longrightarrow \int_G y(x)\, dM_o^h(x) = U_{M_o}(h)$$

nd $\{M^h : M \in K\} \subseteq$ caos (G,H) is uniformly bounded. Therefore by heorem 2 $M_\alpha^h \longrightarrow M_o^h$ weakly for all $h \in H$ and this means $_\alpha \longrightarrow M_o$ weakly. ⌋

orollary. Let $\{M_n : n \in \mathbb{N}\} \subseteq$ spec (G,H) have the Fourier transorms $\{U_n : n \in \mathbb{N}\}$. Suppose that $U_o(y) := \lim U_n(y)$ exists for all $\in \hat{G}$ and that U_o is continuous at the origin. Then there is $_o \in$ spec (G,H) whose Fourier transform is U_o and $M_n \longrightarrow M_o$ eakly.

roof. First of all $U_o(y_1+y_2) = U_o(y_1) \circ U_o(y_2)$ hence U_o is ontinuous on all of $\hat{G}$ and therefore the Fourier transform of ome $M_o \in$ spec (G,H). We have $M_n \longrightarrow M_o$ in the Fourier topology d by Theorem 4 we get $M_n \longrightarrow M_o$ weakly. ⌋

Let now $X_o, X_1, X_2, \ldots$ be a sequence of weakly stationary ochastic processes on $\hat{G}$. Let U_n denote the uniquely determined unitary representation of $\hat{G}$ on $H_n := H_{X_n}$ with $U_n(y)X_n(o) = X_n(y)$, $n \geq 0$, and let finally

$$X_n = X_{M_n} \quad , \quad M_n \in \text{caos}\,(G,H_n)$$

and

$$U_n = U_{\mathcal{M}_n} \quad , \quad \mathcal{M}_n \in \text{spec}\,(G,H_n)\ .$$

We shall assume that X_o is not too degenerate in comparison with $X_1, X_2, \ldots$, i.e. that $H_n \subseteq H_o \quad \forall\, n$. Clearly U_n can be regarded as operator on H_o by setting $U_n(y)(h) = 0$ for $h \in H_o \ominus H_n$. We then have

<u>Theorem 5</u>. From the following four statements

(1) $\quad X_n(y) \longrightarrow X_o(y) \qquad \forall\, y \in \hat{G}$

(2) $\quad M_n \longrightarrow M_o$ weakly

(3) $\quad \mathcal{M}_n \longrightarrow \mathcal{M}_o$ weakly

(4) $\quad U_n(y) \longrightarrow U_o(y) \qquad \forall\, y \in \hat{G}$ (pointwise)

(1) and (2) are equivalent, (3) and (4) are equivalent and (1)-(2) implies (3)-(4).

<u>Proof</u>. Only the last assertion remains to be proved. Suppose $X_n(y) \longrightarrow X_o(y) \quad \forall\, y \in \hat{G}$ and let $h = X_o(y_o)$ be given. Then

$$\begin{aligned}(U_n(y) - U_o(y))(h) &= U_n(y)[X_o(y_o) - X_n(y_o)] + U_n(y)(X_n(y_o)) \\ &\quad - U_o(y)(X_o(y_o)) \\ &= U_n(y)[X_o(y_o) - X_n(y_o)] + X_n(y+y_o) - X_o(y+y_o) \\ &\longrightarrow o\ .\end{aligned}$$

Hence $U_n(y)(h) \longrightarrow U_o(y)(h)$ for all $h \in H_o$. ┘

REFERENCES

[1] AMBROSE: Spectral Resolution of Groups of Unitary Operators. Duke Math.J. 11(1944), 589-595.

[2] GROTHENDIECK: Critères de compacité dans les espaces fonctionnels généraux. Amer.J.Math. 74(1952), 168-186.

[3] RESSEL: Weak Convergence of Certain Vectorvalued Measures. Ann.Prob. 2(1974), 136-142.

[4] RUDIN: Fourier Analysis on Groups. New York, Interscience 1962.

[5] TOPSØE: Topology and Measure. Lecture Notes in Mathematics Vol. 133, Berlin: Springer 1970.

Paul Ressel
Institut für Mathematische Stochastik
der Universität Freiburg
Hermann-Herder-Str. 10
7800 Freiburg

SEMI-GROUPES DE MESURES MATRICIELLES

par

Jean-Pierre ROTH

I. INTRODUCTION .

La notion de semi-groupe de mesures matricielles est naturellement associée à celle de semi-groupe d'opérateurs invariants sur $C_0(G,\mathbb{R}^n)$. $C_0(G,\mathbb{R}^n)$ désigne l'espace des fonctions continues, tendant vers 0 à l'infini, à valeurs dans $\mathbb{R}^n$ et définies sur le groupe localement compact G .

Le but de cet article est de montrer comment les techniques mises au point pour étudier les semi-groupes de mesures complexes (3) s'adaptent ici et permettent d'obtenir un théorème de représentation et un théorème de génération des semi-groupes de mesures matricielles . On généralise de plus les résultats au cas, important pour la pratique, où G n'est qu'un monoïde unitaire localement compact .

II. PRINCIPES DU MAXIMUM .

Dans (2) HIRSCH a énoncé différents principes du maximum dans le cadre d'espaces de Banach généraux . Nous allons en transcrire quelques-uns dans le cas particulier de $C_0(X,\mathbb{R}^n)$ où X est un espace localement compact, $\mathbb{R}^n$ est muni de la norme euclidienne canonique $|.|$, $<.,.>$ et

$$\forall f \in C_0(X,\mathbb{R}^n) \quad , \quad \|f\| = \sup_{x\in X} |f(x)|$$

<u>Proposition 1</u> . <u>A étant un opérateur non partout défini de domaine</u> D(A) <u>dense dans</u> $C_0(X,\mathbb{R}^n)$, <u>les propriétés suivantes sont équivalentes</u> :

1) A <u>est dissipatif</u>, <u>i.e.</u>
$\forall\lambda>0 \quad , \quad \forall f \in D(A) \quad , \quad \|\lambda f - Af\| \geq \|\lambda f\|$,

2) A <u>vérifie le principe faible du maximum de la norme</u>, <u>i.e.</u>
$\forall f \in D(A) \quad , \quad \exists x \in X \quad , \quad |f(x)| = \|f\|$ et $<f(x) \, , \, Af(x)> \leq 0$,

3) A <u>vérifie le principe du maximum de la norme</u>, <u>i.e.</u>
$\forall f \in D(A) \quad , \quad \forall x \in X \quad , \quad |f(x)| = \|f\| \Longrightarrow <f(x) \, , \, Af(x)> \leq 0$,

<u>Si, de plus</u>, Im A <u>est dense dans</u> $C_0(X,\mathbb{R}^n)$ <u>les trois propriétés précédentes sont équivalentes à</u>

4) A <u>vérifie le principe complet du maximum</u>, <u>i.e.</u>
$\forall f \in D(A) \quad , \quad [<s \, , \, f(x)> \leq 1$ sur $\{ (s,x) \in S\times X \, / \, <s \, , \, Af(x)> < 0 \}] \Longrightarrow \|f\| \leq 1$
<u>où</u> S <u>désigne la sphère unité de</u> $\mathbb{R}^n$.

La démonstration de ces équivalences figure dans (2) .

De manière analogue, toujours en suivant HIRSCH, on peut énoncer des co-principes du maximum . Nous ne le faisons pas ici .

III. SEMI-GROUPES D'OPERATEURS INVARIANTS ET SEMI-GROUPES DE MESURES MATRICIELLES .

Dans toute la suite de l'article G est un monoïde topologique localement compact avec un élément neutre e (i.e. $x,y \mapsto xy$ est associative, continue et possède un élément neutre e) .

On suppose sur G l'hypothèse supplémentaire suivante : ou bien G est compact, ou bien, si G n'est pas compact,

$\lim_{y\to\infty} yx = \lim_{y\to\infty} xy = \infty$ uniformément en x sur les compacts de G .

Les exemples de tels G sont nombreux, en particulier tout sous-monoïde fermé d'un groupe localement compact convient .

$[-1\ 1]$, $]0\ 1]$, $[0\ 1]$, $\{z \in \mathbb{C} \ / \ |z| \leq 1\}$, munis de la multiplication,

tout cône convexe fermé de $\mathbb{R}^n$ muni de l'addition,

la boule unité $\mathbb{B}$ fermée de l'espace des endomorphismes de $\mathbb{R}^n$ pour la norme opérateur munie du produit de composition, ...

entrent dans le cadre étudié présentement .

$\mathfrak{M}(G,\mathbb{R})$ désigne l'anneau des mesures réelles bornées sur G muni du produit de convolution suivant

$$\forall f \in C_0(G,\mathbb{R}) \quad , \quad \mu * \nu\ (f) = \int f(xy)\ d\mu(x)\ d\nu(y) \quad .$$

M_n est l'anneau des matrices carrées réelles d'ordre n . On le munit de la norme opérateur issue de la norme sur $\mathbb{R}^n$.

μ est dite une mesure matricielle (à valeurs dans M_n) sur G si μ est une matrice carrée d'ordre n à coefficients dans $\mathfrak{M}(G,\mathbb{R})$.

$\mu = (\mu_{ij})_{i,j = 1,\ldots,n}$ est l'écriture d'une telle mesure .

$\mathfrak{M}(G,M_n)$ désigne l'anneau des mesures matricielles (bornées) sur G muni du produit de convolution suivant

$$\mu = (\mu_{ij})_{i,j} \qquad \nu = (\nu_{ij})_{i,j} \qquad \lambda = (\lambda_{ij})_{i,j}$$

$$\mu * \nu = \lambda \quad \text{si} \quad \forall i,j = 1,\ldots,n \qquad \lambda_{ij} = \sum_k \mu_{ik} * \nu_{kj} \quad .$$

Si $f = (f_i)_{i = 1,\ldots,n}$ est dans $C_0(G,\mathbb{R}^n)$ et si $\mu = (\mu_{ij})_{i,j}$ est une mesure matricielle, on note

$\mu(f) = \int d\mu(x)\ f(x)$ l'élément $a = (a_i)_{i = 1,\ldots,n}$ de $\mathbb{R}^n$ défini par

$$a_i = \sum_j \mu_{ij}(f_j) \quad .$$

On a alors l'expression suivante pour la convolée de deux mesures matricielles μ et ν

$$\forall f \in C_0(G,\mathbb{R}^n) \quad , \quad \mu * \nu\,(f) = \int d\mu(x) \int d\nu(y)\; f(xy) \quad .$$

La norme d'une mesure matricielle μ est définie par

$$\|\mu\| = \operatorname{Sup} \{\ |\mu(f)| \ / \ f \in C_0(G,\mathbb{R}^n) \ , \ \|f\| \leq 1 \ \} \quad .$$

Cette norme vérifie l'inégalité

$$\|\mu * \nu\| \leq \|\mu\|\ \|\nu\| \quad .$$

Si x est dans G et f dans $C_0(G,\mathbb{R}^n)$, la fonction $\tau_x f = f(x.)$ est encore dans $C_0(G,\mathbb{R}^n)$.

Un endomorphisme T de $C_0(G,\mathbb{R}^n)$ est dit invariant (à gauche) si

$$\forall x \in G \quad , \quad \tau_x \circ T = T \circ \tau_x \quad .$$

$\mathcal{L}_i(C_0(G,\mathbb{R}^n))$ désigne l'algèbre des opérateurs bornés invariants sur $C_0(G,\mathbb{R}^n)$.

<u>Proposition 2</u> . <u>L'application qui à tout T de $\mathcal{L}_i(C_0(G,\mathbb{R}^n))$ associe la mesure matricielle μ par la relation</u>

$$\forall f \in C_0(G,\mathbb{R}^n) \quad , \quad \mu(f) = Tf(e) \quad ,$$

<u>est une bijection de $\mathcal{L}_i(C_0(G,\mathbb{R}^n))$ sur $\mathfrak{M}(G,M_n)$ qui vérifie les propriétés</u> suivantes (on note T_μ l'opérateur invariant associé à μ),

1) $T_\mu \circ T_\nu = T_{\mu * \nu}$,

2) $T_{a\mu + b\nu} = aT_\mu + bT_\nu$,

3) $\|T_\mu\| = \|\mu\|$.

<u>Démonstration</u> . T_μ est défini à partir de μ par

$$\forall f \in C_0(G,\mathbb{R}^n) \quad , \quad T_\mu(f)(x) = \int d\mu(y)\; f(xy) \quad .$$

L'hypothèse faite sur G assure que $T_\mu f$ est bien dans $C_0(G,\mathbb{R}^n)$.

Le reste de la démonstration est une simple vérification formelle .

<u>Définition</u> . <u>Un semi-groupe à contraction de mesures matricielles est une famille $(\mu_t)_{t\geq 0} \subset \mathfrak{M}(G,M_n)$ telle que</u>

1) $\forall t \geq 0 \quad , \quad \|\mu_t\| \leq 1$,

2) $\forall s,t \geq 0 \quad , \quad \mu_t * \mu_s = \mu_{t+s}$,

3) $\mu_0 = I\delta_e$ ($I\delta_e$ <u>est la mesure matricielle diagonale dont tous les éléments de la diagonale valent</u> δ_e , <u>mesure de Dirac sur G au point</u> e) ,

4) μ_t <u>converge faiblement vers</u> μ_0 <u>lorsque</u> t <u>tend vers</u> 0 ,

(*i.e.* , $\forall f \in C_0(G,\mathbb{R}^n)$, $\mu_t(f) \to f(e)$ *lorsque* $t \to 0$) .

Proposition 3 . *Soit* $(P_t)_{t\geq 0}$ *un semi-groupe fortement continu à contraction d'opérateurs invariants sur* $C_0(G,\mathbb{R}^n)$. *Les mesures* μ_t *associées forment un semi-groupe à contraction de mesures matricielles* . *La réciproque est vraie* .

Démonstration . Le seul point délicat est celui de la forte continuité de $(P_t)_{t\geq 0}$ dans la réciproque . On le résoud de la manière classique suivante .

Par hypothèse on sait que

$\forall f \in C_0(G,\mathbb{R}^n)$, $\mu_t(f) \to f(e)$ lorsque $t \to 0$.

Donc $P_t(f)$ tend vers f simplement sur G lorsque t tend vers 0 .

Comme pour tout $t \geq 0$ on a $\|P_t(f)\| \leq \|f\|$, on en déduit, d'après le théorème de Lebesgue, que $P_t(f)$ tend vers f faiblement lorsque t tend vers 0 .

Un résultat général affirme que tout semi-groupe d'opérateurs faiblement continu est fortement continu, d'où la conclusion .

IV. REPRESENTATION DES SEMI-GROUPES DE MESURES MATRICIELLES .

Notons $\mathbb{B}$ la boule unité fermée de M_n munie de la multiplication des matrices . $\mathbb{B}$ est un monoïde compact admettant pour élément neutre la matrice identité I de M_n .

Soit $\mathbb{B}$xG le monoïde localement compact d'élément neutre (I,e) dont l'opération est définie par

$(m,x)\ (n,y) = (mn,xy)$.

Si μ et ν sont des éléments de $\mathfrak{M}(\mathbb{B}\times G,\mathbb{R})$ on définit $\mu * \nu$ par

$$\forall g \in C_0(\mathbb{B}\times G,\mathbb{R}) \quad , \quad \mu * \nu\ (g) = \int g(mn,xy)\ d\mu(m,x)\ d\nu(n,y) .$$

Si f est un élément de $C_0(G,\mathbb{R}^n)$ on note Mf l'élément de $C_0(\mathbb{B}\times G,\mathbb{R})$ défini par $Mf(m,x) = mf(x)$ pour tout m dans $\mathbb{B}$ et tout x dans G .

Théorème 1 . *Soit* $(\mu_t)_{t\geq 0}$ *un semi-groupe à contraction de mesures matricielles sur le monoïde* G . *Alors il existe* $(\nu_t)_{t\geq 0}$ *un semi-groupe à contraction de mesures positives sur* $\mathbb{B}$xG *tel que*

$$\forall t \geq 0 \quad , \quad \forall f \in C_0(G,\mathbb{R}^n) \quad , \quad \mu_t(f) = \nu_t(Mf) .$$

La démonstration de ce théorème repose sur les quatre lemmes suivants .

Lemme 1 . Soit μ une mesure matricielle sur G . Il existe une mesure réelle bornée ν sur $\mathbb{B}\times G$ telle que

1) ν est positive ,

2) $\|\nu\| = \|\mu\|$,

3) $\forall f \in C_0(G,\mathbb{R}^n)$, $\int_G d\mu(x)\ f(x) = \int_{\mathbb{B}\times G} mf(x)\ d\nu(m,x)$.

Démonstration . Soit $\mu = (\mu_{ij})_{i,j}$.

Pour φ positive dans $C_0(G,\mathbb{R})$ on pose

$$|\mu|(\varphi) = \operatorname{Sup} \{ |\mu(f)| \ / \ f \in C_0(G,\mathbb{R}^n) \ , \ |f| \leq \varphi \} .$$

$|\mu|$ est additive et positivement homogène sur $C_0^+(G,\mathbb{R})$. Elle se prolonge en une mesure positive de masse $\|\mu\|$ sur G encore notée $|\mu|$.

Toutes les mesures μ_{ij} sont absolument continues par rapport à $|\mu|$ donc $\mu_{ij} = a_{ij}\ |\mu|$, où a_{ij} est une fonction borélienne $|\mu|$-intégrable sur G .

On note $A = (a_{ij})_{i,j}$.

On a alors

$$\forall f \in C_0(G,\mathbb{R}^n) \ , \quad \int d\mu(x)\ f(x) = \int A(x).f(x)\ d|\mu|(x) \ .$$

Montrons que $A(x)$ est dans $\mathbb{B}$ pour $|\mu|$-presque tout x .

Soit φ dans $C_0^+(G,\mathbb{R})$, $\eta = (\eta_i)_i$ et $\xi = (\xi_i)_i$ dans $\mathbb{R}^n$.

On a

$$\mu(\varphi\eta) = \int A(x).\eta\ \varphi(x)\ d|\mu|(x) \ ,$$

$$\int (\tilde{\xi}.A(x).\eta)\ \varphi(x)\ d|\mu|(x) = \tilde{\xi}.\mu(\varphi\eta) \leq |\xi|\,|\eta|\,|\mu|(\varphi) \ .$$

Donc

$$\forall \eta,\xi \in \mathbb{R}^n \ , \ \text{pour } |\mu|\text{-presque tout } x \ , \ \tilde{\xi}.A(x).\eta \leq |\xi|\,|\eta| \ .$$

En prenant des suites $(\eta^p)_{p\in\mathbb{N}}$ et $(\xi^p)_{p\in\mathbb{N}}$ denses dans $\mathbb{R}^n$ on obtient pour $|\mu|$-presque tout x , $A(x) \in \mathbb{B}$.

On modifie A sur un ensemble $|\mu|$-négligeable pour que $A(x)$ soit dans $\mathbb{B}$ pour tout x de G , et on pose , pour g dans $C_0(\mathbb{B}\times G,\mathbb{R})$

$$\nu(g) = \int g(A(x),x)\ d|\mu|(x) \ .$$

ν est une mesure sur $\mathbb{B}\times G$, positive, de masse $\|\nu\| = \||\mu|\| = \|\mu\|$ et de plus, si f est dans $C_0(G,\mathbb{R}^n)$ on a

$$\nu(Mf) = \int A(x)f(x)\ d|\mu|(x) = \mu(f) \ .$$

Lemme 2 . Soient μ_1 , $\mu_2 \in \mathfrak{M}(G,M_n)$ et ν_1 , $\nu_2 \in \mathfrak{M}(\mathbb{B}\times G,\mathbb{R})$ telles que

$$\forall f \in C_0(G,\mathbb{R}^n) \quad , \quad \mu_i(f) = \nu_i(Mf) \text{ pour } i = 1 , 2 .$$

Alors

$$\forall f \in C_0(G,\mathbb{R}^n) \quad , \quad \mu_1 * \mu_2 (f) = \nu_1 * \nu_2 (Mf) .$$

Démonstration . Il s'agit d'une simple vérification sans difficultés .

Lemme 3 . Soit $(\zeta_s)_{s>0}$ une famille résolvante à contraction de mesures matricielles sur G , i.e.

$$\forall s,t > 0 \quad , \quad \zeta_s - \zeta_t = (t - s)\, \zeta_s * \zeta_t \quad \text{et} \quad s\|\zeta_s\| \leq 1 .$$

Il existe $(\beta_s)_{s>0}$, famille résolvante à contraction de mesures positives sur $\mathbb{B}\times G$, telle que

$$\forall s > 0 \quad , \quad \forall f \in C_0(G,\mathbb{R}^n) \quad , \quad \zeta_s(f) = \beta_s(Mf) .$$

Démonstration . Ce lemme est le coeur de la théorie . Sa démonstration est assez longue . Elle est analogue à celle du théorème II.2.3. de (3) compte tenu des lemmes 1et 2 précédents . Nous ne la réécrivons pas ici .

Lemme 4 . Avec les notations du lemme précédent,
si $s\zeta_s$ tend faiblement vers $I\delta_e$ quand s tend vers l'infini,
alors $s\beta_s$ tend faiblement vers $\delta_{(I,e)}$ quand s tend vers l'infini .

Démonstration . Soit γ une valeur d'adhérence de $s\beta_s$ quand s tend vers l'infini . On a

$$\|\gamma\| \leq 1 \quad \text{et} \quad \forall f \in C_0(G,\mathbb{R}^n) \quad , \quad \gamma(Mf) = f(e) .$$

Soit φ dans $C_0(G,\mathbb{R})$ telle que , pour tout x, $0 \leq \varphi(x) \leq \varphi(e) = 1$ et a un élément de $\mathbb{R}^n$ de norme 1 .

$$\gamma(M.a\,\varphi) = a\,\varphi(e) = a \quad ,$$

$$\gamma(\tilde{a}.M.a\,\varphi) = \tilde{a}.a = 1 \quad ,$$

$$0 = \gamma(\tilde{a}.M.a\,\varphi) - 1 \leq \gamma(\tilde{a}.M.a\,\varphi - 1) \leq 0 .$$

Donc $\{ (m,x) / \tilde{a}.m.a\,\varphi(x) = 1 \}$ est un fermé contenant Supp γ , d'où

$$\forall a \in \mathbb{R}^n \quad , \quad \|a\| = 1 \quad , \quad \forall \varphi \in C_0(G,\mathbb{R}) \quad , \quad 0 \leq \varphi \leq \varphi(e) = 1 \quad ,$$

$$\text{Supp}\,\gamma \subset \{ m / \tilde{a}.m.a = 1 \} \times \{ x / \varphi(x) = 1 \} .$$

Donc Supp $\gamma = \{(I,e)\}$ et comme, de plus $\|\gamma\| = 1$, on a bien $\gamma = \delta_{(I,e)}$.

Nous pouvons maintenant démontrer le théorème 1 .

Au semi-groupe $(\mu_t)_{t \geq 0}$ on associe la famille résolvante $(\zeta_s)_{s>0}$ définie par

$$\forall f \in C_0(G,\mathbb{R}^n) \quad , \quad \zeta_s(f) = \int_0^\infty e^{-st}\, \mu_t(f)\, dt \quad .$$

On a les propriétés suivantes pour $(\zeta_s)_{s>0}$,

$\forall s > 0$, $s\|\zeta_s\| \leq 1$ et $s\zeta_s$ tend vers $I\delta_e$ faiblement quand s tend vers l'infini .

D'après le lemme 3 il existe une famille résolvante à contraction positive $(\beta_s)_{s>0}$ sur $\mathbb{B}$xG telle que

$$\forall f \in C_0(G,\mathbb{R}^n) \quad , \quad \zeta_s(f) = \beta_s(Mf) \ .$$

D'après le lemme 4 on a $s\beta_s \to \delta_{(I,e)}$ faiblement quand $s \to \infty$

Un théorème général, proche du théorème de Hille-Yosida (4), permet d'associer à une telle famille résolvante $(\beta_s)_{s>0}$ un semi-groupe à contraction $(\nu_t)_{t\geq 0}$ de mesures positives sur $\mathbb{B}$xG, c'est à dire

$$\forall g \in C_0(\mathbb{B}\mathrm{x}G,\mathbb{R}) \quad , \quad \beta_s(g) = \int_0^\infty e^{-st}\, \nu_t(g)\, dt \quad .$$

Par suite,

$$\forall f \in C_0(G,\mathbb{R}^n) \quad , \quad \forall s > 0 \quad , \int_0^\infty e^{-st}\, \mu_t(f)\, dt = \int_0^\infty e^{-st}\, \nu_t(Mf)\, dt \quad .$$

L'injectivité de la transformation de Laplace implique

$$\forall t \geq 0 \quad , \quad \forall f \in C_0(G,\mathbb{R}^n) \quad , \quad \mu_t(f) = \nu_t(Mf) \quad .$$

Remarque . Le théorème 1 constitue la première étape vers un théorème de représentation du type " Lévy-Kintchine" pour les semi-groupes de mesures matricielles . La seconde étape serait de préciser la forme des semi-groupes de mesures positives sur $\mathbb{B}$xG . C'est le genre de démonstration mise en oeuvre par FARAUT dans (1) pour étudier les semi-groupes de mesures complexes sur $\mathbb{R}^p$.

V. GENERATION DES SEMI-GROUPES DE MESURES MATRICIELLES .

En généralisant, comme dans le paragraphe précédent, les méthodes utilisées dans (3) pour l'étude du cas complexe on aboutit au résultat important suivant .

Théorème 2 . Soit A un opérateur non partout défini de domaine D(A) dense dans $C_0(G,\mathbb{R}^n)$, dissipatif (cf. Proposition 1), et invariant à gauche (i.e.

$$\forall x \in G \quad , \quad \tau_x(D(A)) \subset D(A) \quad \text{et} \quad A \circ \tau_x \supset \tau_x \circ A \;)$$

Alors il existe un semi-groupe à contraction $(\mu_t)_{t \geq 0}$ de mesures matricielles sur G tel que

$$\forall f \in D(A) \quad , \quad \lim_{t \to 0} \frac{1}{t}[\, \mu_t(f) - f(e)\,] = Af(e) \quad .$$

Remarque . Ce théorème recouvre en particulier les résultats de FARAUT (1) concernant les semi-groupes de mesures portées par un cône convexe fermé de $\mathbb{R}^p$.

Remarque . Tout le travail qui vient d'être fait aurait pu être exposé en munissant $\mathbb{R}^n$ d'une autre norme que la norme euclidienne .

Bibliographie .

(1) J. FARAUT , Semi-groupes de mesures complexes et calcul symbolique sur les générateurs infinitésimaux de semi-groupes d'opérateurs, Ann. Inst. Fourier, Grenoble, t. 20, Fasc. 1 (1970), 235 - 301 .

(2) F. HIRSCH , Familles résolvantes, générateurs, cogénérateurs, potentiels, Ann. Inst. Fourier, Grenoble, t. 22, Fasc. 1 (1972), 89 - 210 .

(3) J.P. ROTH , Opérateurs dissipatifs et semi-groupes dans les espaces de fonctions continues, Ann. Inst. Fourier, Grenoble, t. 26, Fasc. 4 (1976), 1 - 97 .

(4) K. YOSIDA , Functional analysis , Springer-Verlag , Berlin .

Jean-Pierre ROTH

I.S.E.A.

4, Rue des Frères Lumière

68093 MULHOUSE CEDEX

FRANCE

GROUPS WITH REPRESENTATIONS OF BOUNDED DEGREE

G. Schlichting

A theorem of Halmos [3] states that ergodic automorphisms τ of compact metrisable abelian groups G have countable Lebesgue spectrum. Kaplansky [6] tried to extend this theorem to compact metrisable groups which are not necessarily abelian. Just as for abelian groups, one has to look for irreducible representations (which are by definition here continuous and unitary). To be more precise one considers the set $\hat{G}$ of all equivalence classes of irreducible representations The point is to show that under the induced action $\pi \to \pi \circ \tau$ of τ on $\hat{G}$ there are infinitely many orbits in $\hat{G}$. For abelian groups, $\hat{G}$ again is a group, and the desired result follows using grouptheoretical considerations [3]. In general (G compact, metrisable but not necessarily abelian), the result is trivial if G has irreducible representations π of arbitrarily high degree n_π (dimension of the representation space), since n_π is invariant under the induced action of τ on $\hat{G}$.

Thus, one is led to study locally compact groups with representations of bounded degree, i.e. groups G satisfying

$$(*) \qquad \sup\{n_\pi ; \pi \in \hat{G}\} < \infty .$$

Among other things Kaplansky [6] proved in 1949 that compact, connected groups satisfying (*) are abelian. On the other hand, Isaacs and Passman [5] proved in 1964 that for discrete groups (*) is equivalent to the existence of an abelian normal subgroup of finite index. This result is also contained in a famous theorem independently proved by Thoma [10] around the same time. In 1967 Großer and Moskowitz [2] showed that the connected component of the identity element in locally compact groups satisfying (*) is necessarily abelian. Finally in 1972 C.C. Moore [7] obtained the general solution: for locally compact groups, (*) is equivalent to the existence of a (closed) abelian normal subgroup of finite index. Of course, from the beginning the main problem was to show the existence part of the theorem.

Our primary concern here is to give an elementary proof of this result. The simple idea of C.C. Moore was to reduce the case of a general locally compact group satisfying (*) to the case of a discrete group by using polynomial identities. But one can even simpler reduce to the case of a compact group and then prove the following

THEOREM. A compact group G satisfying (*) contains an abelian normal subgroup of finite index.

Indeed, by the Raikov theorem, irreducible representations separate points. Therefore locally compact groups satisfying (*) are MAP (maximally almost periodic, cf. [4]). By a general construction, MAP groups can be densely imbedded in a (universal) compact group such that by extension and restriction finite dimensional irreducible representations correspond to each other in a one to one manner. Hence it is sufficient to prove the theorem.

Assume that G is a compact group, μ is the normalized Haar measure on G and $\mu_x (x \in G)$ is the unique probability measure on $K_x = \{y^{-1}xy; y \in G\}$ invariant under the action of inner automorphisms (we remark that K_x can be naturally identified with the coset space of G modulo the centralizer $C(x)$ of $x \in G$ and μ_x with the image of μ on $G/_{C(x)}$. The functional equations for characters $x \to \chi_\pi(x) = \text{trace}\, \pi(x)$ $(\pi \in \hat{G}, x \in G)$ imply

$$(1) \qquad (f * \mu_x, \chi_\pi) = (f, \chi_\pi)\overline{\chi_\pi(x)}n_\pi^{-1} \qquad (x \in G, \pi \in \hat{G}, f \in L^2(G))$$

Furthermore, for the subspace of all central functions $f \in L^2(G)$ (that is $f(x^{-1}yx) = f(y)$ a.e. for all $x \in G$) the characters $\chi_\pi, \pi \in \hat{G}$ give a complete orthonormal system. Hence

$$(2) \qquad (f * \mu_x, f * \mu_x) = \sum_{\pi \in \hat{G}} |(f, \chi_\pi)|^2 |\chi_\pi(x)|^2 n_\pi^{-1}$$

for a central function $f \in L^2(G)$ and $x \in G$. Obviously the right hand side of (2) is an absolutely and uniformly convergent series of continuous (positive definit, central) functions on G. Finally, for central $f \in L^2(G)$ one easily verifies

$$(3) \qquad (f * \mu_x, f * \mu_x) = \int_G \mu_x(zf * f^*)\, d\mu_x(z) = \mu_x(xf * f^*),$$

where $f \to zf$ denotes translation $zf(y) = f(z^{-1}y)$ $(z, y \in G)$ and the involution $*$ is defined by $f^*(x) = \overline{f(x^{-1})}$ $(x \in G)$. Dividing by $(f,f) = f * f^*(e) = \|f * f^*\|_\infty$ and remembering that μ_x gives equal probability to any element in K_x, one obtains $\mu_x(\{x\}) = |K_x|^{-1}$ as supp f approaches the unit $e \in G$. Choosing a directed system of continuous central functions f of this kind and passing to the limit in (2), one gets

$$(4) \qquad \int_G |K_x|^{-1}\, d\mu(x) \geq \inf\{n_\pi^{-2}; \pi \in \hat{G}\}.$$

Notice that $x \to |K_x|$ is lower semicontinuous. Therefore, asuming G satisfies (*) one obtains the conclusion that the σ-compact (normal) subgroup $G_f = \bigcup_{n\in\mathbb{N}} \{x \in G; |K_x| \leq n\}$ has positive Haar measure, hence is an open subgroup of finite index. Furthermore, $x \to |K_x|$ has to be bounded on an open subset of G_f (by Baire's theorem) and therefore must be bounded on G_f, since $|K_{xy}| \leq |K_x||K_y|$ for $x,y \in G$. At this point, we may use a theorem of B.H. Neumann [8] which assures that groups with finite conjugacy classes of bounded lengths have finite commutator subgroups. This in turn implies for compact groups, that the center $Z(G_f)$ of G_f has finite index in G_f and therefore is an abelian normal subgroup of finite index in G. But we can equally well give a direct proof. Consider the compact group $H = G_f/Z(G_f)$ of inner automorphisms of G_f and form the semidirect product $G_f \circledS H$. In the obvious way, there is an equivariant, one to one mapping of $G_f \circledS H/_H$ onto G_f such that the action of H on $G_f \circledS H/_H$ is transformed into the natural action of H on G_f by (inner) automorphisms. Now apply the following Lemma to $G_f \circledS H$ and H.

LEMMA. Let H be a closed subgroup of a compact group G and assume $o_H(x) := |HxH/_H| < \infty$ for all $x \in G$. Then $[H : H_G] < \infty$, where $H_G = \bigcap_{x\in G} xHx^{-1}$ denotes the kernel of the permutation representation of G on $G/_H$.

Proof. We may assume that H_G is trivial. Denote by $H_x = H \cap xHx^{-1}$ the closed subgroup of G fixing $xH \in G/_H$. Since $o_H(x) = [H : H_x] < \infty$ for all $x \in G$, one verifies that $\{H_x; x \in G\}$ defines the basis of a neighbourhood (nbh.) system for a (unique) group topology on G. Define G^* to be the abstract group G equipped with this topology. Evidently G and G^* induce the same topology on H, and in this topology H is a compact open subgroup of G^*. Hence, G^* is locally compact. Fix a Haar measure μ^* on G^* such that $\mu^*(H) = 1$, and denote by δ_{G^*} the modular function on G^*. Then $T_x f(y) = f(x^{-1}yx)\delta_{G^*}^{\frac{1}{2}}(x)$ defines a unitary representation $x \to T_x$ of G^* on $L^2(G^*)$. Therefore $x \to (T_x 1_H, 1_H) = o_H^{-1}(x)\delta_{G^*}^{\frac{1}{2}}(x)$ is a real valued, positive definit, and consequently symmetric function on G^*. Therefore $\delta_{G^*}(x) = o_H(x)o_H(x^{-1})^{-1}$ and $(T_x 1_H, 1_H) = (o_H(x)o_H(x^{-1}))^{\frac{1}{2}}$. Now o_H is lower semicontinuous on $G = \bigcup_{n\in\mathbb{N}} \{x \in G; o_H(x) \leq n\}$, so o_H is bounded on an open subset of G by Baire's theorem and therefore bounded on G, because $o_H(xy) \leq o_H(x)o_H(y)$ for $x,y \in G$. Thus, the closed convex hull of $T_x 1_H, x \in G^*$ in $L^2(G^*)$ contains a (unique) function $f \neq o$ such that $T_x f = f$ for $x \in G^*$. Hence $x \to (xf,f)$ is a central function, $\neq o$, and vanishing at infinity on G^*. This implies the existence of a compact, invariant nbh. U of the identity element in G^*. Denote by $\{V_i\}$ a basis of compact invariant nbh's of the identity element in G. Since the inclusion $G \to G^*$ is continuous, we may conclude that $\{U \cap V_i\}$ is a basis of compact, invariant nbh's of the identity element in G^*. Therefore $\{e\} = \bigcap_{x\in G} xHx^{-1}$ is open in G^*; hence H is a finite subgroup.

REMARKS. We can prove a much more general theorem than the one above, namely a generalisation to locally compact groups G acting on a homogeneous space $X = G/_H$, where H is a compact subgroup of G (cf. [9]): $G/_{H_G}$ contains an abelian normal subgroup of finite index iff all irreducible representations of G weakly contained in $L^2(X)$ are uniformly bounded with respect to their degrees. As an application, we obtain a deep result of R. Baer [1] on the structure of groups of automorphisms acting on abelian groups with finite orbits of uniformly bounded (finite) lengths.

LITERATURE

[1] R. Baer: Automorphismengruppen von Gruppen mit endlichen Bahnen gleichmäßig beschränkter Mächtigkeit. J. reine angew. Math. 262/263, 93-119 (1973)

[2] S. Großer - M. Moskowitz: Representation theory of central topological groups. Trans. AMS 129 (1967) 361-390

[3] P. Halmos: On automorphisms of Compact groups. Bull AMS vol. 49 (1943) 619-624

[4] H. Heyer: Dualität lokalkompakter Gruppen, Springer 1970, Lect. Notes in Math. 150

[5] Isaacs-Passman: Groups with representations of bounded degree. Canad. J. Math. 16 (1964), 299-309

[6] I. Kaplansky: Groups with representations of bounded degree. Canad. J. Math. 1, 105-112 (1949)

[7] C.C. Moore: Groups with finite dimensional irreducible representations. Trans. AMS 166, 401-410 (1972)

[8] B.H. Neumann: Groups with finite classes of conjugate elements. Proc. London Math. Soc. 1, 178-187 (1951)

[9] G. Schlichting: Über Gruppen von beschränktem Darstellungsgrad. Habilitationsschrift TUM-Math-7709 (1977)

10] E. Thoma: Über unitäre Darstellungen abzählbarer diskreter Gruppen. Math. Annalen 153 (1964) 111-138

G. Schlichting
Math. Inst. der TU München
D-8000 München 2, Arcisstr. 21

ol. 551: Algebraic K-Theory, Evanston 1976. Proceedings. Edited M. R. Stein. XI, 409 pages. 1976.

ol. 552: C. G. Gibson, K. Wirthmüller, A. A. du Plessis and J. N. Looijenga. Topological Stability of Smooth Mappings. V, 5 pages. 1976.

ol. 553: M. Petrich, Categories of Algebraic Systems. Vector and ojective Spaces, Semigroups, Rings and Lattices. VIII, 217 pages. 76.

ol. 554: J. D. H. Smith, Mal'cev Varieties. VIII, 158 pages. 1976.

ol. 555: M. Ishida, The Genus Fields of Algebraic Number Fields. I, 116 pages. 1976.

ol. 556: Approximation Theory. Bonn 1976. Proceedings. Edited by Schaback and K. Scherer. VII, 466 pages. 1976.

ol. 557: W. Iberkleid and T. Petrie, Smooth S^1 Manifolds. III, 3 pages. 1976.

ol. 558: B. Weisfeiler, On Construction and Identification of Graphs. V, 237 pages. 1976.

ol. 559: J.-P. Caubet, Le Mouvement Brownien Relativiste. IX, 2 pages. 1976.

ol. 560: Combinatorial Mathematics, IV, Proceedings 1975. Edited L. R. A. Casse and W. D. Wallis. VII, 249 pages. 1976.

ol. 561: Function Theoretic Methods for Partial Differential Equations. armstadt 1976. Proceedings. Edited by V. E. Meister, N. Weck d W. L. Wendland. XVIII, 520 pages. 1976.

ol. 562: R. W. Goodman, Nilpotent Lie Groups: Structure and oplications to Analysis. X, 210 pages. 1976.

l. 563: Séminaire de Théorie du Potentiel. Paris, No. 2. Proceedings 75–1976. Edited by F. Hirsch and G. Mokobodzki. VI, 292 pages. 76.

ol. 564: Ordinary and Partial Differential Equations, Dundee 1976. oceedings. Edited by W. N. Everitt and B. D. Sleeman. XVIII, 551 ges. 1976.

l. 565: Turbulence and Navier Stokes Equations. Proceedings 75. Edited by R. Temam. IX, 194 pages. 1976.

ol. 566: Empirical Distributions and Processes. Oberwolfach 1976. oceedings. Edited by P. Gaenssler and P. Révész. VII, 146 pages. 76.

l. 567: Séminaire Bourbaki vol. 1975/76. Exposés 471–488. IV, 3 pages. 1977.

l. 568: R. E. Gaines and J. L. Mawhin, Coincidence Degree, and onlinear Differential Equations. V, 262 pages. 1977.

l. 569: Cohomologie Etale SGA 4½. Séminaire de Géométrie gébrique du Bois-Marie. Edité par P. Deligne. V, 312 pages. 1977.

ol. 570: Differential Geometrical Methods in Mathematical Physics, nn 1975. Proceedings. Edited by K. Bleuler and A. Reetz. VIII, 6 pages. 1977.

l. 571: Constructive Theory of Functions of Several Variables, berwolfach 1976. Proceedings. Edited by W. Schempp and K. Zel. VI, 290 pages. 1977

ol. 572: Sparse Matrix Techniques, Copenhagen 1976. Edited by A. Barker. V, 184 pages. 1977.

l. 573: Group Theory, Canberra 1975. Proceedings. Edited by A. Bryce, J. Cossey and M. F. Newman. VII, 146 pages. 1977.

l. 574: J. Moldestad, Computations in Higher Types. IV, 203 ges. 1977.

l. 575: K-Theory and Operator Algebras, Athens, Georgia 1975. ited by B. B. Morrel and I. M. Singer. VI, 191 pages. 1977.

l. 576: V. S. Varadarajan, Harmonic Analysis on Real Reductive oups. VI, 521 pages. 1977.

l. 577: J. P. May, E_∞ Ring Spaces and E_∞ Ring Spectra. IV, 8 pages. 1977.

l. 578: Séminaire Pierre Lelong (Analyse) Année 1975/76. Edité P. Lelong. VI, 327 pages. 1977.

l. 579: Combinatoire et Représentation du Groupe Symétrique, rasbourg 1976. Proceedings 1976. Edité par D. Foata. IV, 339 ges. 1977.

Vol. 580: C. Castaing and M. Valadier, Convex Analysis and Measurable Multifunctions. VIII, 278 pages. 1977.

Vol. 581: Séminaire de Probabilités XI, Université de Strasbourg. Proceedings 1975/1976. Edité par C. Dellacherie, P. A. Meyer et M. Weil. VI, 574 pages. 1977.

Vol. 582: J. M. G. Fell, Induced Representations and Banach *-Algebraic Bundles. IV, 349 pages. 1977.

Vol. 583: W. Hirsch, C. C. Pugh and M. Shub, Invariant Manifolds. IV, 149 pages. 1977.

Vol. 584: C. Brezinski, Accélération de la Convergence en Analyse Numérique. IV, 313 pages. 1977.

Vol. 585: T. A. Springer, Invariant Theory. VI, 112 pages. 1977.

Vol. 586: Séminaire d'Algèbre Paul Dubreil, Paris 1975–1976 (29ème Année). Edited by M. P. Malliavin. VI, 188 pages. 1977.

Vol. 587: Non-Commutative Harmonic Analysis. Proceedings 1976. Edited by J. Carmona and M. Vergne. IV, 240 pages. 1977.

Vol. 588: P. Molino, Théorie des G-Structures: Le Problème d'Equivalence. VI, 163 pages. 1977.

Vol. 589: Cohomologie l-adique et Fonctions L. Séminaire de Géométrie Algébrique du Bois-Marie 1965–66, SGA 5. Edité par L. Illusie. XII, 484 pages. 1977.

Vol. 590: H. Matsumoto, Analyse Harmonique dans les Systèmes de Tits Bornologiques de Type Affine. IV, 219 pages. 1977.

Vol. 591: G. A. Anderson, Surgery with Coefficients. VIII, 157 pages. 1977.

Vol. 592: D. Voigt, Induzierte Darstellungen in der Theorie der endlichen, algebraischen Gruppen. V, 413 Seiten. 1977.

Vol. 593: K. Barbey and H. König, Abstract Analytic Function Theory and Hardy Algebras. VIII, 260 pages. 1977.

Vol. 594: Singular Perturbations and Boundary Layer Theory, Lyon 1976. Edited by C. M. Brauner, B. Gay, and J. Mathieu. VIII, 539 pages. 1977.

Vol. 595: W. Hazod, Stetige Faltungshalbgruppen von Wahrscheinlichkeitsmaßen und erzeugende Distributionen. XIII, 157 Seiten. 1977.

Vol. 596: K. Deimling, Ordinary Differential Equations in Banach Spaces. VI, 137 pages. 1977.

Vol. 597: Geometry and Topology, Rio de Janeiro, July 1976. Proceedings. Edited by J. Palis and M. do Carmo. VI, 866 pages. 1977.

Vol. 598: J. Hoffmann-Jørgensen, T. M. Liggett et J. Neveu, Ecole d'Eté de Probabilités de Saint-Flour VI – 1976. Edité par P.-L. Hennequin. XII, 447 pages. 1977.

Vol. 599: Complex Analysis, Kentucky 1976. Proceedings. Edited by J. D. Buckholtz and T. J. Suffridge. X, 159 pages. 1977.

Vol. 600: W. Stoll, Value Distribution on Parabolic Spaces. VIII, 216 pages. 1977.

Vol. 601: Modular Functions of one Variable V, Bonn 1976. Proceedings. Edited by J.-P. Serre and D. B. Zagier. VI, 294 pages. 1977.

Vol. 602: J. P. Brezin, Harmonic Analysis on Compact Solvmanifolds. VIII, 179 pages. 1977.

Vol. 603: B. Moishezon, Complex Surfaces and Connected Sums of Complex Projective Planes. IV, 234 pages. 1977.

Vol. 604: Banach Spaces of Analytic Functions, Kent, Ohio 1976. Proceedings. Edited by J. Baker, C. Cleaver and Joseph Diestel. VI, 141 pages. 1977.

Vol. 605: Sario et al., Classification Theory of Riemannian Manifolds. XX, 498 pages. 1977.

Vol. 606: Mathematical Aspects of Finite Element Methods. Proceedings 1975. Edited by I. Galligani and E. Magenes. VI, 362 pages. 1977.

Vol. 607: M. Métivier, Reelle und Vektorwertige Quasimartingale und die Theorie der Stochastischen Integration. X, 310 Seiten. 1977.

Vol. 608: Bigard et al., Groupes et Anneaux Réticulés. XIV, 334 pages. 1977.

Vol. 609: General Topology and Its Relations to Modern Analysis and Algebra IV. Proceedings 1976. Edited by J. Novák. XVIII, 225 pages. 1977.

Vol. 610: G. Jensen, Higher Order Contact of Submanifolds of Homogeneous Spaces. XII, 154 pages. 1977.

Vol. 611: M. Makkai and G. E. Reyes, First Order Categorical Logic. VIII, 301 pages. 1977.

Vol. 612: E. M. Kleinberg, Infinitary Combinatorics and the Axiom of Determinateness. VIII, 150 pages. 1977.

Vol. 613: E. Behrends et al., L^P-Structure in Real Banach Spaces. X, 108 pages. 1977.

Vol. 614: H. Yanagihara, Theory of Hopf Algebras Attached to Group Schemes. VIII, 308 pages. 1977.

Vol. 615: Turbulence Seminar, Proceedings 1976/77. Edited by P. Bernard and T. Ratiu. VI, 155 pages. 1977.

Vol. 616: Abelian Group Theory, 2nd New Mexico State University Conference, 1976. Proceedings. Edited by D. Arnold, R. Hunter and E. Walker. X, 423 pages. 1977.

Vol. 617: K. J. Devlin, The Axiom of Constructibility: A Guide for the Mathematician. VIII, 96 pages. 1977.

Vol. 618: I. I. Hirschman, Jr. and D. E. Hughes, Extreme Eigen Values of Toeplitz Operators. VI, 145 pages. 1977.

Vol. 619: Set Theory and Hierarchy Theory V, Bierutowice 1976. Edited by A. Lachlan, M. Srebrny, and A. Zarach. VIII, 358 pages. 1977.

Vol. 620: H. Popp, Moduli Theory and Classification Theory of Algebraic Varieties. VIII, 189 pages. 1977.

Vol. 621: Kauffman et al., The Deficiency Index Problem. VI, 112 pages. 1977.

Vol. 622: Combinatorial Mathematics V, Melbourne 1976. Proceedings. Edited by C. Little. VIII, 213 pages. 1977.

Vol. 623: I. Erdelyi and R. Lange, Spectral Decompositions on Banach Spaces. VIII, 122 pages. 1977.

Vol. 624: Y. Guivarc'h et al., Marches Aléatoires sur les Groupes de Lie. VIII, 292 pages. 1977.

Vol. 625: J. P. Alexander et al., Odd Order Group Actions and Witt Classification of Innerproducts. IV, 202 pages. 1977.

Vol. 626: Number Theory Day, New York 1976. Proceedings. Edited by M. B. Nathanson. VI, 241 pages. 1977.

Vol. 627: Modular Functions of One Variable VI, Bonn 1976. Proceedings. Edited by J.-P. Serre and D. B. Zagier. VI, 339 pages. 1977.

Vol. 628: H. J. Baues, Obstruction Theory on the Homotopy Classification of Maps. XII, 387 pages. 1977.

Vol. 629: W. A. Coppel, Dichotomies in Stability Theory. VI, 98 pages. 1978.

Vol. 630: Numerical Analysis, Proceedings, Biennial Conference, Dundee 1977. Edited by G. A. Watson. XII, 199 pages. 1978.

Vol. 631: Numerical Treatment of Differential Equations. Proceedings 1976. Edited by R. Bulirsch, R. D. Grigorieff, and J. Schröder. X, 219 pages. 1978.

Vol. 632: J.-F. Boutot, Schéma de Picard Local. X, 165 pages. 1978.

Vol. 633: N. R. Coleff and M. E. Herrera, Les Courants Résiduels Associés à une Forme Méromorphe. X, 211 pages. 1978.

Vol. 634: H. Kurke et al., Die Approximationseigenschaft lokaler Ringe. IV, 204 Seiten. 1978.

Vol. 635: T. Y. Lam, Serre's Conjecture. XVI, 227 pages. 1978.

Vol. 636: Journées de Statistique des Processus Stochastiques, Grenoble 1977, Proceedings. Edité par Didier Dacunha-Castelle et Bernard Van Cutsem. VII, 202 pages. 1978.

Vol. 637: W. B. Jurkat, Meromorphe Differentialgleichungen. VII, 194 Seiten. 1978.

Vol. 638: P. Shanahan, The Atiyah-Singer Index Theorem, An Introduction. V, 224 pages. 1978.

Vol. 639: N. Adasch et al., Topological Vector Spaces. V, 125 pages. 1978.

Vol. 640: J. L. Dupont, Curvature and Characteristic Classes. X, 175 pages. 1978.

Vol. 641: Séminaire d'Algèbre Paul Dubreil, Proceedings Paris 1976–1977. Edité par M. P. Malliavin. IV, 367 pages. 1978.

Vol. 642: Theory and Applications of Graphs, Proceedings, Michigan 1976. Edited by Y. Alavi and D. R. Lick. XIV, 635 pages. 1978.

Vol. 643: M. Davis, Multiaxial Actions on Manifolds. VI, 141 pages 1978.

Vol. 644: Vector Space Measures and Applications I, Proceedings 1977. Edited by R. M. Aron and S. Dineen. VIII, 451 pages. 1978

Vol. 645: Vector Space Measures and Applications II, Proceedings 1977. Edited by R. M. Aron and S. Dineen. VIII, 218 pages. 1978

Vol. 646: O. Tammi, Extremum Problems for Bounded Univalen Functions. VIII, 313 pages. 1978.

Vol. 647: L. J. Ratliff, Jr., Chain Conjectures in Ring Theory. VIII, 133 pages. 1978.

Vol. 648: Nonlinear Partial Differential Equations and Applications Proceedings, Indiana 1976–1977. Edited by J. M. Chadam. VI, 206 pages. 1978.

Vol. 649: Séminaire de Probabilités XII, Proceedings, Strasbourg 1976–1977. Edité par C. Dellacherie, P. A. Meyer et M. Weil. VIII 805 pages. 1978.

Vol. 650: C*-Algebras and Applications to Physics. Proceedings 1977. Edited by H. Araki and R. V. Kadison. V, 192 pages. 1978.

Vol. 651: P. W. Michor, Functors and Categories of Banach Spaces VI, 99 pages. 1978.

Vol. 652: Differential Topology, Foliations and Gelfand-Fuks-Cohomology, Proceedings 1976. Edited by P. A. Schweitzer. XIV, 25 pages. 1978.

Vol. 653: Locally Interacting Systems and Their Application i Biology. Proceedings, 1976. Edited by R. L. Dobrushin, V. I. Kryuko and A. L. Toom. XI, 202 pages. 1978.

Vol. 654: J. P. Buhler, Icosahedral Golois Representations. III, 14 pages. 1978.

Vol. 655: R. Baeza, Quadratic Forms Over Semilocal Rings. V 199 pages. 1978.

Vol. 656: Probability Theory on Vector Spaces. Proceedings, 1977 Edited by A. Weron. VIII, 274 pages. 1978.

Vol. 657: Geometric Applications of Homotopy Theory I, Proceeding 1977. Edited by M. G. Barratt and M. E. Mahowald. VIII, 459 page 1978.

Vol. 658: Geometric Applications of Homotopy Theory II, Proceeding 1977. Edited by M. G. Barratt and M. E. Mahowald. VIII, 487 page 1978.

Vol. 659: Bruckner, Differentiation of Real Functions. X, 247 page 1978.

Vol. 660: Equations aux Dérivée Partielles. Proceedings, 1977. Edi par Pham The Lai. VI, 216 pages. 1978.

Vol. 661: P. T. Johnstone, R. Paré, R. D. Rosebrugh, D. Schumache R. J. Wood, and G. C. Wraith, Indexed Categories and Their Applica tions. VII, 260 pages. 1978.

Vol. 662: Akin, The Metric Theory of Banach Manifolds. XIX, 30 pages. 1978.

Vol. 663: J. F. Berglund, H. D. Junghenn, P. Milnes, Compact Rig Topological Semigroups and Generalizations of Almost Periodici X, 243 pages. 1978.

Vol. 664: Algebraic and Geometric Topology, Proceedings, 197 Edited by K. C. Millett. XI, 240 pages. 1978.

Vol. 665: Journées d'Analyse Non Linéaire. Proceedings, 197 Edité par P. Bénilan et J. Robert. VIII, 256 pages. 1978.

Vol. 666: B. Beauzamy, Espaces d'Interpolation Réels: Topologie Géometrie. X, 104 pages. 1978.

Vol. 667: J. Gilewicz, Approximants de Padé. XIV, 511 pages. 197

Vol. 668: The Structure of Attractors in Dynamical Systems. Pr ceedings, 1977. Edited by J. C. Martin, N. G. Markley and W. Pe rizo. VI, 264 pages. 1978.

Vol. 669: Higher Set Theory. Proceedings, 1977. Edited by G. Müller and D. S. Scott. XII, 476 pages. 1978.